ADVANCED CALCULUS

ADVANCED CALCULUS
An Introduction to Linear Analysis

Leonard F. Richardson

WILEY-INTERSCIENCE

A JOHN WILEY & SONS, INC., PUBLICATION

Published by John Wiley & Sons, Inc., Hoboken, New Jersey

Published simultaneously in Canada

For general information on our other products and services or for technical support, please contact our Customer Care Department within the United States at (800) 762-2974, outside the United States at (317) 572-3993 or fax (317) 572-4002.

Wiley also publishes its books in a variety of electronic formats. Some content that appears in print may not be available in electronic formats. For more information about Wiley products, visit our web site at www.wiley.com.

Library of Congress Cataloging-in-Publication Data:

Richardson, Leonard F.
   Advanced calculus : an introduction to linear analysis / Leonard F. Richardson.
      p. cm.
   Includes bibliographical references and index.
   ISBN 978-0-470-23288-0 (cloth)
   1. Calculus. I. Title.
   QA303.2.R53 2008
   515--dc22

                          2008007377
Printed in the United States of America

10 9 8 7 6 5 4 3 2 1

To Joan, Daniel, and
Joseph

CONTENTS

PREFACE

Why this Book was Written

The course known as Advanced Calculus (or Introductory Analysis) stands at the summit of the requirements for senior mathematics majors. An important objective of this course is to prepare the student for a critical challenge that he or she will face in the first year of graduate study: the course called Analysis I, Lebesgue Measure and Integration, or Introductory Functional Analysis.

We live in an era of rapid change on a global scale. And the author and his department have been testing ways to improve the preparation of mathematics majors for the challenges they will face. During the past quarter century the United States has emerged as the destination of choice for graduate study in mathematics. The influx of well-prepared, talented students from around the world brings considerable benefit to American graduate programs. The international students usually arrive better prepared for graduate study in mathematics—in particular better prepared in analysis—than their typical U.S. counterparts. There are many reasons for this, including (a) school systems abroad that are oriented toward teaching only the brightest students, and (b) the self-selection that is part of a student taking the step of travel abroad to study in a foreign culture.

The presence of strongly prepared international students in the classroom raises the level at which courses are taught. Thus it is appropriate at the present time, in the early years of the new millennium, for college and university mathematics departments to reconsider their advanced calculus courses with an eye toward preparing graduates for the international environment in American graduate schools. This is a challenge, but it is also an opportunity for American students and international students to learn side-by-side with, and also about, one another. It is more important than ever to teach undergraduate advanced calculus or analysis in such a way as to prepare and reorient the student for graduate study as it is today in mathematics.

Another recent change is that applied mathematics has emerged on a large scale as an important component of many mathematics departments. In applied and numerical mathematics, functional analysis at the graduate level plays a very important role.

Yet another change that is emerging is that undergraduates planning careers in the secondary teaching of mathematics are being required to major in mathematics instead of education. These students must be prepared to teach the next generation of young people for the world in which they will live. Whether or not the mathematics major is planning an academic career, he or she will benefit from better preparation in advanced calculus for careers in the emerging world.

The author has taught mathematics majors and graduate students for thirty-seven years. He has served as director of his department's graduate program for nearly two decades. All the changes described above are present today in the author's department. This book has been written in the hope of addressing the following needs.

1. Students of mathematics should acquire a sense of the unity of mathematics. Hence a course designed for senior mathematics majors should have an integrative effect. Such a course should draw upon at least two branches of mathematics to show how they may be combined with illuminating effect.

2. Students should learn the importance of rigorous proof and develop skill in coherent written exposition to counter the universal temptation to engage in wishful thinking. Students need practice composing and writing proofs of their own, and these must be checked and corrected.

3. The fundamental theorems of the introductory calculus courses need to be established rigorously, along with the traditional theorems of advanced calculus, which are required for this purpose.

4. The task of establishing the rigorous foundations of calculus should be enlivened by taking this opportunity to introduce the student to modern mathematical structures that were not presented in introductory calculus courses.

5. Students should learn the rigorous foundations of calculus in a manner that reorients thinking in the directions taken by modern analysis. The classic theorems should be couched in a manner that reflects the perspectives of modern analysis.

Features of this Text

The author has attempted to address these needs presented above in the following manner.

1. The two parts of mathematics that have been studied by nearly every mathematics major prior to the senior year are introductory calculus, including calculus of several variables, and linear algebra. Thus the author has chosen to highlight the interplay between the calculus and linear algebra, emphasizing the role of the concepts of a vector space, a linear transformation (including a linear functional), a norm, and a scalar product. For example, the customary theorem concerning uniform limits of continuous functions is interpreted as a completeness theorem for $C[a, b]$ as a vector space equipped with the sup-norm. The elementary properties of the Riemann integral gain coherence expressed as a theorem establishing the integral as a bounded linear functional on a convenient function-space. Similarly, the family of absolutely convergent series is presented from the perspective that it is a complete normed vector space equipped with the l_1-norm.

2. Many exercises are offered for each section of the text. These are essential to the course. **An exercise preceded by a dagger symbol † is** *cited* **at some point in the text.** Such citations refer to the exercise by section and number. **An exercise preceded by a diamond symbol ◇ is a** *hard* **problem.** *If a hard problem will be cited later in the text, then there will be a footnote to say precisely where it will be cited.* This is intended to help the professor decide whether or not an exercise should be assigned to a particular class based upon his or her planned coverage for the course. Topics that can be omitted at the professor's discretion without disturbing continuity of the course are so-indicated by means of footnotes.

3. At the end of each chapter there is a brief section called *Test Yourself*, consisting of short questions to test the student's comprehension of the basic concepts and theorems. The *answers* to these short questions, and also to *other selected short questions*, appear in an appendix. There are no proofs provided among those answers to selected questions. The reason is that there are many possible correct proofs for each exercise. Only the professor or the professor's designated assistant will be able to properly evaluate and correct the student's writing in exercises requiring proofs.

4. The *Introduction* to this book is intended to introduce the student to both the importance and the challenges of writing proofs. The guidance provided in the introduction is followed by corresponding illustrative remarks that appear after the first proof in each of the five chapters of Part I of this text.

5. Whether a professor chooses to collect written assignments or to have students present proofs at the board in front of the class, each student must regularly construct and write proofs. The coherence and the presentation of the arguments must be criticized.

6. Most of the traditional theorems of elementary differential and integral calculus are developed rigorously. Since the orientation of the course is toward the role of normed vector spaces, Cauchy completeness is the most natural form of the completeness concept to use. Thus we present the system of real numbers as a Cauchy-complete Archimedean ordered field. The traditional theorems of advanced calculus are presented. These include the elements of the study of integrable and differentiable functions, extreme value theorems, Mean Value Theorems, and convergence theorems, the polynomial approximation theorem of Weierstrass, the inverse and implicit function theorems, Lebesgue's theorem for Riemann integrability, and the Jacobian theorem for change of variables.

7. Students learn in this course such concepts as those of a complete normed vector space (*real* Banach space) and a bounded linear functional. This is *not* a course in functional analysis. Rather the central theorems and examples of advanced calculus are treated as instances and motivations for the concepts of functional analysis. For example, the space of bounded sequences is shown to be the dual space of the space of absolutely summable sequences.

8. The concept of this book is that the student is guided gradually from the study of the topology of the real line to the beginning theorems and concepts of graduate analysis, expressed from a modern viewpoint. Many traditional theorems of advanced calculus list properties that amount to stating that a certain set of functions forms a vector space and that this space is complete with respect to a norm. By phrasing the traditional theorems in this light, we help the student to mentally organize the knowledge of advanced calculus in a coherent and meaningful manner while acquiring a helpful reorientation toward modern graduate-level analysis.

Course Plans that Are Supported by this Book

Part I of this book consists of five chapters covering most of the standard one-variable topics found in two-semester advanced calculus courses. These chapters are arranged in order of dependence, with the later chapters depending on the earlier ones. Though the topics are mainly the ones typically found, they have been reoriented here from the viewpoint of linear spaces, norms, completeness, and linear functionals.

Part II offers a choice of two mutually independent advanced one-variable topics: either Fourier series or Stieltjes integration. It is especially the case in Part II that each professor's individual judgment about the readiness of his or her class should guide what is taught. Some of these topics will not be for the average student, but will make excellent reading material for the student seeking honors credit or writing a senior thesis. Individual reading courses can be employed very effectively to provide advanced experience for the prospective graduate student.

In Chapter 6 the introduction of Fourier series is aided by inclusion of complex-valued functions of a real variable. This is the only chapter in which complex-valued functions appear, and with these the Hermitian inner product is introduced. The chapter includes l^2 and its self-duality, convergence in the L^2-norm,[1] the uniform convergence of Fourier series of smooth functions, and the Riemann localization theorem. The study of a vibrating string is presented to motivate the chapter.

Chapter 7, which is about Stieltjes integration, includes functions of bounded variation and the Riesz Representation Theorem, presenting the dual space of $C[a, b]$ in terms of Stieltjes integration. The latter theorem of F. Riesz is the hardest one presented in this book. It is not required for the later chapters. However, it is an excellent theorem for a promising student planning subsequent doctoral study, and it requires only what has been learned previously in this course. It is a century since the discovery of the Riesz Representation Theorem. The author thinks it is time for it to take its place in an undergraduate text for the twenty-first century.

Part III is about several-variable advanced calculus, including the inverse and implicit function theorems, and the Jacobian theorems for multiple integrals. Where the first two parts place emphasis on infinite-dimensional linear spaces of functions, the third part emphasizes finite-dimensional spaces and the derivative as a linear transformation.

At Louisiana State University, Advanced Calculus is offered as a three-semester *triad* of courses.[2] The first semester is taken by all and is the starting point regardless of the subsequent choices. But the other two semesters can be taken *in either order*. This enables the Department to offer all three semesters each year, with the first semester offered in both fall and spring, and the two other courses being offered with only one of them each semester. These courses are not rushed. One must allow sufficient time for the typical undergraduate mathematics major to learn to prove theorems and to absorb the new concepts. It is the author's experience that all too often, courses in analysis are inadvertently sabotaged by packing too much subject matter into one term. It is best to teach students to take enough time to learn well and learn deeply.

A few words about testing procedures may be helpful too. At the author's institution, and at many others also, it is important to teach Advanced Calculus in a manner that is suitable for *both* those students who are preparing for graduate study in mathematics and those who are not. The author finds that it is appropriate to divide each test into two approximately equal parts: one for short questions of the type represented in the *Test Yourself* sections of this book, and the other consisting of proofs representative of those assigned and collected for homework. Although one would like each student to excel in both, there are many students who excel in one class of question but not the other. And there are indeed many students who do better in proofs than in the concept-testing short questions. Thus tests that combine both types of question provide fuller information about each student and give an opportunity for more students to show what they can do. The author always gives a choice of questions in each of the two categories: typically eight out of twelve for

---

[1]The L^2 norm is used here exclusively with the Riemann integral.

[2]Mathematics majors planning careers in high-school teaching take at least the first semester, while the others must take at least two of the three semesters. Those students who are contemplating graduate study in mathematics are advised strongly to take all three semesters.

the short questions, and two out of three for the proofs, for a one-hour test. The pass rate in these courses is actually high, despite the depth of the subject. Naturally, each professor will need to determine the best approach to testing for his or her own class.

It is most common for colleges and universities to offer either a single semester or else a two-semester sequence in Advanced Calculus or Undergraduate Analysis. Below the author has indicated practical syllabi for a one-semester course, as well as three alternative versions of a two-semester course. It should be understood that, depending on the readiness of the class, it may be possible to do more.

- *Single-semester course:* Sections 1.1–1.8, 2.1–2.4, 3.1–3.3, and 4.1–4.3.

- *Two-semester course leading to Stieltjes integration:*
    1. Chapters 1–3 for the first semester
    2. Chapters 4, 5, and 7 for the second semester

- *Two-semester course leading to Fourier series:*
    1. Chapters 1–3 for the first semester
    2. Chapters 4–6 for the second semester

- *Two-semester course leading to the inverse and implicit function theorems:*
    1. Sections 1.1–1.8, 2.1–2.4, 3.1–3.3, and 4.1–4.3 for the first semester
    2. Sections 8.1–8.3, 9.1–9.3, and 10.1–10.3 for the second semester

- *Three-semester course, with parts 2 and 3 interchangeable in order:*
    1. Chapters 1–3 for the first semester
    2. Either
        (a) Chapters 4–6 for the second semester or
        (b) Chapters 4, 5, and 7 for the second semester
    3. Sections 8.1–8.3, 9.1–9.3, and 10.1–10.3 for the third semester, and with Chapter 11 if there is sufficient time.

No doubt there are other possible combinations. Whatever is the choice made, the author hopes that the whole academic community of mathematicians will devote an increased number of courses to the teaching of analysis to undergraduate mathematics majors.

LEONARD F. RICHARDSON

Baton Rouge, Louisiana
August, 2007

ACKNOWLEDGMENTS

It is a pleasure to thank several colleagues at Louisiana State University who have contributed useful ideas, corrections, and suggestions. They are Professors Jacek Cygan, Mark Davidson, Charles Delzell, Raymond Fabec, Jerome Hoffman, Richard Litherland, Gestur Olafsson, Ambar Sengupta, Lawrence Smolinsky, and Peter Wolenski. Several of these colleagues taught classes using the manuscript that became this book. Of course the errors that remain are entirely my own responsibility, and further corrections and suggestions from the reader will be much appreciated.

In the academic year 1962–1963 I was a student in an advanced calculus course taught by Professor Frank J. Hahn at Yale University. His inclusion in that course of the Riesz Representation Theorem and its proof was a highlight of my undergraduate education. Though I didn't realize it at the time, that course likely was the source of the idea for this book.

Professor Hahn was a young member of the Yale faculty when I was a student in his advanced calculus course that included the Riesz theorem. He was an extraordinary and generous teacher. I became his PhD student, but his death intervened about a year later. Then Professor George D. Mostow adopted me as his student. Professor Mostow took an interest in improving undergraduate education in mathematics, having co-authored a book [14] that had as one of its goals the earlier inclusion and integration of abstract algebra into the undergraduate curriculum. I have been very fortunate with regard to my teachers. They taught lessons that grow over time like branches, integral parts of one tree. I am grateful for the opportunity to record my gratitude and indebtedness to them.

My book is intended to facilitate the integration of linear spaces, functionals and transformations, both finite- and infinite-dimensional, into Advanced Calculus. It is not a new idea that mathematics should be taught to undergraduate students in a manner that demonstrates the overarching coherence of the subject. As mathematics grows, in both pure and applied directions, the need to emphasize its unity remains a pressing objective.

Questions and observations from students over the years have resulted in numerous exercises and explanatory remarks. It has been a privilege to share some of my favorite mathematics with students, and I hope the experience has been a good one for them.

I am grateful to John Wiley & Sons for the opportunity to offer this book, as well as the course it represents and advocates, to a wider audience. I appreciate especially the role of Ms. Susanne Steitz-Filler, the Mathematics and Statistics Editor of John Wiley & Sons, in making this opportunity available. She and her colleagues provided valued advice, support, and technical assistance, all of which were needed to transform a professor's course notes into a book.

L. F. R.

INTRODUCTION

Why Advanced Calculus is Important

What is the meaning of knowledge? And what is the meaning of learning? The author believes these are questions that must be addressed in order to grasp the purpose of advanced calculus. In primary and secondary education, and also in some introductory college courses, we are asked to accept many statements or claims and to remember them, perhaps to apply them. Individuals vary greatly in temperament and are more willing or less willing to acquiesce in the acceptance of what is taught. But whether or not we are inclined to do so, we must ask responsible questions about the basis upon which knowledge rests.

Here are a few examples.

- Have we been taught accurate renditions of the history of our civilization? Is there nothing to indicate that history is presented sometimes in a biased or misleading way?

- Were we taught correct claims about the nature of the physical or biological world? Are there not examples of famous claims regarding the natural sciences, endorsed ardently, yet proven in time to be false?

- How do we know what is or is not true about mathematics? Is there no record of error or disagreement? Is there an infallible expert who can be trusted to tell correctly the answers to all questions?

- *If* there are authorities who can be trusted without doubt to instruct us correctly, what will be our fate when these authorities, perhaps older than ourselves, die? Can we not learn for ourselves to determine the difference between truth and falsehood, between valid reason and error?

In the serious study of history, one must learn how to search for records or evidence and how to appraise its reliability. In the natural sciences, one must learn to construct sound experiments or to conduct accurate observations so as to distinguish between truth and wishful thinking. And in the study of mathematics it is through logical proof by deductive reasoning that we can check our thinking or our guesswork. Learning how to confirm the foundations of our knowledge transforms us from receptacles for the claims made by others into stewards for the knowledge mankind has acquired through millennia of exertion. It is both our right as human beings and our responsibility to assume this role.

Throughout our lives, we find ourselves with the need to resolve the conflict between opposing forces. On the one hand, the human mind is impulsive, eager to leap from one spot to another that may have a clearer view. This spark is an engine of creativity. We would not be human in its absence. It is also our Achilles' heel. Training and self-discipline are required that we may distinguish the worthwhile leaps of imagination from the faulty ones.

A vital aspect of the self-discipline that must be learned by each student of mathematics is that *proofs must be written down, scrutinized step-by-step, and re-written wherever there is doubt.* In a proof the reasoning must be solid and secure from start to finish. There is no one among us who can reliably devise a proof mentally, leaving it unwritten and unscrutinized. Indeed, mankind's capacity for wishful thinking is boundless. Discipline in the standard of logical proof is severe, and it is essential to our task.

Mathematics is not a spectator sport. It can be learned only by doing. It is necessary but never sufficient to watch proofs being constructed by an experienced practitioner. The latter activity (which includes attendance in class and active participation, as well as careful study of the text) can help one to learn good technique. But only the effort of writing our own proofs can teach each of us by trial and error how to do it. See this as not only a warning but also good news that strenuous effort in this work is effective. From more than three decades of teaching as well as personal experience, the author can assure each student that this is so. It is possible also to assure the student that through vigorous effort in mathematics the student may come to enjoy this subject very much and to relish the light that it can shed. Even a seemingly small question can be a portal to a whole world of unforeseen surprise and wonder. In this spirit it is a pleasure to welcome the student and the reader to advanced calculus.

Learning to Write Proofs: A Guide for the Perplexed Student

I want to do my proof-writing homework, but I don't know how to begin! It is an oft-heard lament. In elementary mathematics courses, the student is provided customarily with a set of instructions, or algorithms, that will lead upon implementation to the solution of certain types of problems. Thus many conscientious students have requested instructions for writing proofs. All sets of instructions for writing proofs, however, suffer from one defect: They do not work. Yet one can learn to write proofs, and there are many living mathematicians and successful mathematics students whose existence proves this point. The author believes that learning to write proofs is not a matter of following theorem-proving instructions. The answer lies rather in learning *how to study* advanced calculus. The student, having been in school for much of his or her life, may bridle at the suggestion that he or she has not learned how to study. Yet in the case of studying theoretical mathematics, that is very likely to be true. Every single theorem and every single proof that is presented in this book, or by the student's professor in class, is a vivid example of theorem-proving technique. But to benefit from these fine examples, the student must learn how to study. Mathematicians find that *the best way to read mathematics is with paper and pencil!* This means that it is the reader's task to figure out how to think about the theorem and its proof and to *write it down coherently.*

In reading the proofs of theorems in this text, or in the study of proofs presented by one's teacher in class, the student must understand that what is written is much more than a body of facts to be remembered and reproduced upon demand. Each proof has a story that guided the author in its writing. There is a beginning (the hypotheses), a challenge (the objective to be achieved), and a plan that might, with hard work, skill, and good fortune, lead to the desired conclusion. It will take time and a concerted effort for the student to learn to think about the statements and proofs of the presented theorems in this light. Such practice will cultivate the ability to read the exercises as well in a fruitful manner. With experience at recognizing the story of the proof or problem at hand, the student will be in a position to develop technique through the work done in the exercises.

The first step, before attempting to read a proof, is to read the statement of the theorem carefully, trying to get an overall picture of its content. The student should make sure he or she knows precisely the definition of each term used in the statement of the theorem. Without that information, it is impossible to understand even the claim of the theorem, let alone its proof. If a term or a symbol in the statement of a theorem or exercise is not recognized, look in the index! Write on paper what you find.

After clarifying explicitly the meaning of each term used, if the student does not see what the theorem is attempting to achieve, it is often helpful to write down a few examples to see what difficulties might arise, leading to the need for the theorem. *Working with examples is the mathematical equivalent of laboratory work for a natural scientist.* At this point the student will have read the statement of the theorem at least twice, and probably more often than that, accumulating written notes on a scratch pad along the way. Read the theorem again! Remember that in constructing a building or a bridge, it is not a waste of time to dwell upon the foundation. The author has assured many students, from freshman to doctoral level, that the way to make faster progress is to slow down—especially at the outset. If you were planning a grand two-week backpacking trip in a national park, would you simply run out of the house? Of course not—you would plan and make preparations for the coming adventure.

At this point we suppose the reader understands the statement of the theorem and wishes next to learn why the claimed conclusion is true. How does the author or teacher in class overcome the obstacles at hand? Read the whole proof a first time, *taking written notes* as to what combination of steps the author has chosen to proceed from the hypotheses to the conclusions. This first reading of the proof itself can be likened to one's first look at a road map drawn for a cross-country trip. It will give one an overall sense of the journey ahead. But taking the trip, or walking the walk, is another matter. Having noted that the journey ahead can be divided into segments, much like a trip with several overnight stops, the student should begin in earnest at the beginning. For each leg of the journey, it is important to understand thoroughly, and to write on paper, the logical justification of each individual step. There must be no magical disappearance from point A and reappearance at point B! No external authority can be substituted for the student's own understanding of each step taken. It is the both the right and the responsibility of the student to understand in full detail.[3]

By studying the theorems in this book in the manner explained above, the student will cultivate the modes of thinking that will enable him or her to write the proofs that are required in the exercises.

The exercises are a vital part of this course, and the proof exercises are the most important of all. There is an answer section for *selected short-answer* exercises among the appendices of this book. It includes all the answers to the Test Yourself self-tests at the ends of the chapters. But the student will not find solutions to the proof exercises there. That is because it is not satisfactory merely to copy a written proof. Many correct proofs are possible. Only an experienced teacher can judge the correctness and the quality of the proofs you write. The student can and must depend upon his or her professor or the professor's designated assistant to read and correct proofs written as exercises.

One of the ways that a teacher can help a student is by explaining that he or she has been where the student stands. The student is not alone and can meet the challenges ahead much as his or her teacher has done before. When the author was young, he had long walks to and from school: about twenty minutes each way at a brisk pace. It was a favorite pastime during these walks to review mentally the logical structure of advanced calculus–reconstructing the proofs of theorems about Riemann integrals or uniform convergence from the axioms of the real number system. Many colleagues within mathematics, and some from theoretical physics, have shared with the author similar experiences from their own lives. It is the active engagement with a subject

---

[3]The student should reread this introduction before reading Remark 1.1.1, which appears after the proof of the first theorem in this book. Corresponding remarks appear following the first proof in each of the five chapters of Part I of this book.

that builds firm understanding and that incorporates the knowledge gained into ones own mind.

Experiences in life can be enjoyed only once for the first time. The student is about to embark on a mathematical adventure with advanced calculus for his or her first time. Neither the author nor your teacher can do this again. But we can wish you a wonderful journey, and we do.

PART I

ADVANCED CALCULUS IN ONE VARIABLE

CHAPTER 1

REAL NUMBERS AND LIMITS OF SEQUENCES

1.1 THE REAL NUMBER SYSTEM

During the 19th century, as applications of the differential and integral calculus in the physical sciences grew in importance and complexity, it became apparent that intuitive use of the concept of limit was inadequate. Intuitive arguments could lead to seemingly correct or incorrect conclusions in important examples. Much effort and creativity went into placing the calculus on a rigorous foundation so that such problems could be resolved. In order to see how this process unfolded, it is helpful to look far back into the history of mathematics.

Approximately 2000 years ago, Greek mathematicians placed Euclidean geometry on the foundations of deductive logic. Axioms were chosen as assumptions, and the major theorems of geometry were proven, using fairly rigorous logic, in an orderly progression. These ancient mathematicians also had concepts of numbers. They used *natural numbers*, known also as *counting numbers*, the set of which is denoted by

$$\mathbb{N} = \{1, 2, 3, \ldots, n, n+1, \ldots\}.$$

This is the endless sequence of numbers beginning with 1 and proceeding without end by adding 1 at each step. Also used were *positive rational numbers*, which we

Advanced Calculus: An Introduction to Linear Analysis. By Leonard F. Richardson
Copyright © 2008 John Wiley & Sons, Inc.

denote as

$$\mathbb{Q}^+ = \left\{ \frac{p}{q} \,\middle|\, p, q \in \mathbb{N} \right\}.$$

These numbers were regarded as representing proportions of positive whole numbers.

Members of the Pythagorean school of geometry discovered that there was no ratio of positive whole numbers that could serve as a square root for 2. (See Exercise 1.11.) This was disturbing to them because it meant that the side and the diagonal of a square must be *incommensurable*. That is, the side and the diagonal of a square cannot both be measured as a whole number multiple of some other line segment, or *unit*. So great was these geometers' consternation over the failure of the set of rational numbers to provide the proportion between the side and the diagonal of a square that confidence in the logical capacity of algebra was diminished. Mathematical reasoning was phrased, to the extent possible, in terms of geometry.

For example, today we would express the area of a circle algebraically as $A = \pi r^2$. We could express this common formula alternatively as $A = \frac{\pi}{4}d^2$, where d is the diameter of the circle. But the ancient Greeks put it this way: The areas of two circles are in the same proportion as the areas of the *squares on their diameters*. The squares were constructed, each with a side coinciding with the diameter of the corresponding circle, and the areas of the squares were in the same proportion as the areas of the circles. Much later, in the 17th century, Isaac Newton continued to be influenced by this perspective. In his celebrated work on the calculus, *Principia Mathematica*, we can see repeatedly that where we would use an algebraic calculation, he used a geometrical argument, even if greater effort is required. The reader interested in the history of mathematics may enjoy the book *The Exact Sciences in Antiquity* by Otto Neugebauer [15] and the one by Carl Boyer [3], *The History of the Calculus*.

It took until the 19th century for mathematicians to liberate themselves from their misgivings regarding algebra. It came to be understood that the *real numbers*, the numbers that correspond to the points on an endless geometrical line, could be placed on a systematic logical foundation just as had been done for geometry nearly two thousand years earlier. Most of the axioms that were needed to prove the properties of the real number system were already quite familiar from the arithmetic of the rational numbers. There was one crucial new axiom needed: the *Completeness Axiom of the Real Number System*. Once this axiom had been added, the theorems of the calculus could be proven rigorously, and future development of the subject of *Mathematical Analysis* in the 20th century was facilitated.

Although we will not attempt the laborious task of rigorously proving every familiar property of the real number system, we will sketch the axioms that summarize familiar properties, and we will explain carefully the completeness axiom. With the latter axiom in hand, we will develop the theory of the calculus with great care. Students interested in studying the full and formal development of the real number system are referred to J. M. H. Olmsted's book [16], or to a stylistically distinctive classic by E. Landau [12].

In addition to the set $\mathbb{N}$ of natural numbers, we will consider the set $\mathbb{Z}$ of *integers*, or whole numbers. Thus

$$\mathbb{Z} = \{0, \pm 1, \pm 2, \ldots\} = \{\pm n \mid n \in \mathbb{N}\} \cup \{0\}.$$

We need also the full set of rational numbers:

$$\mathbb{Q} = \left\{ \frac{p}{q} \,\middle|\, p, q \in \mathbb{Z}, q \neq 0 \right\}.$$

We list in Table 1.1 the axioms for a general *Archimedean Ordered Field* $\mathbb{F}$. You will observe that the set $\mathbb{Q}$ is an Archimedean ordered field. However, the set $\mathbb{R}$ of *real numbers*, which we will define in Section 1.3, will obey all the axioms for an Archimedean ordered field together with one more axiom, called the *Completeness Axiom*, which is *not* satisfied by $\mathbb{Q}$.

Table 1.1 Archimedean Ordered Field

An **Archimedean Ordered Field** $\mathbb{F}$ is a set with two operations, called addition and multiplication. There is also an *order relation*, denoted by $a < b$. These satisfy the following properties:

1. *Closure:* If a and b are elements of $\mathbb{F}$, then $a + b \in \mathbb{F}$ and $ab \in \mathbb{F}$.

2. *Commutativity:* If a and b are elements of $\mathbb{F}$, then $a + b = b + a$ and $ab = ba$.

3. *Associativity:* If a, b, and c are elements of $\mathbb{F}$, then $a + (b + c) = (a + b) + c$ and $a(bc) = (ab)c$.

4. *Distributivity:* If a, b, and c are elements of $\mathbb{F}$, then $a(b + c) = ab + ac$.

5. *Identity:* There exist elements 0 and 1 in $\mathbb{F}$ such $0 + a = a$ and $1a = a$, for all $a \in \mathbb{F}$. Moreover, $0 \neq 1$.

6. *Inverses:* If $a \in \mathbb{F}$, then there exists $-a \in \mathbb{F}$ such that $-a + a = 0$. Also, for all $a \neq 0$, then there exists $a^{-1} = \frac{1}{a} \in \mathbb{F}$ such that $a\frac{1}{a} = 1$.

7. *Transitivity:* If $a < b$ and $b < c$, then $a < c$.

8. *Preservation of Order:* if $a < b$ and if $c \in \mathbb{F}$, then $a + c < b + c$. Moreover, if $c > 0$, then $ac < bc$.

9. *Trichotomy:* For all a and b in $\mathbb{F}$, exactly one of the following three statements will be true: $a < b$, or $a = b$, or $a > b$ (which means $b < a$).

10. *Archimedean Property:* If $\epsilon > 0$ and if $M > 0$, then there exists $n \in \mathbb{N}$ such that $n\epsilon > M$. (In this general context, $\mathbb{N}$ is defined as the smallest subset of $\mathbb{F}$ that contains 1 and is closed under addition.)

There is an old adage that loosely paraphrases the Archimedean Property found in the table: If you save a penny a day, eventually you will become a millionaire (or a billionaire, etc.).

From the axioms for an Archimedean ordered field, many familiar properties of the real numbers can be deduced. In particular, the behavior of all the operations used in solving equations and inequalities follows directly, with the exception that we have not established yet that roots of positive numbers, such as square roots, exist. Here we will concentrate on those properties that received less emphasis in elementary mathematics courses.

The order axioms are particularly useful for analysis. In this connection, it is important to make the following definition.

Definition 1.1.1 *We define*

$$|a| = \begin{cases} a & \text{if } a \geq 0, \\ -a & \text{if } a < 0. \end{cases}$$

We think of $|a|$ as representing the *distance* of a from 0 on the number line. Note that $|a|$ is always nonnegative. The absolute value satisfies a vital inequality known as the *Triangle Inequality*.

Theorem 1.1.1 *For all a and b in $\mathbb{R}$, $|a + b| \leq |a| + |b|$.*

Proof: Observe that

$$-|a| \leq a \leq |a|,$$

and

$$-|b| \leq b \leq |b|,$$

so that

$$-(|a| + |b|) \leq a + b \leq |a| + |b|. \tag{1.1}$$

Thus, if $a + b \geq 0$,

$$|a + b| = a + b \leq |a| + |b|.$$

But if $a + b < 0$, then from the first inequality in Equation (1.1), we obtain

$$|a + b| = -(a + b) \leq |a| + |b|.$$

We see that whether $a + b$ is negative or nonnegative, we have in either case that $|a + b| \leq |a| + |b|$. ∎

Remark 1.1.1 If the student has not yet *read the Introduction*, including the discussion of *Learning to Write Proofs* on page xxiii, this should be done now. It was explained that in order to learn to write proofs, the student must learn first how to study the theorems and proofs that are presented in this book. Let us note how the remarks made there apply to the short proof of the first theorem in this book.

First we read carefully the statement of Theorem 1.1.1. We note that this is a theorem about absolute values, so we reread Definition 1.1.1 to insure that we know the meaning of this concept. Since the absolute value of a number a depends upon the sign of a, we should test the claimed inequality in the theorem with several pairs of numbers: two positive numbers, two negative numbers, and two numbers of opposite sign. The reader should *do this*, with examples of his or her choice of numbers, noting that the triangle inequality in real application gives either *equality*, if the two numbers have the same sign, or else strict inequality, if the two numbers have opposite sign. This gives us an intuitive appreciation that the triangle inequality ought to be true. Now how do we prove it? Testing more examples will not suffice, because infinitely many pairs are possible. Many correct proofs can be given, but we will discuss the one chosen by the author.

The next step in writing a proof requires some playfulness or inquisitiveness on the part of the student. In theoretical mathematics we are discouraged from following rote procedures in the hope of finding an answer without thought. To bypass thought would be to bypass mathematics itself. The student should not even consider such a route, just as he or she should not substitute a pill for a good meal.

We see by playing with the definition of absolute value that $|a|$ must be *equal* to either a or $-a$. This reminds us of what we observed when checking pairs of specific numbers of the same or opposite sign, as explained above. The playfulness appears when we choose to write this as $-|a| \leq a \leq |a|$ for all a, even though the truth of this double inequality hinges upon a being equal to either the left side or the right side. Then we do the same for b, recognizing that a and b do play symmetrical roles in the statement of the theorem. Then we add the two double inequalities, obtaining Equation (1.1). The remainder of the proof unfolds from considering that the value of $|a + b|$ hinges upon the sign of $a + b$.

This analysis of the proof of the triangle inequality is representative of what the student should do with each proof in this book, and with each proof presented in class by his or her professor. Take a fresh sheet of paper and write out a full analysis of the proof, including the perceived rationale for the course that it takes. Work on this until you are sure you understand correctly. If in doubt, ask your teacher! This is the way to learn advanced mathematics, and it is what the student must do to learn to prove theorems.

EXERCISES

1.1   Let $\epsilon > 0$. Determine how large $n \in \mathbb{N}$ must be to ensure that the given inequality is satisfied, and use the Archimedean Property to establish that such n exist.

    a) $\frac{1}{n} < \epsilon$?

    b) $\frac{1}{n^2} < \epsilon$?

    c) $\frac{1}{\sqrt{n}} < \epsilon$? (Assume that $\sqrt{n}$ exists in $\mathbb{R}$.)

1.2   Prove the uniqueness of the additive inverse $-a$ of a. (Hint: Suppose that

$$x + a = 0 = y + a$$

and prove that $x = y$.)

1.3   Use the Axiom of Distributivity to prove that $a0 = 0$ for all $a \in \mathbb{R}$, and use this to prove that $(-1)(-1) = 1$.

1.4   Prove that $(-1)a = -a$ for all $a \in \mathbb{R}$.

1.5   Prove the uniqueness of the multiplicative inverse a^{-1} of a for all $a \neq 0$ in $\mathbb{R}$.

1.6   Prove: For all a and b in $\mathbb{R}$, $|ab| = |a||b|$. (Hint: Consider the three cases a and b both nonnegative, a and b both negative, and a and b of opposite sign.)

1.7   Prove: For all a, b, c in $\mathbb{R}$,

$$|a - c| \leq |a - b| + |b - c|.$$

(Hint: Use the triangle inequality.)

1.8   Let $\epsilon > 0$. Find a number $\delta > 0$ small enough so that $|a-b| < \delta$ and $|c-b| < \delta$ implies $|a - c| < \epsilon$.

$|a-b-(c-b)| = |a-b+b-c| \leq |a-b| + |b-c|$

$\leq 2\delta$

1.9   † Prove: For all a and b in $\mathbb{R}$,

let $\delta = \frac{\epsilon}{2}$

$$||a| - |b|| \leq |a - b|.$$

Intuitively, this says that $|a|$ and $|b|$ cannot be farther apart than a and b are. (Hint: Write $|a| = |(a - b) + b|$ and use the triangle inequality. Then do the same thing for $|b|$.)

1.10   Prove or give a counterexample:
 a) If $a < b$ and $c < d$, then $a - c < b - d$.
 b) If $a < b$ and $c < d$, then $a + c < b + d$.

1.11   † This exercise leads in three parts to a proof that there is no rational number the square of which is 2. The reader will need to know from another source that each rational number can be written in the form $\frac{m}{n}$ in *lowest terms*. This means that m and n have no common factors other than ± 1.
 a) If $m \in \mathbb{Z}$ is odd, prove that m^2 is odd.
 b) If $m \in \mathbb{Z}$ is such that m^2 is even, prove that m is even.
 c) Suppose there exists $\frac{m}{n} \in \mathbb{Q}$, expressed in *lowest terms*, such that

$$\left(\frac{m}{n}\right)^2 = 2.$$

Prove that m and n are both even, resulting in a contradiction.
   (Hint: For this problem, if the student has not taken any class in number theory, the following definitions may be helpful. A number n is called *even* if and only if it can be written as $n = 2k$ for some integer k. A number n is called *odd* if and only if it can be written as $n = 2k - 1$ for some integer k.)

1.2   LIMITS OF SEQUENCES & CAUCHY SEQUENCES

By a *sequence* x_n of elements of a set S we mean that to each natural number $n \in \mathbb{N}$ there is assigned an element $x_n \in S$. Unless otherwise stated, we will deal with sequences of real numbers. We can think of a sequence as an endless list of real numbers, or we could equivalently think of a sequence as being a *function* whose domain is $\mathbb{N}$ and whose range lies in $\mathbb{R}$. It is very important to define the concept of the *limit* of a sequence. Intuitively, we say that x_n *approaches* the *real* number L as n *approaches infinity*, written $x_n \to L \in \mathbb{R}$ as $n \to \infty$, provided we can force $|x_n - L|$ to become as small as we like just by making n sufficiently big. This is also written with the symbols $\lim_{n \to \infty} x_n = L$. The advantage of writing the definition symbolically as follows is that this definition provides inequalities that can be solved to determine whether or not $x_n \to L$.

Definition 1.2.1 *A sequence* $x_n \to L \in \mathbb{R}$ *as* $n \to \infty$ *if and only if for all* $\epsilon > 0$, *there exists* $N \in \mathbb{N}$ *corresponding to* ϵ *such that*

$$n \geq N \Rightarrow |x_n - L| < \epsilon.$$

If there exists a number L such that $x_n \to L$, *we say x_n is* convergent. *Otherwise we say that x_n is* divergent.

See Exercise 1.12.

$\left| \frac{1}{n} - 0 \right| < \varepsilon$, $\frac{1}{n} < \varepsilon$, $1 < \varepsilon n$, $\frac{1}{\varepsilon} < n$

Pick $N > \frac{1}{\varepsilon}$

■ **EXAMPLE 1.1**

We claim that if $x_n = \frac{1}{n}$, then $x_n \to 0$.

Proof: Let $\epsilon > 0$. We need $N \in \mathbb{N}$ such that $n \geq N$ implies

$$\left| \frac{1}{n} - 0 \right| < \epsilon.$$

That is, we need to solve the inequality $\frac{1}{n} < \epsilon$. Multiplying both sides of this inequality by the positive number $\frac{n}{\epsilon}$, we see that $\frac{1}{\epsilon} < n$. That is, if we pick $N \in \mathbb{N}$ such that $N > \frac{1}{\epsilon}$, then

$$n \geq N \implies \frac{1}{n} \leq \frac{1}{N} < \epsilon.$$

We know that such an N exists in $\mathbb{N}$ since ϵ and 1 are both positive. Thus there exists $N \in \mathbb{N}$ such that $N1 = N > \frac{1}{\epsilon}$ by the Archimedean Principle.   ■

The student should note that the value of N does indeed correspond to ϵ. If $\epsilon > 0$ is made smaller, then N must be chosen larger.

■ **EXAMPLE 1.2**

Let $|r| < 1$. We claim that $r^n \to 0$ as $n \to \infty$.
   Let $\epsilon > 0$. We need to find $N \in \mathbb{N}$ such that $n \geq N$ implies

$$|r^n - 0| = |r|^n < \epsilon. \qquad \frac{1}{|r|^n} > \frac{1}{\varepsilon}$$

In the special case in which $r = 0$, it would suffice to take $N = 1$. So suppose $r \neq 0$. Then we need to solve

$$\left(\frac{1}{|r|}\right)^n > \frac{1}{\epsilon}.$$

Note that we do not proceed by taking nth roots of both sides of this inequality, since we have not yet established the existence of such roots for all positive real numbers. Since $|r| < 1$, $\frac{1}{|r|} = 1 + p > 1$ for some $p > 0$. Thus

$$\left(\frac{1}{|r|}\right)^n = (1 + p)^n$$
$$= (1 + p)(1 + p) \cdots (1 + p)$$
$$= 1^n + np + \cdots + p^n$$
$$> np.$$

By transitivity of inequalities, it would suffice to find $N \in \mathbb{N}$ such that $Np > \frac{1}{\epsilon}$. Such integers N exist because of the Archimedean property. So pick $N \in \mathbb{N}$ such $Np > \frac{1}{\epsilon}$ and we find that $n \geq N$ implies $np \geq Np > \frac{1}{\epsilon}$ so that $|r^n - 0| = |r|^n < \epsilon$.

Notice that if x_n is convergent, then after some finite number N of terms, all subsequent terms are bunched very close to one another: in fact, within ϵ of some number L. This motivates the following definition and theorem.

Definition 1.2.2 *A sequence x_n is called a Cauchy sequence if and only if, for all $\epsilon > 0$, there exists $N \in \mathbb{N}$, corresponding to ϵ, such that n and $m \geq N$ implies $|x_n - x_m| < \epsilon$.*

Theorem 1.2.1 *If x_n is any convergent sequence of real numbers, then x_n is a Cauchy sequence.*

Proof: Suppose x_n is convergent: say $x_n \to L$. Let $\epsilon > 0$. Then, since $\frac{\epsilon}{2} > 0$ as well, we see there exists $N \in \mathbb{N}$, *corresponding to ϵ,* such that $n \geq N$ implies $|x_n - L| < \frac{\epsilon}{2}$. Then, if n and $m \geq N$, we have

$$|x_n - x_m| = |(x_n - L) + (L - x_m)|$$
$$\leq |x_n - L| + |L - x_m|$$
$$< \frac{\epsilon}{2} + \frac{\epsilon}{2} = \epsilon.$$

∎

Remark 1.2.1 We make some remarks here to help the student to write his or her own detailed analysis of the proof of Theorem 1.2.1, as recommended in the introduction, on page xxiii. The student should begin with the intuitive understanding that if $x_n \to L$, then x_n will be very close to L for all sufficiently big n. The point is that we want both x_n and x_m to be so close to L that x_n and x_m must be within ϵ of one another. The student should use visualization to recognize that since x_n and x_m can be on opposite sides of L, we will need both x_n and x_m to be within $\frac{\epsilon}{2}$ of L. Then the triangle inequality for real numbers assures that x_n and x_m are no more than ϵ apart. The student should write a careful analysis of every proof in this course, whether proved in the text or by the professor in class.

■ **EXAMPLE 1.3**

We claim the sequence $x_n = (-1)^{n+1}$ is divergent.
In fact, if x_n were convergent, then x_n would have to be Cauchy. But $|x_n - x_{n+1}| \equiv 2$, for all n. Thus, if $0 < \epsilon \leq 2$, it is impossible to find $N \in \mathbb{N}$ such that n and $m > N$ implies $|x_n - x_m| < \epsilon$.

Definition 1.2.3 *A sequence x_n is called* bounded *if and only if there exists $M \in \mathbb{R}$ such that $|x_n| \leq M$, for all $n \in \mathbb{N}$.*

Theorem 1.2.2 *If x_n is Cauchy, then x_n must be bounded.*

Remark 1.2.2 Observe that if x_n is convergent, then it is Cauchy, so this theorem implies that every convergent sequence is bounded.

Cauchy holds $\forall \epsilon > 0$

Proof: We will show that every Cauchy sequence is bounded. In fact, taking $\epsilon = 1$, we see that there exists $N \in \mathbb{N}$ such that n and $m \geq N$ implies $|x_n - x_m| < 1$. In particular, $n \geq N$ implies

$|x_n - x_m| < \epsilon$

$$|x_n| - |x_N| \leq \left||x_n| - |x_N|\right| \leq |x_n - x_N| < 1 \qquad \Rightarrow \text{ Take } m = N,$$

$|x_n - x_N| < 1$

so that $|x_n| < 1 + |x_N|.$ If we let

$\forall n \geq N$

$$M = \max\left\{|x_1|, \ldots, |x_{N-1}|, 1 + |x_N|\right\},$$

making M the largest element of the indicated set of N numbers, then $|x_n| \leq M$ for all $n \in \mathbb{N}$.   ■

■ **EXAMPLE 1.4**

If $x_n = n$, then x_n is *not* convergent.
If x_n were convergent, then x_n would be bounded. But for all $M > 0$, there exists $n \in \mathbb{N}$, corresponding to M, such that $n > M$ by the Archimedean Property. So x_n is not bounded.
It is also convenient to define the concepts $x_n \to \infty$ and $x_n \to -\infty$. However, ∞ is not a real number, so we have not defined anything like $|x_n - \infty|$ and thus cannot prove such a difference is less than ϵ. (Compare this with the discussion on page 9.) We adopt the following definition.

Definition 1.2.4 *We write* $x_n \to \infty$ *if and only if for all* $M > 0$ *there exists* $N \in \mathbb{N}$ *such that* $n \geq N$ *implies* $x_n > M$. *Similarly, we write* $x_n \to -\infty$ *if and only if for all* $m < 0$ *there exists* $N \in \mathbb{N}$ *such that* $n \geq N$ *implies* $x_n < m$.

EXERCISES

1.12   † Use Definition 1.2.1 to prove that the limit of a convergent sequence x_n is unique. That is, prove that if $x_n \to L$ and $x_n \to M$ then $L = M$.

1.13   Let

$$x_n = \begin{cases} 0 & \text{if } n < 100, \\ 1 & \text{if } n \geq 100. \end{cases}$$

Prove that x_n converges and find $\lim x_n$.

1.14   Let $x_n = \frac{n-1}{n}$. Prove x_n converges and find the limit.

1.15   Let $x_n = \frac{(-1)^n}{\sqrt{n}}$. Prove x_n converges and find the limit.

1.16   Let $x_n = \frac{1}{n^2}$. Prove x_n converges and find the limit.

1.17   Let $x_n = \frac{n^2-n}{n}$. Does x_n converge or diverge? Prove your claim.

1.18   Let $x_n = \frac{(-1)^n+1}{n}$. Does x_n converge or diverge? Prove your claim.

1.19   † Prove: If $s_n \leq t_n \leq u_n$ for all n and if both $s_n \to L$ and $u_n \to L$ then $t_n \to L$ as $n \to \infty$ as well. (This is sometimes called the *squeeze theorem* or the *sandwich theorem* for sequences.)

1.20   Prove or give a counterexample:
    a) $x_n + y_n$ converges if and only if both x_n and y_n converge.
    b) $x_n y_n$ converges if and only if both x_n and y_n converge.
    c) If $x_n y_n$ converges, then $\lim x_n y_n = \lim x_n \lim y_n$.

1.21   Let $x_n = \frac{\sin n}{n}$. Prove x_n converges, and find the limit.

1.22   † Suppose $a \leq x_n \leq b$ for all n and suppose further that $x_n \to L$. Prove: $L \in [a, b]$. (Hint: If $L < a$ or if $L > b$, obtain a contradiction.)

1.23   Suppose $s_n \leq t_n \leq u_n$ for all n, $s_n \to a < b$, and $u_n \to b$. Prove or give a counterexample: $\lim_{n\to\infty} t_n \in [a, b]$.

1.24   For each of the following sequences:

    i. Determine whether or not the sequence is Cauchy and explain why.

    ii. Find $\lim_{n\to\infty} |x_{n+1} - x_n|$.

       a) $x_n = (-1)^n n$
       b) $x_n = n + \frac{1}{n}$
       c) $x_n = \frac{1}{n^2}$

d) x_n is described as follows:

$$0, 1, \frac{1}{2}, 0, \frac{1}{3}, \frac{2}{3}, 1, \frac{3}{4}, \frac{1}{2}, \frac{1}{4}, 0, \frac{1}{5}, \frac{2}{5}, \frac{3}{5}, \frac{4}{5}, 1, \dots .$$

1.25 † Prove: The sequence x_n is *Cauchy* if and only if for all $\epsilon > 0$ there exists $N \in \mathbb{N}$ such that for all $k \geq N$, we have $|x_k - x_N| < \epsilon$.

1.26 Prove that if $x_n \to \infty$ then x_n is not Cauchy.

1.27 Let $x_n \neq 0$, for all $n \in \mathbb{N}$. Prove: $|x_n| \to \infty$ if and only if $\frac{1}{|x_n|} \to 0$.

1.3 THE COMPLETENESS AXIOM AND SOME CONSEQUENCES

Consider the following sequence of decimal approximations to $\sqrt{2}$:

$$x_1 = 1, \ x_2 = 1.4, \ x_3 = 1.41, \ x_4 = 1.414, \dots .$$

Each x_k is a rational number, having only finitely many nonzero decimal places. For each k, the last nonzero decimal digit of x_k is selected in such a way that $x_k^2 < 2$ yet if that last digit were one bigger the square would be larger than 2. The number x_k^2 cannot equal 2, since there is no $\sqrt{2}$ in the rational number system. Naturally we hope for x_k to converge and for $\lim x_k = \sqrt{2}$. Indeed, x_k is a Cauchy sequence. We can see this by observing that if m and n are greater than or equal to N, then

$$|x_m - x_n| < \frac{1}{10^{N-1}}.$$

Since the sequence of successive powers of $\frac{1}{10}$ converges to 0, if $\epsilon > 0$ we can pick N large enough to ensure that $\frac{1}{10^{N-1}} < \epsilon$.

Since there is no $\sqrt{2}$ in $\mathbb{Q}$, there are Cauchy sequences in $\mathbb{Q}$ that have no limit in the set $\mathbb{Q}$ of rational numbers. It is reasonable, knowing from geometrical considerations that there *should* be a $\sqrt{2} \in \mathbb{R}$, to select the following axiom as the final axiom for the real number system.

Completeness Axiom of $\mathbb{R}$. Every Cauchy sequence of real numbers has a limit in the set $\mathbb{R}$ of real numbers.

In Example 1.10 we will see that in fact the completeness axiom does imply that there exists a $\sqrt{2}$ in $\mathbb{R}$.

Remark 1.3.1 In books that use a different but equivalent version of the Completeness Axiom, the statement that every Cauchy sequence of real numbers converges to a real number is called the *Cauchy Criterion for sequences*.

Definition 1.3.1 *The set $\mathbb{R}$ of real numbers is an Archimedean ordered field satisfying the Completeness Axiom.*

Thus a sequence of real numbers converges if and only if it is Cauchy. We remark that it can be proven, although we will not do so here, that any two complete

Archimedean ordered fields must be *isomorphic* in the sense of algebra. The interested reader can find a proof in the book [16] by Olmsted. On the other hand, the reader can find an *explicit construction* of a set having all the properties of a complete Archimedean ordered field, beginning from the natural numbers, in the book [12] by Landau.

In the next chapter, after studying the Intermediate Value Theorem, we will see easily that $\mathbb{R}$, with the Completeness Axiom, does possess an $\sqrt[n]{p}$ for each $p > 0$ and for all $n \in \mathbb{N}$. Most of the current chapter, however, will deal with other consequences of completeness, that we will begin exploring right now.

Definition 1.3.2 *A number M is called an* upper bound *for a set $A \subset \mathbb{R}$ if and only if for all $a \in A$ we have $a \leq M$. Similarly, a number m is called a* lower bound *for A if and only if for all $a \in A$ we have $a \geq m$. A set A of real numbers is called* bounded *provided that it has both an upper bound and a lower bound. A* least upper bound *for a set A is an upper bound L for A with the property that no number $L' < L$ is an upper bound of A. A least upper bound is denoted by* $\text{lub}(A)$.

Note that not every subset of $\mathbb{R}$ has an upper or a lower bound. For example, $\mathbb{N}$ has no upper bound, and $\mathbb{Z}$ has neither an upper nor a lower bound. It is important to bear in mind also that many *bounded* sets of real numbers have neither a largest nor a smallest element. For example, this is true for the set of numbers in the *open* interval $(0,1)$. The reader should prove this claim as an informal exercise.

Theorem 1.3.1 *If a nonempty set S has an upper bound, then S has a least upper bound L.*

Remark 1.3.2 If S has an upper bound, then its least upper bound is denoted by $\text{lub}(S)$. If $\text{lub}(S)$ exists, then it must have a unique value L. The reader should prove that no number greater or smaller than L could satisfy the definition of $\text{lub}(S)$.

Proof: Since $S \neq \emptyset$, there exists $s \in S$. Select any number $a_1 < s$ so that a_1 is too small to be an upper bound for S. Let b_1 be any upper bound of S. We will use a process known as *interval halving*, in which we will cut the interval $[a_1, b_1]$ in half again and again without end. The midpoint between a_1 and b_1 is $\frac{a_1+b_1}{2}$.

   i. If $\frac{a_1+b_1}{2}$ is an upper bound for S, then let $b_2 = \frac{a_1+b_1}{2}$ and let $a_2 = a_1$.

   ii. But if $\frac{a_1+b_1}{2}$ is *not* an upper bound for S, then let $a_2 = \frac{a_1+b_1}{2}$ and let $b_2 = b_1$.

Thus we have chosen $[a_2, b_2]$ to be one of the two half-intervals of $[a_1, b_1]$, and we have done this in such a way that b_2 is again an upper bound of S and a_2 is too small to be an upper bound for S. Now we cut $[a_2, b_2]$ in half and select a half-interval of it to be $[a_3, b_3]$ in the same way we did for $[a_2, b_2]$. Note that

$$|b_N - a_N| = \frac{|b_1 - a_1|}{2^{N-1}} \to 0$$

as $N \to \infty$. Thus if $\epsilon > 0$, there exists $N \in \mathbb{N}$, *corresponding to* ϵ, such that $|b_N - a_N| < \epsilon$. But, if $n \geq N$, then a_n and $b_n \in [a_N, b_N]$, so n and $m \geq N$ implies $|a_n - a_m| < \epsilon$ and also $|b_n - b_m| < \epsilon$. Thus a_n and b_n are Cauchy sequences. Hence $a_n \to a$ and $b_n \to b$, for some real numbers a, b. By Exercise 1.22, a and b are in $[a_N, b_N]$, for all N. Thus $0 \leq |a - b| < \epsilon$, for all $\epsilon > 0$. Thus $|a - b| = 0$ and $a = b$. We claim that the number $L = a = b$ is the least upper bound of S. Note that for each k we have $a_k \leq L \leq b_k$, since for all $j \geq k$ we have a_j and $b_j \in [a_k, b_k]$.

First, observe that if $s \in S$, then $s \leq L$. In fact, if we did have $s > L$, then, since $b_k \to L$, for some big enough value of k we would have $|b_k - L| < |s - L|$ and so $b_k < s$. But this is impossible, since b_k is an upper bound of S. Thus $s \leq L$ and L is an upper bound of S.

Finally, we claim L is the least upper bound of S. In fact, suppose $L' < L$. Then since $a_k \to L$, there exists k such that $L' < a_k$. But a_k is not an upper bound of S. Thus L' cannot be an upper bound of S. ∎

Remark 1.3.3 The proof of Theorem 1.3.1 is the most difficult proof presented thus far in this book. It proceeds by the method of *interval-halving*. This method can be likened to the way that a first baseman and a second baseman in a baseball game will attempt to tag a base-runner out by throwing the ball back and forth between them, steadily reducing the distance between them until one baseman is close enough to tag the runner. Interval halving is a very useful method of calculating roots of equations with a computer, provided it is possible to tell from the endpoints of each half-interval which half would need to contain the root. The student should take careful note of how the method of interval-halving produces two natural Cauchy sequences, a_n and b_n, corresponding to the left and right endpoints of the selected half-intervals.

Corollary 1.3.1 *If S is any* nonempty *set of real numbers that has a lower bound, then S has a greatest lower bound.*

For the *proof* see Exercise 1.28 in this section.

Remark 1.3.4 If S has a lower bound, then its greatest lower bound is denoted by $\text{glb}(S)$.

Since not every subset $S \subset \mathbb{R}$ has either an upper or a lower bound, least upper bounds and greatest lower bounds do not exist in every case. Thus we introduce the concepts of the *supremum* and the *infimum* of an arbitrary set $S \subset \mathbb{R}$.

Definition 1.3.3 *Let S be any nonempty subset of $\mathbb{R}$. Define the* supremum *of S, denoted* $\sup(S)$, *to be the least upper bound of S if S is bounded above and define* $\sup(S) = \infty$ *if S has no upper bound. Similarly, define the* infimum *of S, denoted* $\inf(S)$, *to be the greatest lower bound of S if S is bounded below, and define* $\inf(S) = -\infty$ *if S has no lower bound.*

Thus

$$\sup(S) = \begin{cases} \text{lub}(S) & \text{if S is bounded above,} \\ \infty & \text{if S is not bounded above} \end{cases}$$

and

$$\inf(S) = \begin{cases} \text{glb}(S) & \text{if S is bounded below,} \\ -\infty & \text{if S is not bounded below.} \end{cases}$$

■ **EXAMPLE 1.5**

Let $S = \{x \mid 0 < x < 1\} = (0,1)$. Then $\sup(S) = 1$ and $\inf(S) = 0$.

Proof: Clearly, 1 is an upper bound of S. But if $M < 1$, then there exists $x \in S \cap (M,1)$. Thus M cannot be an upper bound of S. Hence 1 is the least upper bound of S. The argument for $\inf(S)$ is similar.   ■

■ **EXAMPLE 1.6**

Observe that $\sup(\mathbb{N}) = \infty$ and $\inf(\mathbb{N}) = 1$. This follows because $\mathbb{N}$ has no upper bound, but $\mathbb{N}$ does have a least element, namely 1.

Definition 1.3.4 *We call a sequence x_n increasing provided $x_n \leq x_{n+1}$ for all $n \in \mathbb{N}$, and then we write this symbolically as*

$$x_n \nearrow .$$

Similarly, we call x_n a decreasing sequence if $x_n \geq x_{n+1}$ for all $n \in \mathbb{N}$, which we denote by

$$x_n \searrow .$$

In either case, we call x_n a monotone sequence. Similarly, if $x_n < x_{n+1}$ for all $n \in \mathbb{N}$, we write

$$x_n \uparrow$$

and call x_n strictly monotone increasing. And if $x_n > x_{n+1}$ for all $n \in \mathbb{N}$, we write

$$x_n \downarrow$$

and call x_n strictly monotone decreasing.

 For $\mathbb{R}^\ast$ up

Theorem 1.3.2 *If x_k is an increasing sequence, then $x_k \to \sup\{x_n\}$. Similarly, if x_k is a decreasing sequence, then $x_k \to \inf\{x_n\}$.*

Remark 1.3.5 If $\sup\{x_n\} = L$, a real number, then this theorem says the increasing sequence $x_n \to L$ and this is an instance of convergence. But if $\sup\{x_n\} = \infty$, we write $x_n \to \infty$, but this is called *divergence to infinity*. We do not consider the latter circumstance as convergence because we cannot make $|x_n - \infty| < \epsilon$. In fact, $x_n - \infty$ is *meaningless*, since ∞ is not a real number and the arithmetic operations of real numbers are not defined for ∞. Similar remarks apply if x_n is a decreasing sequence.

Proof: Consider the case of x_n increasing. If $\{x_n\}$ is not bounded above, so that the supremum is infinite, we see that for all $M \in \mathbb{R}$ there exists N such that $x_N > M$. Then, $n \geq N$ implies $x_n \geq x_N > M$ too, and we call this divergence of x_n to infinity, denoted by $x_n \to \infty$.

Now suppose x_n is bounded above, so $\sup\{x_n\} = L$ is the least upper bound of the set of numbers $\{x_n\}$. We must show that $x_n \to L$. Let $\epsilon > 0$. Since $L - \epsilon < L$, $L - \epsilon$ cannot be an upper bound of $\{x_n\}$, so there exists N such that $L \geq x_N > L - \epsilon$. Thus for all $n \geq N$ we have

$$L - \epsilon < x_N \leq x_n \leq L,$$

so $n \geq N$ implies $|x_n - L| < \epsilon$; that is, $x_n \to L$.

The case in which x_n decreases is Exercise 1.29. ∎

Corollary 1.3.2 *A monotone sequence converges if and only if it is bounded.*

Proof: Exercise 1.30.

One inconvenience in the concept of limit is that $\lim x_n$ does not exist for every sequence x_n. One may not be sure in advance whether a given sequence is convergent or divergent. However, there are two related concepts called the *Limit Superior*[4] and the *Limit Inferior* which are always defined.

Definition 1.3.5 *Let x_n be any sequence of real numbers. Denote $T_n = \{x_k \mid k \geq n\}$, which we call the nth* tail *of the sequence x_n.*

Note that
$$T_1 \supseteq T_2 \supseteq \ldots \supseteq T_n \supseteq \ldots.$$

Define
$$i_n = \inf(T_n) \text{ and } s_n = \sup(T_n).$$

It is easy to see that $i_n \leq s_n$, for all n. Moreover, as n increases, the set T_n of which one takes sup or inf shrinks to a subset of what it was the step before. Thus i_n increases and s_n decreases. Consequently, $i_k \to \sup\{i_n \mid n \in \mathbb{N}\}$ and $s_k \to \inf\{s_n \mid n \in \mathbb{N}\}$. Recall that this horizontal-arrow notation means convergence if the sequence is approaching a real number, but it indicates a special type of divergence if the sequence is approaching plus or minus infinity.

Definition 1.3.6 *We define the* limit superior *of x_n by*

$$\limsup x_n = \inf\{s_n \mid n \in \mathbb{N}\} = \inf\{\sup(T_n) \mid n \in \mathbb{N}\}$$

and we define the limit inferior *of x_n*

$$\liminf x_n = \sup\{i_n \mid n \in \mathbb{N}\} = \sup\{\inf(T_n) \mid n \in \mathbb{N}\},$$

---

[4]The lim sup and lim inf appear only occasionally in this book, but the concepts are presented because they are intrinsically interesting. Also they are very useful to know for further study in graduate courses. On the other hand, the sup, inf, lub, and glb appear often and are needed throughout this book.

where T_n is the nth tail of the sequence x_n.

Of course, lim sup and lim inf may be real numbers or they may be $\pm\infty$.

Theorem 1.3.3 *Let $L \in \mathbb{R}$ and let x_n be a sequence of real numbers. Then $x_n \to L$ if and only if* $\limsup x_n = L = \liminf x_n$.

Proof: First, suppose $x_n \to L$. Thus if $\epsilon > 0$ there exists $N \in \mathbb{N}$ such that $n \geq N$ implies $|x_n - L| < \epsilon/2$, which implies $s_n = \sup(T_n) \leq L + \epsilon/2$ and $i_n = \inf(T_n) \geq L - \frac{\epsilon}{2}$. Thus $-\epsilon/2 < x_n - L < \epsilon/2$

$$L - \frac{\epsilon}{2} \leq i_n \leq s_n \leq L + \frac{\epsilon}{2}$$

which implies that $|s_n - L| \leq \frac{\epsilon}{2} < \epsilon$ and $|i_n - L| \leq \frac{\epsilon}{2} < \epsilon$, for all $\epsilon > 0$. Thus $s_n \to L = \limsup x_n$ and $i_n \to L = \liminf x_n$.
   For the opposite implication, suppose $\limsup x_n = \liminf x_n = L \in \mathbb{R}$. Thus there exists N_1 such that $n \geq N_1$ implies $\sup(T_n) \leq L + \frac{\epsilon}{2}$ and there exists N_2 such that $n \geq N_2$ implies $\inf(T_n) \geq L - \frac{\epsilon}{2}$. Let $N = \max\{N_1, N_2\}$, and $n \geq N$ implies $|x_n - L| \leq \epsilon/2 < \epsilon$. Thus $x_n \to L$.   ■

EXERCISES

1.28   † Prove Corollary 1.3.1. (Hint: Let $-S = \{-s \mid s \in S\}$. Which theorem can you apply to the set $-S$?)

1.29   † Prove the case in which x_n decreases in Theorem 1.3.2.

1.30   † Prove Corollary 1.3.2.

1.31   Find $\sup(S)$ and $\inf(S)$ for each set S below, and justify your conclusions.
   a) $S = \{(-1)^n \mid n \in \mathbb{N}\}$.
   b) $S = \{(-1)^n n \mid n \in \mathbb{N}\}$.
   c) $S = \{x \in \mathbb{R} \mid x^2 < 1\}$.

1.32   Suppose A and B are subsets of $\mathbb{R}$, both nonempty, with the special property that $a \leq b$ for all $a \in A$ and for all $b \in B$. Prove: $\sup(A) \leq \inf(B)$. (Hint: Every b is an upper bound of A. So how does the $\sup(A)$ relate to each $b \in B$?)

1.33   Prove that every real number $M \in \mathbb{R}$ is both an upper bound and a lower bound of the empty set, $\emptyset$.

1.34   Let $x_n = \frac{n-1}{n}$. Show that x_n is convergent and find $\lim x_n$. Justify your conclusions.

1.35   Let $x_n = (1.5)^n$, for all $n \in \mathbb{N}$. Find $\sup(T_n)$ and $\inf(T_n)$, where T_n is the nth *tail* of the sequence, and explain. Find $\liminf x_n$ and $\limsup x_n$.

1.36   Prove or give a counterexample: If x_n increases and y_n increases, then $(x_n + y_n)$ is monotone.

1.37   Prove or give a counterexample: if x_n increases and y_n increases then $(x_n - y_n)$ is monotone.

1.38   Prove or give a counterexample: if x_n increases and y_n increases then the product $(x_n y_n)$ is monotone.

1.39   Prove: If x_n is a *constant* sequence if and only if x_n is *both* monotone increasing and monotone decreasing.

1.40   Let $x_n = \frac{(-1)^n}{n}$. Find $\inf(T_n)$, $\sup(T_n)$, $\limsup x_n$, and $\liminf x_n$. Does $\lim_{n \to \infty} x_n$ exist? (Hint: T_n is the nth *tail* of the sequence x_n.)

1.41   Let $x_n = (-1)^n + \frac{1}{n}$. Find $\limsup x_n$ and $\liminf x_n$. Does $\lim_{n \to \infty} x_n$ exist?

1.42   Give an example of a sequence $x_n \to \infty$ for which x_n is *not monotone*.

1.43   Let x_n be any sequence of real numbers. Prove: x_n diverges to ∞ if and only if $\liminf x_n = \limsup x_n = \infty$.

1.44   Let x_n be any sequence of real numbers. Prove: x_n diverges to $-\infty$ if and only if $\liminf x_n = \limsup x_n = -\infty$.

1.45   Prove that $\liminf x_n \leq \limsup x_n$, for every *bounded* sequence x_n of real numbers. (Hint: The result of problem 5 may help.)

1.46   Let x_n be any *unbounded* sequence of real numbers. Let s_n and i_n be defined as in the proof of Theorem 1.3.3.

    a) If $\{x_n \mid n \in \mathbb{N}\}$ has no upper bound, prove $s_n = \infty$ for all n, so that $\limsup x_n = \infty$.

    b) If $\{x_n \mid n \in \mathbb{N}\}$ has no lower bound, prove $i_n = -\infty$ for all n, so that $\liminf x_n = -\infty$.

    c) In either of the two cases above, conclude that

$$\liminf x_n \leq \limsup x_n.$$

1.4   ALGEBRAIC COMBINATIONS OF SEQUENCES

If s_n is some algebraic combination of other sequences, then we may be able to determine whether or not s_n converges if we know the behavior of the other sequences of which s_n is composed.

Theorem 1.4.1 *Suppose x_n and y_n both converge, with $x_n \to L$ and $y_n \to M$ as $n \to \infty$. Then*

    i. $x_n + y_n \to L + M$.

    ii. $x_n - y_n \to L - M$.

    iii. $x_n y_n \to LM$.

iv. $\frac{x_n}{y_n} \to \frac{L}{M}$, provided that $M \neq 0$ and $y_n \neq 0$, for all $n \in \mathbb{N}$.

In order to prove this four-part theorem, it is helpful first to introduce the following definition and the two lemmas that follow it.

Definition 1.4.1 *A sequence that converges to zero is called a* null *sequence.*

Lemma 1.4.1 *The sequence $x_n \to L \in \mathbb{R}$ if and only if $x_n - L \to 0$.*

Proof of Lemma. We remark that in words we are proving that $x_n \to L$ if and only if $x_n - L$ is a null sequence. By definition, $x_n \to L \in \mathbb{R}$ if and only if for all $\epsilon > 0$ there exists $N \in \mathbb{N}$, *corresponding to* ϵ, such that $n \geq N$ implies $|x_n - L| < \epsilon$. This is equivalent to $|(x_n - L) - 0| < \epsilon$, which is equivalent to the statement that $(x_n - L) \to 0$, since $|x_n - L| = |(x_n - L) - 0|$. ∎

Lemma 1.4.2 *If $s_n \to 0$ and if t_n is bounded, then $s_n t_n \to 0$.*

Proof: We are proving that a null sequence times a bounded sequence must be a null sequence. There exists $M > 0$ such that $|t_n| \leq M$, for all $n \in \mathbb{N}$. Let $\epsilon > 0$. Since $s_n \to 0$, there exists N such that $n \geq N$ implies $|s_n - 0| = |s_n| < \frac{\epsilon}{M}$. Now, $n \geq N$ implies

$$|s_n t_n - 0| = |s_n t_n| = |s_n||t_n| < \frac{\epsilon}{M} M = \epsilon.$$

∎

With the preceding definition and two lemmas in hand, we proceed to the main task of proving the theorem.

Proof:

i. Let $\epsilon > 0$. There exists N_1 such that $n \geq N_1$ implies $|x_n - L| < \epsilon/2$, and there exists N_2 such that $n \geq N_2$ implies $|y_n - M| < \epsilon/2$. Now let $N = \max\{N_1, N_2\}$. Then $n \geq N$ implies

$$|(x_n + y_n) - (L + M)| \leq |x_n - L| + |y_n - M| < \epsilon.$$

ii. This proof is almost identical to the preceding case.

iii. Since y_n converges, y_n is bounded. And

$$
\begin{aligned}
x_n y_n - LM &= x_n y_n - L y_n + L y_n - LM \\
&= (x_n - L) y_n + L(y_n - M) \\
&\to 0 + 0 = 0
\end{aligned}
$$

using the two lemmas and the first part, proven above.

iv. Because of the third part, proven above, it suffices to prove that $\frac{1}{y_n} \to \frac{1}{M}$. But

$$\left| \frac{1}{y_n} - \frac{1}{M} \right| = \frac{|y_n - M|}{|y_n M|} = |y_n - M| \frac{1}{|y_n M|}.$$

Since $|y_n - M| \to 0$, it suffices to show $\frac{1}{|y_n M|}$ is bounded. There exists N such that $n \geq N$ implies $|y_n - M| < \frac{|M|}{2}$. Thus $|y_n| > \frac{|M|}{2}$ and $\frac{1}{|y_n M|} < \frac{2}{|M|^2}$. Thus $\frac{1}{|y_n M|}$ is bounded by $\max \left\{ \frac{1}{|y_1 M|}, \ldots, \frac{1}{|y_{N-1} M|}, \frac{2}{|M|^2} \right\}$.

∎

EXERCISES

1.47 Give examples of divergent sequences x_n and y_n such that $x_n + y_n$ converges.

1.48 Let $a \in \mathbb{R}$ be arbitrary. Give examples of sequences $x_n \to \infty$ and $y_n \to \infty$ such that $x_n - y_n \to a$.

1.49 Give examples of divergent sequences x_n and y_n such that $x_n y_n$ converges.

1.50 Let the real number $a \geq 0$ be arbitrary. Give examples of sequences $x_n \to \infty$ and $y_n \to \infty$ such that $\frac{x_n}{y_n} \to a$.

1.51 Prove or else give a counterexample: If $x_n + y_n$ converges and if $x_n - y_n$ converges, then x_n converges and y_n converges.

1.52 Prove or else give a counterexample: If $ad - bc \neq 0$ and if

$$a x_n + b y_n \to L \quad \text{and} \quad c x_n + d y_n \to M$$

as $n \to \infty$, then x_n converges and y_n converges.

1.53 Suppose for all $n \in \mathbb{N}$ we have $y_n \neq 0$. Prove or else give a counterexample: If both $x_n y_n$ and $\frac{x_n}{y_n}$ converge, then x_n converges and y_n converges.

1.54 Prove or else give a counterexample:
  a) A bounded sequence times a convergent sequence must be convergent.
  b) A null sequence times a bounded sequence must be a null sequence.

1.55

  a) If $q(n) = b_k n^k + b_{k-1} n^{k-1} + \cdots + b_1 n + b_0$ is a polynomial in the variable $n \in \mathbb{N}$ with $b_k \neq 0$, show that there exists $N \in \mathbb{N}$ such that $n \geq N$ implies $q(n) \neq 0$.
  b) Show that

$$\lim_{n \to \infty} \frac{a_k n^k + a_{k-1} n^{k-1} + \cdots + a_1 n + a_0}{b_k n^k + b_{k-1} n^{k-1} + \cdots + b_1 n + b_0} = \frac{a_k}{b_k},$$

  provided that $b_k \neq 0$ and k is a positive integer.

1.56   ◇ †[5] Define the nth Cesàro mean of a sequence x_n by

$$\sigma_n = \frac{1}{n}(x_1 + \ldots + x_n)$$

for all $n \in \mathbb{N}$.

    a) Suppose $x_n \to L$ as $n \to \infty$. Prove: $\sigma_n \to L$ as $n \to \infty$. (Hint: Write $|\sigma_n - L| = \left|\sum_{k=1}^{n} \frac{x_k - L}{n}\right|$.)

    b) Give an example of a *divergent* sequence x_n for which σ_n *converges*.

1.57   Let x_n and y_n be any two bounded sequences of real numbers. Prove that

$$\limsup(x_n + y_n) \le \limsup x_n + \limsup y_n.$$

Give an example in which strict inequality occurs.

1.58   Let x_n and y_n be any two bounded sequences of real numbers. Prove that

$$\liminf(x_n + y_n) \ge \liminf x_n + \liminf y_n.$$

Give an example in which strict inequality occurs.

1.5   THE BOLZANO–WEIERSTRASS THEOREM

A *subsequence* of a sequence x_n is a sequence consisting of some (but not necessarily all) of the terms of the sequence x_n. The terms appear in the same order as they appeared in x_n, but with omissions. We formalize this concept in the following definition.

Definition 1.5.1 *Let n_k be any* strictly *increasing sequence of natural numbers, so that*

$$n_1 < n_2 < \cdots < n_k < \cdots .$$

Then we call x_{n_k} a subsequence of x_n.

We remark that since $n_1 \ge 1$, it follows that $n_2 \ge 2$, ..., and $n_k \ge k$, for all k. An alternative way to think about and to notate subsequences is to write that $n_k = \phi(k)$, where $\phi : \mathbb{N} \to \mathbb{N}$ is a *strictly increasing* function, in the sense that $j < k \implies \phi(j) < \phi(k)$. Then we could alternatively write x_{n_k} as $x_{\phi(k)}$.

■ **EXAMPLE 1.7**

Let $x_n = n^2$, for all $n \in \mathbb{N}$. If $n_k = 2k$, then $x_{n_k} = (2k)^2$ is the sequence of squares of even natural numbers.

---

[5]This exercise is used to develop the Fejer kernel for Fourier series in Exercise 6.47.

Theorem 1.5.1 *If x_n converges to the limit L as $n \to \infty$, then every subsequence $x_{n_k} \to L$ as $k \to \infty$.*

Proof: Let $\epsilon > 0$. There exists $N \in \mathbb{N}$ such that $n \geq N$ implies $|x_n - L| < \epsilon$. Since $n_k \geq k$ for all k, it follows that $k \geq N \implies |x_{n_k} - L| < \epsilon$. ∎

Corollary 1.5.1 *If x_n has two subsequences that converge to different limits, then x_n is not convergent.*

Theorem 1.5.1 should be compared carefully with the following example.

■ **EXAMPLE 1.8**

Let $x_n = (-1)^{n+1}$. The sequence x_n is bounded but is not convergent. The subsequences $x_{2k-1} \to 1$ and $x_{2k} \to -1$ as $k \to \infty$,
We have learned previously that every convergent sequence is bounded. Although the student has seen several examples of bounded sequences that are *not* convergent, we do have the following very important theorem.

Theorem 1.5.2 (Bolzano–Weierstrass) *Let x_n be any* bounded *sequence of real numbers, so that there exists $M \in \mathbb{R}$ such that $|x_n| \leq M$ for all n. Then there exists a convergent subsequence x_{n_k} of x_n. That is, there exists a subsequence x_{n_k} that converges to some $L \in [-M, M]$.*

Proof: We will use the method of interval-halving introduced previously to prove the existence of least upper bounds. Let $a_1 = -M$ and $b_1 = M$. So $x_n \in [a_1, b_1]$, for all $n \in \mathbb{N}$. Let $x_{n_1} = x_1$. Now divide $[a_1, b_1]$ in half using the midpoint $\frac{a_1+b_1}{2} = 0$.

i. If there exist ∞-many values of n such that $x_n \in [a_1, 0]$, then let $a_2 = a_1$ and $b_2 = 0$.

ii. But if there do not exist ∞-many such terms in $[a_1, 0]$, then there exist ∞-many such terms in $[0, b_1]$. In that case let $a_2 = 0$ and $b_2 = b_1$.

Now since there exist ∞-many terms of x_n in $[a_2, b_2]$, pick any $n_2 > n_1$ such that $x_{n_2} \in [a_2, b_2]$. Next divide $[a_2, b_2]$ in half and pick one of the halves $[a_3, b_3]$ having ∞-many terms of x_n in it. Then pick $n_3 > n_2$ such that $x_{n_3} \in [a_3, b_3]$. Observe that

$$|b_k - a_k| = \frac{2M}{2^{k-1}} \to 0$$

as $k \to \infty$. So if $\epsilon > 0$, there exists K such that $k \geq K$ implies $|b_k - a_k| < \epsilon$. Thus if j and $k \geq K$, we have $|x_{n_j} - x_{n_k}| < \epsilon$ as well. Hence x_{n_k} is a Cauchy sequence and must converge. Since $[-M, M]$ is a closed interval, we know from a previous exercise that $x_{n_k} \to L$ as $k \to \infty$ for some $L \in [-M, M]$. ∎

EXERCISES

1.59   Give an example of a bounded sequence that does not converge.

1.60   Use Corollary 1.5.1 to prove that the sequence $x_n = (-1)^n + \frac{1}{n}$ does not converge.

1.61   Suppose $x_n \to \infty$. Prove that every subsequence $x_{n_k} \to \infty$ as $k \to \infty$ as well. (Hint: The sequence x_n is divergent, so it is not enough to quote Theorem 1.5.1.)

1.62   Use the following steps to prove that the sequence x_n has no convergent subsequences if and only if $|x_n| \to \infty$ as $n \to \infty$.

  a) Suppose that the sequence x_n has no convergent subsequences. Let $M > 0$. Prove that there exist at most finitely many values of n such that $x_n \in [-M, M]$. Explain why this implies $|x_n| \to \infty$ as $n \to \infty$.

  b) Suppose $|x_n| \to \infty$ as $n \to \infty$. Show that x_n has no convergent subsequence. (Hint: Exercise 1.61 may help.)

1.63   Give an example in which $y_j > 0$ for all j and $y_j \to 0$ yet y_j is *not* monotone.

1.64   The following questions provide an easy, alternative proof of the Bolzano–Weierstrass Theorem.

  a) Use the following steps to prove that *every* sequence x_n of real numbers has a monotone subsequence. Denote the nth tail of the sequence by $T_n = \{x_j \mid j \geq n\}$.

  (i) Suppose the following special condition is satisfied: For each $n \in \mathbb{N}$, T_n has a smallest element. Prove that there exists an increasing subsequence x_{n_j}.

  (ii) Suppose the condition above fails, so that there exists $N \in \mathbb{N}$ such that T_N has no smallest element. Prove that there exists a decreasing subsequence x_{n_j}.

  b) Give an easy alternative proof of the Bolzano–Weierstrass Theorem.

1.65   Prove: A sequence $x_n \to L \in \mathbb{R}$ if and only if *every* subsequence x_{n_i} possesses a sub-subsequence $x_{n_{i_j}}$ that converges to L as $j \to \infty$. (Hint: To prove the *if* part, suppose false and write out the logical negation of convergence of x_n to L.)

1.66   *Prove or Give a Counterexample*: A sequence $x_n \in \mathbb{R}$ converges if and only if *every* subsequence x_{n_i} possesses a sub-subsequence $x_{n_{i_j}}$ that converges as $j \to \infty$.

1.6   THE NESTED INTERVALS THEOREM

Having used the method of interval-halving twice already, it is natural to consider the following theorem.

Theorem 1.6.1 (Nested Intervals Theorem) *Suppose*

$$[a_1, b_1] \supseteq [a_2, b_2] \supseteq \cdots \supseteq [a_k, b_k] \supseteq \cdots$$

is a decreasing nest *of closed finite intervals. Suppose also that*

$$b_k - a_k \to 0 \text{ as } k \to \infty.$$

Then there exists exactly one *point* $L \in \bigcap_{k=1}^{\infty} [a_k, b_k]$. *Moreover,* $a_k \to L$ *and* $b_k \to L$ *as* $k \to \infty$.

Proof: Let $\epsilon > 0$. Then there exists K such that $k \geq K$ implies $|b_k - a_k| < \epsilon$. But, for all $k \geq K$, $a_k \in [a_K, b_K]$. Thus $j, k \geq K$ implies $|a_j - a_k| < \epsilon$. Hence the sequence a_k is a Cauchy sequence so there exists a point L such that $a_k \to L$. Since $k \geq n$ implies for all n that $a_k \in [a_n, b_n]$, it follows that $L \in [a_n, b_n]$ for all n and that

$$L \in \bigcap_{k=1}^{\infty} [a_k, b_k].$$

Now, if

$$L' \in \bigcap_{k=1}^{\infty} [a_k, b_k]$$

also, then $|L - L'| \leq |b_k - a_k| \to 0$, which implies $L = L'$. Hence the point L is unique. Observe that $|b_k - L| \leq |b_k - a_k| \to 0$ so that $b_k \to L$ as claimed. ∎

The reader is aware that there are real numbers that are not rational. For example, we will prove that there is a square root of 2 in $\mathbb{R}$ in Example 1.10. Yet we know that no rational number can be a square root of 2 as was shown in Exercise 1.11. Despite the fact that not every real number is rational, every finitely long decimal expansion represents a rational number, and common sense tells us that we may approximate any real number as closely as we wish by using a suitable but finitely long decimal expansion. This observation gives rise to the following definition of what it means for a subset $S \subseteq \mathbb{R}$ to be *dense* in $\mathbb{R}$.

Definition 1.6.1 *A subset* $S \subseteq \mathbb{R}$ *is called* dense *in* $\mathbb{R}$ *if and only if for all* $x \in \mathbb{R}$, *there exists a sequence* s_k *of elements of* S *such that* $s_k \to x$.

■ **EXAMPLE 1.9**

We will show that $\mathbb{Q}$ is dense in $\mathbb{R}$.

Proof: Let $x \in \mathbb{R}$. If $x \in \mathbb{Q}$, we could simply let $s_k \equiv x$ so that $s_k \to x$, being a constant sequence.

So suppose $x \notin \mathbb{Q}$, so that x is irrational. Then there exists $n \in \mathbb{Z}$ such that $n < x < n + 1$. Let $a_1 = n$ and $b_1 = n + 1$, both rational numbers. Then the midpoint is also a rational number, and x must lie in one half-interval but not the other. Let $[a_2, b_2]$ be the half-interval containing x. Now cut $[a_2, b_2]$ in half again and select $[a_3, b_3]$ containing x again. Note that

$$|b_k - a_k| = \frac{1}{2^{k-1}} \to 0.$$

Since $x \in [a_k, b_k]$ for all k, $|a_k - x| \to 0$, so $a_k \to x$, and $a_k \in \mathbb{Q}$ for all k. Thus we have a sequence of rational numbers converging to x in this case as well. (Note that b_k would have served just as well as a_k.)   ∎

Remark. Because $\mathbb{R}$ is complete and because the set $\mathbb{Q} \subset \mathbb{R}$ is dense in $\mathbb{R}$, it follows that any set of numbers that contains limits for all its Cauchy sequences and that contains $\mathbb{Q}$ must also contain $\mathbb{R}$. For this reason $\mathbb{R}$ is called the *completion* of $\mathbb{Q}$.

■ **EXAMPLE 1.10**

We will show that $\sqrt{2}$ exists in $\mathbb{R}$.

Proof: Recall that in the first paragraph of Section 1.3 we constructed an increasing sequence x_k as follows:

$$
\begin{aligned}
x_1 &= 1 \\
x_2 &= 1.4 \\
x_3 &= 1.41 \\
x_4 &= 1.414 \\
&\vdots
\end{aligned}
$$

Here x_k is the largest *k-digit* decimal greater than 1 such that $x_k^2 < 2$. We could have constructed also a decreasing sequence y_k by letting y_k be the smallest k-digit decimal such that $y_k^2 > 2$. Thus

$$|y_k - x_k| \leq \frac{1}{10^{k-1}} \to 0$$

as $k \to \infty$ We see that the intervals $[x_k, y_k]$ satisfy the hypotheses of Theorem 1.6.1. Thus there exists a unique

$$L \in \bigcap_{k=1}^{\infty} [x_k, y_k]$$

such that $x_k \to L$ and $y_k \to L$. Hence $x_k^2 \to L^2$, so that $L^2 \leq 2$, and $y_k^2 \to L^2$, so that $L^2 \geq 2$. Thus $L^2 = 2$ and $L = \sqrt{2}$ exists in $\mathbb{R}$.   ∎

EXERCISES

1.67   Give an example of a decreasing nest of nonempty open finite intervals

$$(a_1, b_1) \supseteq (a_2, b_2) \supseteq \cdots$$

such that $\bigcap_{k=1}^{\infty}(a_k, b_k) = \emptyset$, the *empty set*.

1.68 Give an example of a decreasing nest of open intervals

$$(a_1, b_1) \supseteq (a_2, b_2) \supseteq \cdots$$

such that $b_k - a_k \to 0$ yet $\bigcap_{k=1}^{\infty}(a_k, b_k) \neq \emptyset$.

1.69 Give an example of a decreasing nest of *infinite* intervals with empty intersection.

1.70 Prove or give a counterexample: If $a_n \uparrow$, $b_n \downarrow$, and (a_n, b_n) is a decreasing nest of finite open intervals, then there exists $L \in \mathbb{R}$ such that

$$\bigcap_{n=1}^{\infty}(a_n, b_n) = \{L\}.$$

1.71 Show that every open interval $(a, b) \subset \mathbb{R}$, with $0 < b - a$ but no matter how small, must contain a rational number. (Hint: Apply Example 1.9.)

1.72 † Let I denote the set of all irrational numbers. The following steps will lead to the conclusion that I is dense in $\mathbb{R}$. (You may assume it is known that $\sqrt{2} \in I$.) Let $x \in \mathbb{R}$. We must show there exists a sequence s_k of elements of I converging to x.

    a) Show that if $\frac{m}{n}$ is any nonzero rational number then $\frac{m}{n}\sqrt{2}$ is irrational. (Hint: Suppose the claim is false, and deduce a contradiction.)

    b) Now suppose x is any real number. Explain why there exists a sequence t_k of *nonzero* elements of $\mathbb{Q}$ converging to $\frac{x}{\sqrt{2}}$. Define a sequence s_k of elements of I converging to x.

1.73 Show that every open interval (a,b), with $b - a > 0$ but no matter how small, must contain an irrational number. (Hint: Use the result of Exercise 1.72.)

1.74 Is the set $\left\{\frac{m}{2^n} \,\middle|\, m \in \mathbb{Z}, n \in \mathbb{N}\right\}$ dense in $\mathbb{R}$? Prove your conclusion.

1.75 ◇ Let $D \neq \emptyset$ be a subset of the set of strictly positive real numbers, and let $S = \{nd \mid n \in \mathbb{Z}, d \in D\}$. Prove: S is dense in $\mathbb{R}$ if and only if $\inf(D) = 0$.

1.7 THE HEINE–BOREL COVERING THEOREM

Although the study of continuous functions belongs to the next chapter, let us think in advance on an intuitive level about this concept. A function $f : \mathbb{R} \to \mathbb{R}$ is said to be everywhere continuous provided that for each point $p \in \mathbb{R}$, $f(x)$ remains very close to $f(p)$ provided that x is kept sufficiently close to p. For example, the set

$$S = \{x \,|\, |f(x) - f(p)| < \epsilon\}$$

should contain some sufficiently small open interval around p, although S may also include points far away from p.

Consider next an open interval (a, b) that is contained in the set of values achieved by f. Let $O = \{x \mid f(x) \in (a, b)\}$. For each $p \in O$ there will be a corresponding small number $\epsilon > 0$ such that

$$(f(p) - \epsilon, f(p) + \epsilon) \subseteq (a, b).$$

Because f is continuous at p, there will be a small open interval around p that is contained in O. This example motivates the concept of an *open* set, which generalizes the familiar notion of an open interval.

Definition 1.7.1 *A set $O \subseteq \mathbb{R}$ is called an* open *subset of $\mathbb{R}$ provided for each $x \in O$ there exists $r_x > 0$ such that*

$$(x - r_x, x + r_x) \subseteq O.$$

Thus O is called open provided that each $x \in O$ has some (perhaps very small) open interval of *radius $r_x > 0$* around it that is entirely in O.

■ **EXAMPLE 1.11**

We claim that every open interval (a, b) is an open set. In fact, if $x \in (a, b)$, then $a < x < b$ and we can let

$$r_x = \min\{|x - a|, |x - b|\}.$$

Then $(x - r_x, x + r_x) \subseteq (a, b)$.

Theorem 1.7.1 *Every open subset $O \subseteq \mathbb{R}$ is a union of (perhaps infinitely many) open intervals. Moreover, every union of open sets is an open set.*

Proof: Let $O \subseteq \mathbb{R}$ be open. Then, using the notation of Definition 1.7.1, you will show in Exercise 1.78 that

$$O = \bigcup_{x \in O} (x - r_x, x + r_x).$$

To prove the second conclusion, let $O = \bigcup_{\alpha \in A} O_\alpha$ be any union[6] of open sets. Let $x \in O$. We know there exists $\alpha_0 \in A$ such that $x \in O_{\alpha_0}$, which is open. Thus there exists $r_x > 0$ such that $(x - r_x, x + r_x) \subseteq O_{\alpha_0} \subseteq O$. Thus O is open.   ■

Definition 1.7.2 *An* open cover *of a set $S \subseteq \mathbb{R}$ is a collection*

$$\mathcal{O} = \{O_\alpha \mid \alpha \in A\}$$

---

[6]When denoting an *arbitrary* union of open sets, it is customary to use a so-called *index set*, such as the set A used here. One should think of A as being a set of labels, or names, used to tag, or identify the sets of which the union is being formed. One cannot always index sets by means of natural numbers, because there exist sets so large that they cannot be uniquely indexed by natural numbers. Even the infinite set $\mathbb{N}$ is too small. The reader will learn more about this in Theorem 1.15.

of (perhaps infinitely many) open sets O_α, where α ranges over some index set A, such that $S \subseteq \bigcup_{\alpha \in A} O_\alpha$.

In analysis, it is often necessary to try to control small-scale local variations of some structure defined on a domain D. Under suitable conditions, one can control variations by restricting ones view to a very small open set surrounding each given point of D. Then in the large we *cover* the whole domain D with a family $\mathcal{O}$ of these (possibly small) open sets whose *union* contains D. Usually $\mathcal{O}$ will have infinitely many open sets as members, or elements of itself. Within each one of the open sets that are elements of $\mathcal{O}$ the fine structure varies only slightly. We hope for the availability of a finite *sub*cover, consisting of only finitely many of the open sets belonging to $\mathcal{O}$, so as to produce *uniform* controls on fine-scale variations for the entire large domain D. Below, we show an example of an open covering of a set for which there is no finite subcover. This will motivate the Heine–Borel Theorem which follows.

■ **EXAMPLE 1.12**

"bounded interval"... ie finite length, not finite number of elements

Consider the set $S = (0, 2)$, a finite open interval. We claim that

$$S \subseteq \bigcup_{n=1}^{\infty} \left(\frac{1}{n}, 2 \right).$$

ie, $x \in S \Rightarrow x \in \bigcup_{n=1}^{\infty} \mathcal{O}$
so, $S \subseteq \bigcup_{n=1}^{\infty} \mathcal{O}$

since for all $n \in \mathbb{N}$ will contain
over all $n \in \mathbb{N}$

In fact, for each $x \in (0, 2)$ there exists $n \in \mathbb{N}$ such that $x \in \left(\frac{1}{n}, 2 \right)$. (Make sure you see why this is so.) Thus $\mathcal{O} = \left\{ \left(\frac{1}{n}, 2 \right) \mid n \in \mathbb{N} \right\}$ is an open cover of S. However, it is impossible to select any *finite* subset of $\mathcal{O}$ that covers S. The reason is that any finite subset of $\mathcal{O}$ would have a largest value n_0 of n for which $\frac{1}{n}$ would be the left hand endpoint of an interval belonging to the chosen finite subset of $\mathcal{O}$. Thus the finite subset would fail to cover any points to the left of $\frac{1}{n_0}$.

Remark 1.7.1 Note that the term *finite interval* means an interval of finite *length*. Any finite interval with strictly positive length has infinitely many distinct points within it. Thus the word *finite* in *finite interval* means the same thing as *bounded*. ★ On the other hand, a *finite set* means a set with *finitely many elements*. In Example 1.12, a finite subset of a set of intervals means a collection of finitely many of those intervals. This does *not* mean that the intervals in question have finitely many points.

The Heine–Borel theorem is one of the most important in advanced calculus. But it is the most abstract theorem presented thus far in this book, and the reader will need time and experience to absorb fully its significance. It is recommended to consider Exercise 1.80 below after reading the statement of the theorem.

Theorem 1.7.2 (Heine–Borel) *Suppose the closed finite interval*

$$[a, b] \subseteq \bigcup_{\alpha \in A} O_\alpha,$$

where $\mathcal{O} = \{O_\alpha \mid \alpha \in A\}$ is an open *cover of* $[a, b]$. *Then there exists a finite set* $F = \{\alpha_1, \ldots, \alpha_n\} \subseteq A$ *such that*

$$[a, b] \subseteq \bigcup_{\alpha \in F} O_\alpha = \bigcup_{i=1}^{n} O_{\alpha_i}.$$

The collection $\{O_{\alpha_1}, \ldots, O_{\alpha_n}\} \subseteq \mathcal{O}$ *is called a* finite subcover *of* $[a, b]$.

Proof: We suppose the theorem were false. We will deduce a logical self-contradiction from that supposition. This will prove the theorem. So suppose the Heine–Borel theorem were false: Thus we can assume the given cover does not admit a finite subcover of [a,b].

Let $a_1 = a$ and $b_1 = b$, and let $c = \frac{a_1 + b_1}{2}$. Then each of the intervals $[a_1, c]$ and $[c, b_1]$ is covered by $\bigcup_{\alpha \in A} O_\alpha$. If *both* of these half-intervals had finite subcovers, then the whole interval [a,b] would have a finite subcover since the union of two finite families is still finite. Since we are supposing [a,b] has no finite subcover, pick a half-interval $[a_2, b_2]$ that has *no* finite subcover. Now cut $[a_2, b_2]$ in half and reason the same way for $[a_2, b_2]$ as we did for $[a_1, b_1]$. We obtain a decreasing next of intervals

$$[a_1, b_1] \supseteq \cdots \supseteq [a_k, b_k] \supseteq \cdots$$

such that each $[a_k, b_k]$ is covered by $\bigcup_{\alpha \in A} O_\alpha$ but has no finite subcover.

However,

$$|b_k - a_k| = \frac{b - a}{2^{k-1}} \to 0$$

as $k \to \infty$. By the nested intervals theorem, there exists a unique

$$x \in \bigcap_{k=1}^{\infty} [a_k, b_k] \subseteq [a, b].$$

Since $x \in [a, b]$, there exists $\alpha \in A$ such that $x \in O_\alpha$. So there exists $r_x > 0$ such that $(x - r_x, x + r_x) \subseteq O_\alpha$. Now pick k big enough so that $b_k - a_k < r_x$. Thus

$$x \in [a_k, b_k] \subset (x - r_x, x + r_x) \subseteq O_\alpha$$

and we have covered $[a_k, b_k]$ with a *single* open set O_α from the original cover. This is a (very small) finite subcover. This contradicts the statement that $[a_k, b_k]$ could not have a finite subcover. This contradiction proves the Heine–Borel theorem.   ∎

EXERCISES

1.76   Show that a closed finite interval [a,b] is *not* an open set.

1.77   Show that a half-closed finite interval (a,b] is *not* an open set.

1.78   Let O be any open subset of $\mathbb{R}$, and for each $x \in O$ let r_x be defined as in the proof of Theorem 1.7.1. Complete the proof of that theorem by showing that $O = \bigcup_{x \in O} (x - r_x, x + r_x)$.

1.79 The empty set $\emptyset$ satisfies the definition of being open. Explain.

1.80 Find an open cover of the interval $(-1, 1)$ that has no finite subcover. Justify your claims.

1.81 Find an open cover of the interval $(-\infty, \infty)$ that has no finite subcover. Justify your claims.

1.82 Let $E \subseteq \mathbb{R}$ be any *unbounded* set. Find an open cover of E that has no finite subcover. Prove that you have chosen an open cover and that it has no finite subcover.

1.83 Let $E = \left\{ \frac{1}{n} \,\middle|\, n \in \mathbb{N} \right\}$. Find an open cover $\mathcal{O} = \{O_n \mid n \in \mathbb{N}\}$ of E that has no finite subcover, and prove that $\mathcal{O}$ is an open cover and that $\mathcal{O}$ has no finite subcover.

1.84 $\diamond$ We call p a *cluster point* of E, provided that for all $\epsilon > 0$ there exists $e \in E$ such that

$$0 < |e - p| < \epsilon.$$

(See Defintion 2.1.1.) Let $E \subset \mathbb{R}$ be any set with the property that there is a *cluster point p of E such that $p \notin E$*. Show that there exists an open cover of E that has no finite subcover. Justify your claims. (Note: Exercise 1.83 is an example of the claim of this exercise.)

1.85 True or False: Finitely many of the open sets in the collection

$$\left\{ \left(\frac{x}{2}, \frac{3x}{2} \right) \,\middle|\, x \in [0, 1] \right\}$$

would suffice to cover $[0, 1]$.

1.86 Prove or give a counterexample: Every open cover of a finite subset of $\mathbb{R}$ has a finite subcover. (Note: For the real line, the phrase *finite subset* does *not* mean the same thing as *finite interval*.)

1.8 COUNTABILITY OF THE RATIONAL NUMBERS

Definition 1.8.1 *A set S is called* countable *if it is an* infinite *set for which it is possible to arrange all the elements of S into a sequence. That is, S is countable if $S = \{s_1, s_2, \ldots, s_k, \ldots\}$ with each element of S listed exactly once in the sequence.*

Equivalently, we may say that S is countable if and only if there exists a *function* $s : \mathbb{N} \to S$ that is both one-to-one, which is also called *injective*, and onto S. Onto maps are often called *surjective*. The term s_n in the definition above would be $s(n)$ in this notation.

■ **EXAMPLE 1.13**

Let E denote the set of all *even* natural numbers. Thus $E \subsetneq \mathbb{N}$. We claim that E is countable. In fact, the elements of E can be arranged into a sequence by means of a function $s : n \rightarrow 2n$ that is both an injection and a surjection of E onto $\mathbb{N}$. That is, the sequence is given by $s_n = 2n$. It may surprise the reader that the elements of an infinite set can be paired one-to-one with those of a *proper* subset.

■ **EXAMPLE 1.14**

We will prove the surprising and useful fact that the set $\mathbb{Q}$ of all rational numbers is countable. It is important to understand that if a sequence s_n is to include *all* the rational numbers, then these numbers cannot be listed in *size places*. That is, if $s_n < s_{n+1}$, both in $\mathbb{Q}$, then $\frac{s_n + s_{n+1}}{2}$ lies between them and is again rational. Hence there is no *next smallest* rational number after s_n.

We can explain how to list the rational numbers in a sequence, disregarding the order relation, as follows. We are going to consider a table of numbers with infinitely many rows. The entry in the mth row and nth column will be the fraction $\frac{n}{m}$. Here $m \in \mathbb{N}$ and $n \in \mathbb{Z}$. Thus there will be a first row, in which each denominator is understood to be 1, but no last row. Each row will extend endlessly to left and to the right. We can draw only part of this table below.

$\cdots$	-4	-3	-2	-1	**0**	1	2	3	4	$\cdots$
$\cdots$	$-\frac{4}{2}$	$-\frac{3}{2}$	$-\frac{2}{2}$	$-\frac{1}{2}$	$\frac{0}{2}$	$\frac{1}{2}$	$\frac{2}{2}$	$\frac{3}{2}$	$\frac{4}{2}$	$\cdots$
$\cdots$	$-\frac{4}{3}$	$-\frac{3}{3}$	$-\frac{2}{3}$	$-\frac{1}{3}$	$\frac{0}{3}$	$\frac{1}{3}$	$\frac{2}{3}$	$\frac{3}{3}$	$\frac{4}{3}$	$\cdots$
$\cdots$	$-\frac{4}{4}$	$-\frac{3}{4}$	$-\frac{2}{4}$	$-\frac{1}{4}$	$\frac{0}{4}$	$\frac{1}{4}$	$\frac{2}{4}$	$\frac{3}{4}$	$\frac{4}{4}$	$\cdots$
$\vdots$	$\vdots$	$\vdots$	$\vdots$	$\vdots$	$\vdots$	$\vdots$	$\vdots$	$\vdots$	$\vdots$	

We will describe a systematic *expanding search pattern* that reaches each term on the infinite table after some finite number of terms in the sequence described below. We will list side-by-side those terms $\frac{n}{m}$ for which

$$|m| + |n| = k$$

beginning with $k = 1$, $k = 2$, and so on. If parentheses are placed around a number, we are skipping that number because it was already listed previously. Here is the resulting list:

$$0, -1, \left(\frac{0}{2} = 0\right), 1, -2, -\frac{1}{2}, \left(\frac{0}{3} = 0\right), \frac{1}{2}, 2, -3, \left(-\frac{2}{2} = -1\right), -\frac{1}{3},$$

$$\left(\frac{0}{4} = 0\right), \frac{1}{3}, \left(\frac{2}{2} = 1\right), 3, \ldots .$$

It is clear that this expanding search pattern eventually reaches any rational number $\frac{n}{m}$ that one might choose, and each rational number is listed exactly once in the resulting endless sequence. The first several terms of the sequence s_n, corresponding to $k = 0, 1, 2, 3, 4$, are

$$0, -1, 1, -2, -\frac{1}{2}, \frac{1}{2}, 2, -3, -\frac{1}{3}, \frac{1}{3}, 3, \dots .$$

We will see several applications of the countability of $\mathbb{Q}$ in this book. However, for now we describe a startling example.

■ EXAMPLE 1.15

We will describe a set O that is *both open* and *dense* in $\mathbb{R}$, yet which is *quite small*.

Let $\epsilon > 0$, a small positive number. Consider the line segment $[0, \epsilon]$ of length ϵ. We will construct a sequence of intervals (a_k, b_k), each of length $\frac{\epsilon}{2^k}$. That is, the first interval, (a_1, b_1) will have length $\frac{\epsilon}{2}$. This leaves half of $[0, \epsilon]$ remaining. But for (a_2, b_2) we will use only half that remainder: namely, $\frac{\epsilon}{4}$. $b_3 - a_3$ will be taken to be $\frac{\epsilon}{8}$, or half of the remaining $\frac{\epsilon}{4}$ from the original interval $[0, \epsilon]$.

Let $\mathbb{Q} = \{s_1, s_2, \dots, s_k, \dots\}$, which can be arranged since $\mathbb{Q}$ is countable, as explained above. Let (a_1, b_1) be centered around s_1, (a_2, b_2) centered around s_2, and in general (a_k, b_k) will be centered around the point s_k. For any finite subcollection of the intervals (a_k, b_k), $k = 1, 2, 3, \cdots$, the sum of the lengths of each of the finitely many intervals chosen must be less than ϵ. That is because the whole infinite sequence of intervals is chosen by cutting ϵ in half again and again without end.

Now consider that $\mathbb{Q}$ is dense in $\mathbb{R}$. But if we let $O = \bigcup_{k=1}^{\infty} (a_k, b_k)$, then $\mathbb{Q} \subset O$ and so O is also dense in $\mathbb{R}$. Moreover, O is *open* by Theorem 1.7.1.

We claim that O is a small set in the following sense. Let [a,b] be *any* closed finite interval of length $\geq \epsilon$. We claim it is impossible for $[a, b] \subseteq O$. In fact, if [a,b] were a subset of O, then

$$[a, b] \subseteq \bigcup_{k=1}^{\infty} (a_k, b_k),$$

an open cover of [a,b]. By the Heine–Borel theorem, there must be a finite number of intervals from among the (a_k, b_k)'s that cover [a,b]. Yet the sum of the lengths of these finitely many intervals must be less than $\epsilon \leq b - a$. This is impossible.

It is interesting to compare the preceding example with Exercise 1.91. The interested student can learn much more about surprising subsets of the line in the book by Gelbaum and Olmsted [7].

It is natural to wonder at this point whether or not perhaps every infinite set is countable. The answer is *no*, as is shown by the following surprising theorem.

Theorem 1.8.1 (Cantor) *The set $\mathbb{R}$ of real numbers is* uncountable. *(That is, it is impossible to include all the real numbers in a sequence.)*

Proof: We begin by noting that the (possibly endless) decimal expansions of real numbers are not unique, because an infinite *tail* of 9's can always be replaced by an expansion ending in an infinite *tail* of 0's. For example,

$$0.999\ldots = 1.000\ldots$$

This is understood in the sense that if $x_n = 0.999\ldots9$ with n 9's then

$$|x_n - 1| = \frac{1}{10^n} \to 0$$

as $n \to \infty$. But if we agree not to allow endless tails of 9's, then decimal expansions of real numbers are unique. Moreover, *every infinite decimal representation corresponds to a real number*. The reason for this fact is as follows. Consider any infinite decimal expression. It could be written in terms of a whole number K in the form

$$K + 0.d_1 d_2 d_3 \ldots d_n \ldots,$$

where d_n is the nth digit to the right of the decimal point. Then let

$$x_n = K + 0.d_1 d_2 \ldots d_n.$$

It follows that if m and n are both greater than N, then

$$|x_n - x_m| < \frac{1}{10^N} \to 0$$

as $N \to \infty$. Hence the sequence x_n of *truncations* of the endless decimal expression to n digits is itself a Cauchy sequence. By the *completeness axiom* this sequence x_n must converge to a limit $x \in \mathbb{R}$. That is why we say that the endless decimal *represents* x.

Now we suppose that Cantor's theorem were false and deduce a contradiction. Suppose therefore that all real numbers could be placed into a sequence. Then there would be a subsequence x_n containing *all* the real numbers in [0,1). We denote the decimal expansions of the numbers x_n in a vertical column below.

$$
\begin{aligned}
x_1 &= .d_{11}d_{12}d_{13}\ldots d_{1k}\ldots\\
x_2 &= .d_{21}d_{22}d_{23}\ldots d_{2k}\ldots\\
x_3 &= .d_{31}d_{32}d_{33}\ldots d_{3k}\ldots\\
&\;\;\vdots\\
x_n &= .d_{n1}d_{n2}d_{n3}\ldots d_{nk}\ldots\\
&\;\;\vdots
\end{aligned}
$$

Now we obtain a contradiction by constructing a number $x \in [0,1)$ that is *not* in the sequence x_n. We define x by the digits d_k in its decimal expansion. If $d_{11} \neq 0$, we let $d_1 = 0$. If $d_{11} = 0$, let $d_1 = 1$. If $d_{22} \neq 0$, we let $d_2 = 0$. If $d_{22} = 0$, we let $d_2 = 1$. In general, if $d_{kk} \neq 0$, we let $d_k = 0$, but if $d_{kk} = 0$, then we let $d_k = 1$. We observe that $x = .d_1 d_2 d_3 \ldots d_k \ldots \in [0,1)$, yet $x \notin \{x_n\}$ since for all n, x differs from x_n in the nth decimal digit. ∎

EXERCISES

1.87 †
    a) If A and B are each countable sets, show that $A \cup B$ is countable. (Hint: For each set, consider a sequence of all elements, and show how to *splice* the sequences together to make one sequence. Remember that the sets need not be disjoint.)
    b) Prove that the union of countably many finite sets is either countable or finite.

1.88 † If A_n is a countable set for each $n \in \mathbb{N}$, show that

$$\mathcal{A} = \bigcup_{n=1}^{\infty} A_n,$$

is again a countable set. (Hint: Explain why each set A_n can be written in the form

$$A_n = \{a_{nk} \mid k \in \mathbb{N}\}$$

but these sets need not be *disjoint* from one another. Consider an array similar to that displayed in the proof of Cantor's Theorem in this section, but reason in a manner similar to that in Example 1.14.)

1.89 Is the set $\mathbb{Z}$ of integers countable? Why or why not? How about the set of all odd positive integers? Even integers?

1.90 Show that the set I of all *irrational* numbers must be uncountable. (Hint: Use Exercise 1.87.)

1.91 A subset $E \subset \mathbb{R}$ is called *closed* if and only if its complement $\mathbb{R} \setminus E$ is *open*. (For example, $\mathbb{R}$ itself is a closed set since $\mathbb{R} \setminus \mathbb{R} = \emptyset$ is an open set.) *Prove* that a closed set E that is also *dense* in $\mathbb{R}$ must be all of $\mathbb{R}$. (Hint: Suppose the claim were false, so that $\mathbb{R} \setminus E$ is a nonempty open set. Deduce a contradiction.)

1.92 Referring to the definition in Exercise 1.91, answer the following questions.
    a) Prove that every closed finite interval $[a, b]$ is a closed set.
    b) Give an example of subset $E \subset \mathbb{R}$ for which E is *neither* open *nor* closed. Justify your example.
    c) Give an example of a set $S \subseteq \mathbb{R}$ that is *both* open and closed.

1.93 ◇ Prove that every open set $S \subseteq \mathbb{R}$ can be expressed as the union of a *countable* set of open intervals. *Hint*: Let $S \cap \mathbb{Q} = \{q_n \mid n \in \mathbb{N}\}$ be a sequence listing all the rational numbers in S. Let

$$r_n = \sup\{r \mid (q_n - r, q_n + r) \subseteq S\}$$

1.94 Prove that every subset E of $\mathbb{R}$ is the union of some family of *closed* sets. Can every subset E of $\mathbb{R}$ be the union of a family of *open* sets? Prove your answer.

1.95   Let $S = \mathbb{Q} \cap [0, 1]$. Then S is countable, so we can write

$$S = \{s_n \mid n \in \mathbb{N}\}.$$

We follow the model of Example 1.15 using $\epsilon = 1/2$. Thus, for each n, (a_n, b_n) is an open interval centered about s_n and $b_n - a_n = \frac{1}{2^{n+1}}$.

    a) Show that $O = \bigcup_{n=1}^{\infty}(a_n, b_n)$ is an *open* subset of $\mathbb{R}$ and that every point of $[0, 1]$ is the limit of a sequence of points from O.

    b) Use the Heine–Borel Theorem to prove that $\mathcal{O} = \{(a_n, b_n) \mid n \in \mathbb{N}\}$ is not an open cover of $[0, 1]$.

1.96 $\Diamond$ A real number a is called an *algebraic number* provided there exists a polynomial equation $p(x) = 0$ with integer coefficients such that $p(a) = 0$.

    a) Let $P_{N,n}$ denote the set of all polynomials with *integer* coefficients of the form $p(x) = a_n x^n + \cdots + a_1 x + a_0$ for which the sum of the absolute values of the coefficients is bounded by N. That is

$$\sum_{k=0}^{n} |a_k| \leq N.$$

       Show that $P_{N,n}$ is a finite set.

    b) Prove that the set of algebraic numbers is countable. (Hint: Consider first the set of those numbers that are roots of a polynomial equation of degree n with integer coefficients.)

1.97   A real number is called *transcendental* provided that it is not algebraic. Prove that the set of all transcendental numbers is uncountable.

Remark 1.8.1 The method of proof employed in Cantor's theorem is known as the *Cantor diagonalization process* after its inventor, Georg Cantor (1845–1918). The discovery that some infinite sets are significantly larger than others, as uncountable sets are larger than countable ones, led to the invention of the subject of transfinite arithmetic. The student who is curious to learn more about this may enjoy the classic book by E. Kamke [11].

    It is interesting to note that Cantor embarked upon his study of transfinite sets with particular applications to analysis in mind. So-called trigonometric series, or Fourier series, are representations of suitable functions as sums of perhaps infinitely many sine and cosine waves of various periods. Such representations had been shown by Fourier to be very useful for the solution of the heat equation in physics. There were, however, major difficulties regarding the uniqueness of these representations and the actual pointwise convergence of the sums of sine and cosine waves to the function under study. In the long run, it turned out that a different development undertaken by Henri Lebesgue (the Lebesgue integral) was more effective than set theory for this application. However, Cantor's research cast a new light upon the whole of mathematics, far beyond the applications that motivated the initial study. This is a good example of how investigation of an interesting question can lead to vast and totally unanticipated branches of mathematical knowledge.

The interested reader can find this and many other historical topics in Mathematics at the website of the MacTutor History of Mathematics archive[7] at the University of St. Andrews in Scotland.

1.9 TEST YOURSELF

Test Yourself sections, found at the end of each chapter, contain short questions to check your understanding of basic concepts and examples. Proofs are not tested in these sections, since proofs must be read individually by the student's teacher or teaching assistant.

EXERCISES

1.98 $\epsilon = \frac{1}{100}$. Find a *number* $\delta > 0$ small enough so that $|a-b| < \delta$ and $|c-b| < \delta$ implies $|a - c| < \epsilon$.

1.99 The sequence x_n begins as follows: $\mathbf{0}, \mathbf{1}, \frac{3}{2}, \mathbf{2}, \frac{7}{3}, \frac{8}{3}, \mathbf{3}, \frac{13}{4}, \frac{14}{4}, \frac{15}{4}, \mathbf{4}, \ldots$ and continues according to the same pattern.
  a) True or False: $\lim_{n \to \infty} |x_n - x_{n+1}| = 0$. T
  b) True or False: x_n is a Cauchy sequence. F

1.100 Give an example of two sequences, x_n and $y_n \neq 0$ such that $x_n y_n$ converges, $\frac{x_n}{y_n}$ converges, but *neither* x_n *nor* y_n converges. $x_n = (-1)^n \quad y_n = (-1)^{n+1}$

1.101 Let $x_n = ((-1)^n + 1) + \frac{1}{2^n}$ for all $n \in \mathbb{N}$. Find both $\liminf x_n$ and $\limsup x_n$. $\liminf x_n = 0$, $\limsup x_n = 2$

1.102 Give an example of *two sequences* of *real numbers* x_n and y_n for which $\liminf(x_n + y_n) = 0$ but $\liminf x_n = -\infty = \liminf y_n$.

1.103 State *True* or *Give a Counterexample*: If x_n is an *unbounded* sequence, then x_n has no convergent subsequences.

1.104 Give an example of a *decreasing nest* of *nonempty* open intervals (a_n, b_n) such that $b_n - a_n \to 0$ but $\bigcap_{i=1}^{\infty} (a_n, b_n) = \emptyset$.

1.105 True or False: The set $S = \left\{ \frac{m}{2^n} \mid m \in \mathbb{Z}, n \in \mathbb{N} \right\}$ is dense in $\mathbb{R}$.

1.106 Let $E = \left\{ \frac{1}{n} \mid n \in \mathbb{N} \right\}$. Find an open cover $\mathcal{O} = \{O_n \mid n \in \mathbb{N}\}$ of E that has no finite subcover.

1.107 True or False: The set $\mathbb{Q}$ is closed in the real line $\mathbb{R}$.

1.108 True or False: The set $S = \{0.d_1 d_2 \ldots d_n \mid n \in \mathbb{N}\}$ of all *finitely long* decimal expansions (with each d_i an integer between 0 and 9) is countable.

1.109 True or False: The set $S = \left\{ \frac{p}{q} \sqrt{2} \mid \frac{p}{q} \in \mathbb{Q} \right\}$ is *uncountable*.

---

[7]http://www-history.mcs.st-andrews.ac.uk/history/

1.110   Let $x_n = 1 + \frac{(-1)^n}{\sqrt{n}}$. If $\epsilon > 0$ find a $N \in \mathbb{N}$ sufficiently big so that $n \geq N$ implies $|x_n - 1| < \epsilon$.

1.111   True or Give a Counterexample: A bounded sequence times a convergent sequence must be a convergent sequence.

1.112   Find $\bigcap_{n=1}^{\infty}(-\infty, -n]$.

1.113   Give an example of an open cover $\mathcal{O} = \{O_n \mid |n \in \mathbb{N}\}$ of the set

$$S = \left\{\frac{1}{n} \,\middle|\, n \geq 2, \, n \in \mathbb{N}\right\}$$

such that S has no finite subcover from $\mathcal{O}$.

1.114   Let

$$E = \left\{\frac{1}{n} \,\middle|\, n \in \mathbb{N}\right\} \cup \{0\}.$$

True or False: The set $\mathbb{R} \setminus E$, that is the complement of E, is an open subset of $\mathbb{R}$.

CHAPTER 2

CONTINUOUS FUNCTIONS

2.1 LIMITS OF FUNCTIONS

During the 19th century, many mathematicians worked to identify those classes of functions to which useful techniques, such as differentiation, integration, and decomposition into infinite series could be applied correctly. Mathematicians and physical scientists had worked with many types of functions described by formulas. These included polynomials, quotients of polynomials, and combinations of these with roots and powers. Sines, cosines, tangents, exponential and logarithmic functions played a role too. But in order to discover a coherent body of theorems that would explain to which functions what techniques could be properly applied, it was necessary to identify the underlying properties that functions would need to have for certain theorems to work. Those properties could be shared by functions that might be described in terms of very different-looking formulas.

The formal concept of a function includes all the familiar examples and many others that are not so easily described. A function f assigns a numerical value $f(x)$ to each point x in the domain D_f on which f is defined. The range of f is the set

$$R_f = \{f(x) \mid x \in D_f\}. \subseteq \ codomain$$

Advanced Calculus: An Introduction to Linear Analysis. By Leonard F. Richardson
Copyright © 2008 John Wiley & Sons, Inc.

Since derivatives, integrals, and infinite series are all defined in terms of limits, we need to define the concept of the *limit* of $f(x)$ as $x \to a$. In order to understand the subtleties of the definition that is required, we look ahead to one of the most important applications of this concept, which will be the definition of the derivative of a function. The student will recall that in elementary calculus the derivative of f at a point x is defined as the limit as $h \to 0$ of the *difference quotient*

$$Q(h) = \frac{f(x+h) - f(x)}{h}.$$

Thus $f'(x) = \lim_{h \to 0} Q(h)$. It is important to note that we wish to define the limit of $Q(h)$ as $h \to 0$, although Q *is not even defined at 0*. Thus we are forced to define $\lim_{x \to a} f(x)$ in such a way that it will be irrelevant whether or not the function f is actually defined at a. On the other hand, for $\lim_{x \to a} f(x)$ to make sense, it will have to be possible for x to become as close as we like to a *without* x being a itself. Thus we formulate the following preliminary concept.

Definition 2.1.1 *A point a is called a* cluster point *of the set D if and only if for all $\delta > 0$ there exists $x \in D$ such that $0 < |x - a| < \delta$.*

Thus a cluster point a of a set D has the property that it is always possible to find points $x \in D$ for which $x \neq a$ and yet x is as close to a as we like.

■ **EXAMPLE 2.1**

Let $D = [0, 1)$, the interval closed at 0 but open at 1. Then the set of cluster points of D is the interval $[0,1]$. See Exercise 2.1.

■ **EXAMPLE 2.2**

Let $D = \{\frac{1}{n} \mid n \in \mathbb{N}\}$. Then D has only one cluster point: namely, the point 0. Note that in this example $0 \notin D$. See Exercise 2.2.

We are ready to define the concept of the limit of a function, which the reader should compare with Exercise 2.18.

Definition 2.1.2 *Let a be any cluster point of the domain D_f of a function f. Then*

$$\lim_{x \to a} f(x) = L$$

if and only if for each $\epsilon > 0$ there exists $\delta > 0$ such that

$$x \in D_f \text{ and } 0 < |x - a| < \delta \implies |f(x) - L| < \epsilon.$$

■ EXAMPLE 2.3

We present four useful examples of limits.

(1.) We claim that for all m and b in $\mathbb{R}$ we have

$$\lim_{x \to a} (mx + b) = ma + b.$$

We understand implicitly that the domain of $mx + b$ is the whole real line $\mathbb{R}$. Now, let $\epsilon > 0$. We need $\delta > 0$ such that $0 < |x - a| < \delta$ implies

$$|(mx + b) - (ma + b)| < \epsilon.$$

But the latter inequality is equivalent to

$$|m(x - a)| = |m||x - a| < \epsilon$$

to which the solution is

$$|x - a| < \frac{\epsilon}{|m|},$$

provided that $m \neq 0$. (If $m = 0$, then *all* x would satisfy the required inequality.)

(2.) We claim that

$$\lim_{x \to 1} \frac{x^2 - 1}{x - 1} = 2.$$

Since the concept of limit does not permit $x = 1$ we can say that for all $x \neq 1$ we have $\frac{x^2-1}{x-1} = x + 1 \to 2$ as $x \to 1$.

(3.) Let

$$f(x) = \begin{cases} 0 & \text{if } x \neq 0, \\ 1 & \text{if } x = 0. \end{cases}$$

Then $\lim_{x \to 0} f(x) = 0 \neq f(0)$.

(4.) Let

$$f(x) = \begin{cases} 1 & \text{if } x \geq 0, \\ 0 & \text{if } x < 0. \end{cases}$$

Then $\lim_{x \to 0} f(x)$ does not exist. In fact, no matter how small we make $\delta > 0$, the inequality $0 < |x - 0| < \delta$ will be satisfied by points x for which $f(x)$ can be either 1 or 0, and we cannot force both 1 and 0 to be simultaneously within arbitrarily small $\epsilon > 0$ of any one number L.

In mathematics it is very important to have *equivalent forms* of definitions. For some purposes, one form of a definition is most convenient to use; for other purposes, another form is more suitable.

Theorem 2.1.1 *Suppose a is a cluster point of the domain D_f of f. Then the following two statements are logically equivalent.*

  i. *$\lim_{x \to a} f(x) = L$.*

 ii. *For every sequence of points $x_n \in D_f \setminus \{a\}$ such that $x_n \to a$, we have the sequence $f(x_n) \to L$.* (This is called the Sequential Criterion for Limits.)

Proof: Let us prove first that (i) implies (ii). So suppose (i) and consider any sequence of points $x_n \in D_f \setminus \{a\}$ such that $x_n \to a$. We know for all $\epsilon > 0$ there exists $\delta > 0$ such that $x \in D_f$ and $0 < |x - a| < \delta$ implies $|f(x) - L| < \epsilon$. But since $x_n \to a$, there exists N such that $n \geq N$ implies $0 < |x_n - a| < \delta$. This implies $|f(x_n) - L| < \epsilon$, so that $f(x_n) \to L$.

Next we prove (ii) implies (i). So suppose (ii) is true. We need to show $\lim_{x \to a} f(x) = L$. We will show that if this conclusion *were false* then a self-contradiction would result. But if it *were* false that $\lim_{x \to a} f(x) = L$, then there *would exist* $\epsilon > 0$ such that for all $\delta > 0$ there exists $x \in D_f$ such that $0 < |x-a| < \delta$ and yet $|f(x) - L| \geq \epsilon$. In particular, if we let $\delta_n = \frac{1}{n}$, then we get x_n such that

$$0 < |x_n - a| < \frac{1}{n} \text{ and } |f(x_n) - L| \geq \epsilon.$$

Now, $x_n \in D_f \setminus \{a\}$ and $x_n \to a$ yet $f(x_n) \not\to L$. This contradicts (ii) which was assumed to be true.   ∎

Remark 2.1.1 Since this is the first theorem of Chapter 2, we remind the reader of the importance of writing out a full analysis of each proof in the reader's own words and to his or her own satisfaction. This has been discussed in the Introduction on page xxiii. For the first implication, (i) implies (ii), the reasoning follows from a careful reading of the relevant definitions. The opposite implication is trickier and benefits much from the choice of an indirect method of proof. Without the indirect proof, it would be very difficult to establish that there is a suitable δ corresponding to each $\epsilon > 0$, just from knowing that $x_n \to a$ implies $f(x_n) \to f(a)$.

If we have two functions f and g we can form the sum, difference, and product of these functions with the domain of this combination being $D_f \cap D_g$. On the other hand, the domain $D_{\frac{f}{g}}$ of the quotient will be only those points of $D_f \cap D_g$ for which $g(x) \neq 0$.

Corollary 2.1.1 *Suppose a is a cluster point of the intersection $D_f \cap D_g$ of the domains of f and g, and suppose further that*

$$\lim_{x \to a} f(x) = L \text{ and } \lim_{x \to a} g(x) = M.$$

Then

  i. *$\lim_{x \to a} (f \pm g)(x) = L \pm M$*

ii. $\lim_{x \to a} (fg)(x) = LM$

iii. $\lim_{x \to a} \frac{f}{g}(x) = \frac{L}{M}$ *provided that* $M \neq 0$ *and that* a *is a cluster point of* $D_{\frac{f}{g}}$.

Proof: Consider first conclusion (i). Consider a sequence of points

$$x_n \in D_f \cap D_g \setminus \{a\}$$

such that $x_n \to a$. Then

$$(f \pm g)(x_n) = f(x_n) \pm g(x_n) \to L \pm M$$

by the corresponding theorem for limits of sequences. The proof of (ii) is very similar. The proof of (iii) is Exercise 2.3.   ∎

Remark 2.1.2 It is possible to make up many variations on the definition of limit of a function, for assorted specialized uses. Here are two examples, in which we adapt the sequential condition of Theorem 2.1.1 to generalize the concept of limit. The reader should compare the definitions below with the statements in Exercise 2.17.

Definition 2.1.3 *We define* limits at infinity *and also* one-sided limits *as follows.*

i. *If the domain* D_f *is not bounded above, we say* $f(x) \to L$ *as* $x \to \infty$ *provided that for all sequences* $x_n \in D_f$ *such that* $x_n \to \infty$ *we have* $f(x_n) \to L$.

ii. *If* a *is a cluster point of* $D_f \cap (a, \infty)$*, we say* $\lim_{x \to a+} f(x) = L$*, provided that for all sequences* $x_n \in D_f \cap (a, \infty)$ *such that* $x_n \to a$ *we have* $f(x_n) \to L$. *In this case, we write* $f(a+) = \lim_{x \to a+} f(x)$.

In both cases, L can be a real number, in which case we speak of *convergence* of $f(x)$ to L. However, if L is $\pm\infty$, then we speak of the *divergence* of $f(x)$ to L.

EXERCISES

2.1   † Prove the claim of Example 2.1.

2.2   † Prove the claim of Example 2.2.

2.3   † Prove part (iii) of Corollary 2.1.1.

2.4   Find all the cluster points of the set $\mathbb{Q}$ of all rational numbers, and justify your conclusion.

2.5   Find all the cluster points of the set $\mathbb{Z}$ of all integers, and justify your conclusion.

2.6   Show that if

$$f(x) = \begin{cases} \sin \frac{1}{x} & \text{if } 0 < x \leq 1, \\ 0 & \text{if } x = 0, \end{cases}$$

then $\lim_{x \to 0+} f(x)$ does not exist. (See Fig. 2.2.)

2.7

    a) Suppose $f(x) \le g(x) \le h(x)$ for all $x \in (a - \delta, a + \delta) \setminus \{a\}$. Suppose also that $\lim_{x \to a} f(x)$ and $\lim_{x \to a} h(x)$ both exist and equal $L \in \mathbb{R}$. Prove that $\lim_{x \to a} g(x)$ exists and equals L. (This statement is sometimes called the *squeeze theorem* or the *sandwich theorem* for functions.)

    b) Use the squeeze theorem for functions to prove that

$$\lim_{x \to 0} \frac{\sin x}{x}$$

    exists and equals 1. (Hint: You may assume your prior knowledge that $\lim_{x \to 0} \sec x = 1$.)

2.8   † A function f is called monotone increasing, denoted by $f \nearrow$, provided whenever $x_1 < x_2$ in D_f we have $f(x_1) \le f(x_2)$. Prove: If f is monotone increasing on $\mathbb{R}$, then for all $a \in \mathbb{R}$,

$$\lim_{x \to a+} f(x)$$

exists and is a real number. The latter limit is called the *limit from the right*, and it is denoted by $f(a+)$. (Hint: Let $S = \{f(x) \mid x > a\}$ and show that $\inf(S)$ is a real number L. Then show that for every sequence $x_n \to a+$ we must have $f(x_n) \to L$.)

2.9   † Adapt Definition 2.1.3 to formulate the concept of a *limit from the left*, $\lim_{x \to a-} f(x)$, and prove that if f is increasing on $\mathbb{R}$ then for all $a \in \mathbb{R}$ we have $\lim_{x \to a-} f(x) = f(a-)$ exists and is a real number.

2.10

    a) Suppose $f : D \to \mathbb{R}$ and a is a cluster point of $D \cap (a, \infty)$. Prove that $\lim_{x \to a+} f(x)$ exists and equals $L \in \mathbb{R}$ if and only if for each $\epsilon > 0$ there exists $\delta > 0$ such that $x \in D \cap (a, a + \delta)$ implies $|f(x) - L| < \epsilon$.

    b) Suppose $f : D \to \mathbb{R}$ and a is a cluster point of $D \cap (-\infty, a)$. Prove that $\lim_{x \to a-} f(x)$ exists and equals $L \in \mathbb{R}$ if and only if for each $\epsilon > 0$ there exists $\delta > 0$ such that $x \in D \cap (a - \delta, a)$ implies $|f(x) - L| < \epsilon$.

2.11   Use the Definition you constructed in Exercise 2.9 to prove that if a is a cluster point of both $D_f \cap (a, \infty)$ and $D_f \cap (-\infty, a)$, then $\lim_{x \to a} f(x)$ exists and equals L if and only if $\lim_{x \to a+} f(x)$ and $\lim_{x \to a-} f(x)$ both exist and equal L. (Hint: The result of Exercise 2.10 may be helpful.)

2.12   † A function f is said to have a *jump discontinuity* at a if and only if $\lim_{x \to a+} f(x) \neq \lim_{x \to a+} f(x)$ although both one-sided limits exist. Generalize Exercises 2.8 and 2.9 to show that if a monotone function f has a discontinuity at a point then that discontinuity must be a jump discontinuity.

2.13   Find

a)

$$\lim_{x \to a} \frac{x^2 - a^2}{x - a}.$$

b)

$$\lim_{x \to a} \frac{x^n - a^n}{x - a},$$

where $n \in \mathbb{N}$.

2.14   Show

$$\lim_{x \to \infty} \frac{a_n x^n + \cdots a_1 x + a_0}{b_n x^n + \cdots b_1 x + b_0} = \frac{a_n}{b_n},$$

provided that $b_n \neq 0$.

2.15   Suppose there exists $a \in \mathbb{R}$ such that $x > a$ implies $f(x) > 0$. Prove:

$$\lim_{x \to \infty} f(x) = \infty \Leftrightarrow \lim_{x \to \infty} \frac{1}{f(x)} = 0.$$

2.16   Prove the Cauchy Criterion for Limits of Functions: Suppose a is a cluster point of the domain D_f of f. Then $\lim_{x \to a} f(x)$ exists if and only if for all $\epsilon > 0$ there exists $\delta > 0$ such that x and y in

$$(D_f \setminus \{a\}) \cap (a - \delta, a + \delta)$$

implies $|f(x) - f(y)| < \epsilon$. (Hint: For one direction the Completeness Axiom or the Cauchy Criterion for sequences of real numbers is helpful.)

2.17   † Use Definition 2.1.3 to prove the following statements.
   a) Suppose $f(x) \to L \in \mathbb{R}$ as $x \to \infty$. Prove that for each $\epsilon > 0$ there exists $M > 0$ such that for all $x \in D_f \cap (M, \infty)$ we have $|f(x) - L| < \epsilon$.
   b) Suppose D_f is not bounded above, and suppose that for all $\epsilon > 0$ there exists $M > 0$ such that $x \in D_f \cap (M, \infty) \implies |f(x) - L| < \epsilon$. Prove that $f(x) \to L \in \mathbb{R}$ as $x \to \infty$.
   c) Suppose that a is a cluster point of $D_f \cap (a, \infty)$ and $\lim_{x \to a+} f(x)$ exists and equals $L \in \mathbb{R}$. Prove that for each $\epsilon > 0$ there exists $\delta > 0$ such that $x \in D_f \cap (a, a + \delta)$ implies $|f(x) - L| < \epsilon$.
   d) Suppose a is a cluster point of $D_f \cap (a, \infty)$ and that for each $\epsilon > 0$ there exists $\delta > 0$ such that $x \in D_f \cap (a, a + \delta)$ implies $|f(x) - L| < \epsilon$. Prove that $\lim_{x \to a+} f(x)$ exists and equals $L \in \mathbb{R}$.

2.18   Let $f : E \to \mathbb{R}$ and $\epsilon > 0$ be arbitrary. If p is *not* a cluster point of E and if $L \in \mathbb{R}$ is *arbitrary*, prove that there exists $\delta > 0$ with the following property: for all $e \in E$ such that $0 < |e - p| < \delta$ we have $|f(e) - L| < \epsilon$. (This exercise explains why the concept of limit of a function is meaningless at points that are not cluster points of the domain E of f.)

2.2 CONTINUOUS FUNCTIONS

The intuitive concept is that a function is continuous (on an interval) provided its graph can be drawn *without lifting the pencil from the paper.* The formal concept is meant to embody the requirement that there can be no abrupt jumps or gaps in the values of the function, but it says more than this.

Definition 2.2.1 *A function f is called* continuous *at a point* $a \in D$, *the domain of f, provided that for all $\epsilon > 0$ there exists $\delta > 0$ such that*

$$x \in D \cap (a - \delta, a + \delta) \implies |f(x) - f(a)| < \epsilon.$$

If f is continuous at every point $a \in D$, we say $f \in C(D)$, the set of all continuous functions on D.

or, $\forall \epsilon > 0, \ \exists \delta > 0$ s.th. $x \in D$, with $|x-a| < \delta \Rightarrow |f(x)-f(a)| < \epsilon$

We remark that in the special case in which D is an interval, such as $[a, b] \subset \mathbb{R}$, we denote $C(D)$ by $C[a, b]$.

There is a surprising degenerate case of continuity. If there exists $\delta > 0$ such that $D_f \cap (a - \delta, a + \delta) = \{a\}$, then the stipulation in the definition of continuity is *true,* in that there are no points in the designated intersection other than a itself, and of course $|f(a) - f(a)| < \epsilon$.

Definition 2.2.2 *We call a point $a \in D$ an* isolated *point of $D \subseteq \mathbb{R}$, provided that there exists $\delta > 0$ such that* $D \cap (a - \delta, a + \delta) = \{a\}$. i.e, if there exists $N_\delta(a)$ that contains no points other than a

In other words, if D_f has any isolated points a, then f is automatically continuous at a. The more interesting case, however, is that of a nonisolated point $a \in D_f$. Note that $a \in D_f$ is a nonisolated point, provided that $a \in D_f$ and that a is a *cluster point* of D_f.

Theorem 2.2.1 *A function f is continuous at a cluster point $a \in D_f$ if and only if $\lim_{x \to a} f(x)$ exists and*

$$\lim_{x \to a} f(x) = f(a).$$

Proof: First suppose f is continuous at $a \in D_f$. Then for all $\epsilon > 0$ there exists $\delta > 0$ such that $|x - a| < \delta$ and $x \in D_f$ implies $|f(x) - f(a)| < \epsilon$. Hence for all $x \in D_f$ such that $0 < |x - a| < \delta$ we have $|f(x) - f(a)| < \epsilon$, which implies $\lim_{x \to a} f(x)$ exists and $\lim_{x \to a} f(x) = f(a)$.

Now suppose $\lim_{x \to a} f(x)$ exists and is $f(a)$. Then for all $\epsilon > 0$ there exists $\delta > 0$ such that $0 < |x - a| < \delta$ and $x \in D_f$ implies $|f(x) - f(a)| < \epsilon$. This implies for all $x \in D_f \cap (a - \delta, a + \delta)$ we have $|f(x) - f(a)| < \epsilon$ so that f is continuous at a. ∎

Corollary 2.2.1 *A function $f : D \to \mathbb{R}$ is continuous at $a \in D$ if and only if it satisfies the following* Sequential Criterion for Continuity: *For every sequence of points $x_n \in D$ such that $x_n \to a$ we have $f(x_n) \to f(a)$.*

Proof:

i. Suppose a is an isolated point of D. In this case there is little to prove, since f is automatically continuous at a and since x_n as in the Sequential Criterion must have the property that there exists $N \in \mathbb{N}$ such that $n \geq N \implies x_n = a$ and $f(x_n) = f(a)$, which implies that $f(x_n) \to f(a)$.

ii. Suppose a is a cluster point of D. We will apply Theorem 2.1.1 and Corollary 2.2.1. If f is continuous at a, then $\lim_{x \to a} f(x)$ exists and equals $f(a)$. Hence the Sequential Criterion for limits implies that the sequential criterion for Continuity is satisfied. Conversely, if the Sequential Criterion for continuity is satisfied, then $\lim_{x \to a} f(x)$ exists and equals $f(a)$.

■

Theorem 2.2.2 *Suppose both f and g are continuous at $a \in D_f \cap D_g$. Then*

i. *$f \pm g$ is continuous at a.*

ii. *fg is continuous at a.*

iii. *$\frac{f}{g}$ is continuous at a provided $g(a) \neq 0$.*

iv. *If, moreover, h is continuous at $f(a)$, then the composition $h \circ f$ is continuous at a, where $h \circ f(x) = h(f(x))$.*

Proof: We will apply the Sequential Criterion for Continuity (Corollary 2.2.1) throughout.

i. For the first case, bear in mind that $D_{f \pm g} = D_f \cap D_g$. If $x_n \to a$ with all $x_n \in D_f \cap D_g$, then $f(x_n) \to f(a)$ and $g(x_n) \to g(a)$. We conclude that $\lim_{x \to a} (f \pm g)(x)$ exists and equals

$$\lim_{x \to a} f(x) \pm \lim_{x \to a} g(x) = f(a) \pm g(a).$$

ii. This part has a nearly identical proof.

iii. This part is left to the reader in Exercise 2.19.

iv. We note that $D_{h \circ f} = \{x \in D_f \mid f(x) \in D_h\}$. Now let $x_n \in D_{h \circ f}$ such that $x_n \to a$. Then letting $y_n = f(x_n)$, we have

$$\lim_{n \to \infty} h(f(x_n)) = \lim_{n \to \infty} h(y_n) = h(f(a))$$

since $y_n \to f(a)$ as $n \to \infty$ since f and h are continuous at a and $f(a)$, respectively.

■

■ **EXAMPLE 2.4**

Examples of Continuity and Discontinuity.

1. Let $i(x) = x$ for all $x \in \mathbb{R}$. Then i is continuous at every point $a \in \mathbb{R}$. In fact, if $x_n \to a$ we have $i(x_n) \equiv x_n \to a = i(a)$ for all $a \in \mathbb{R}$.

2. Let $f(x) = c$, a constant. It is easy to check that f satisfies the Sequential Criterion for Continuity at every $a \in \mathbb{R}$ so that $f \in \mathcal{C}(\mathbb{R})$.

3. Let $f(x) = x^2$, for all $x \in \mathbb{R}$. By Theorem 2.2.2, $f \in \mathcal{C}(\mathbb{R})$. This generalizes easily to show that $g(x) = x^n$ is in $\mathcal{C}(\mathbb{R})$ for all $n \in \mathbb{N}$. Now apply Theorem 2.2.2 to see that every polynomial $p(x) = a_n x^n + \cdots a_1 x + a_0$ is continuous on all of $\mathbb{R}$ as well. Moreover, every *rational* function $Q(x) = \frac{p(x)}{q(x)}$, where p and q are polynomials, is continuous wherever defined.

4. Let
$$f(x) = \begin{cases} 1 & \text{if } x \geq 0, \\ 0 & \text{if } x < 0. \end{cases}$$

Then f is *dis*continuous at 0 but continuous everywhere else. Why?

5. Let
$$f(x) = \begin{cases} 1 & \text{if } x \geq 0, \\ 0 & \text{if } x = -1. \end{cases}$$

Then f is continuous wherever it is defined. Note that -1 is an *isolated* point of D_f.

6. Let
$$f(x) = \begin{cases} x^2 & \text{if } x \in \mathbb{Q}, \\ -x^2 & \text{if } x \notin \mathbb{Q}. \end{cases}$$

We claim that that f is continuous at $x = p$ if and only if $p = 0$. Perhaps the easiest way to prove this is to utilize Corollary 2.2.1. Consider two sequences $\bar{x}_n \in \mathbb{Q}$ and $\tilde{x}_n \notin \mathbb{Q}$ such that $\bar{x}_n \to p$ and $\tilde{x}_n \to p$. If f is continuous at p, then

$$f(\bar{x}_n) = \bar{x}_n^2 \to p^2 = f(p) \text{ and } f(\tilde{x}_n) = -\tilde{x}_n^2 \to -p^2 = f(p).$$

If f is continuous at p, then $p^2 = -p^2$ and $p = 0$.

It remains to prove that f actually *is* continuous at 0. Suppose now that x_n is any sequence converging to 0. We need to prove that $f(x_n) \to f(0) = 0$. Let $\epsilon > 0$. Note that the sequences x_n^2 converges to 0, as does the sequence $-x_n^2$. Moreover, there exists $N \in \mathbb{N}$, *corresponding to* ϵ, such that $n \geq N$ implies

$$\left| x_n^2 - 0 \right| = x_n^2 < \epsilon.$$

It follows also that $n \geq N$ implies $\left|-x_n^2 - 0\right| < \epsilon$. Now let $n \geq N$. If $x_n \in \mathbb{Q}$, then $|f(x_n) - 0| = x_n^2 < \epsilon$. On the other hand, if $x_n \notin \mathbb{Q}$, then

$$|f(x_n) - 0| = \left|-x_n^2 - 0\right| = x_n^2 < \epsilon.$$

Thus $n \geq N$ implies $|f(x_n) - f(0)| < \epsilon$ so that $f(x_n) \to f(0) = 0$.

EXERCISES

2.19   † Prove part (iii) of Theorem 2.2.2. The domain $D_{\frac{f}{g}}$ is that subset of $D_f \cap D_g$ consisting of points x for which $g(x) \neq 0$.

2.20   † Prove that if $a(x) = |x|$, the absolute value function, then $a \in \mathcal{C}(\mathbb{R})$. (Use Exercise 1.9.)

2.21   † Let $Q(x) = \sqrt{x}$, which is defined for all $x \geq 0$. Prove: $Q \in \mathcal{C}[0, \infty)$. (Hint: If $a \geq 0$, and $\epsilon > 0$, we seek $\delta > 0$ such that $x \geq 0$ and $|x - a| < \delta$ implies $|Q(x) - Q(a)| < \epsilon$. Begin by showing that $\left|\sqrt{x} - \sqrt{a}\right|^2 \leq |x - a|$.)

2.22   Let f be continuous at a, and $\epsilon > 0$. Show there exists $\delta > 0$ such that $x_1, x_2 \in D_f \cap (a - \delta, a + \delta)$ implies $|f(x_1) - f(x_2)| < \epsilon$.

2.23   Let

$$f(x) = \begin{cases} 1 - x^2 & \text{if } x \in \mathbb{Q}, \\ 0 & \text{if } x \notin \mathbb{Q}. \end{cases}$$

Find *all* points p at which f is *continuous*. Prove your conclusion.

2.24   Let

$$f(x) = \begin{cases} 1 - x & \text{if } x \in \mathbb{Q}, \\ 1 - x^2 & \text{if } x \notin \mathbb{Q}. \end{cases}$$

Prove that f is continuous at p if and only if $p \in \{0, 1\}$.

2.25   † Let

$$f(x) = \begin{cases} \frac{x^n - a^n}{x - a} & \text{if } x \neq a, \\ c & \text{if } x = a, \end{cases}$$

where $n \in \mathbb{N}$. Find the value of c that makes $f \in \mathcal{C}(\mathbb{R})$.

2.26   ◇ Let f be monotone increasing on $[a, b]$. For all $p \in (a, b)$, we denote $\lim_{x \to p+} f(x) = f(p+)$ and $\lim_{x \to p-} f(x) = f(p-)$ both of which exist. Let

$$j(p) = f(p+) - f(p-)$$

denote the *height* of the *jump* at p. Let

$$j(b) = f(b) - f(b-) \quad \text{and} \quad j(a) = f(a+) - f(a)$$

and let

$$E_n = \left\{ x \,\middle|\, j(x) \geq \frac{1}{n} \right\}$$

for all $n \in \mathbb{N}$.

    a) Prove: f is continuous at $x \in [a, b] \Leftrightarrow j(x) = 0$.

    b) Show that the set E of points of *discontinuity* of f is given by

$$E = \bigcup_{n \in \mathbb{N}} E_n.$$

    c) If $a = z_0 \leq x_1 < z_1 < x_2 < z_2 < \cdots < x_k \leq z_k = b$, prove that $j(x_i) \leq f(z_i) - f(z_{i-1})$ for all i.

    d) Prove that E_n is finite for all n. Hint: If $p_1 < \ldots < p_k$ all lie in E_n, show that

$$\frac{k}{n} \leq \sum_{i=1}^{k} j(p_i) \leq f(b) - f(a).$$

    e) Prove: The set E is either *countable* or else finite.

2.27   Let $f : \mathbb{R} \to \mathbb{R}$ be any function such that $f(x + y) \equiv f(x) + f(y)$. (Such a function is called a *homomorphism* of the additive group $\mathbb{R}$ in abstract algebra.) Suppose also that f is *continuous*, and let $c = f(1)$.

    a) For each $n \in \mathbb{N}$, prove that $f\left(\frac{1}{n}\right) = \frac{c}{n}$.

    b) Let $x \in \mathbb{Q}$, so that x can be written as $\frac{p}{q}$. Prove that $f(x) = cx$.

    c) Use the continuity of f to prove that $f(x) = cx$ for all $x \in \mathbb{R}$.

2.28 $\dagger \diamond$ [8] Let $f : D \to \mathbb{R}$ and suppose $p \in D$.

    a) Let $s(\delta) = \sup\{f(x) \mid x \in D \cap (p - \delta, p + \delta)\}$. Prove that s is a monotone increasing function of $\delta \in (0, \infty)$.

    b) Let $i(\delta) = \inf\{f(x) \mid x \in D \cap (p - \delta, p + \delta)\}$. Prove that i is a monotone decreasing function of $\delta \in (0, \infty)$.

    c) Define the *oscillation* $o(p) = \lim_{\delta \to 0+}(s(\delta) - i(\delta))$ and prove that $o(p)$ exists.

    d) Prove that f is continuous at a point $p \in D$ if and only if $o(p) = 0$.

2.3   SOME PROPERTIES OF CONTINUOUS FUNCTIONS

Lemma 2.3.1 (Preservation of Sign) *Suppose f is continuous at p and suppose $f(p) > 0$. Then there exists $\delta > 0$ such that $x \in D_f \cap (p - \delta, p + \delta)$ implies $f(x) > 0$.*

(No matter how close to 0 we take p.)

[8]This exercise is used only for the proof Theorem 11.3.1, which is Lebesgue's criterion for Riemann integrability.

Proof: Let $\epsilon = \frac{f(p)}{2}$, which is positive. Then there exists $\delta > 0$ such that

$$x \in D_f \cap (p - \delta, p + \delta) \implies |f(x) - f(p)| < \epsilon,$$

$$-\varepsilon < f(x) - f(p) < \varepsilon$$

which implies

$$0 < \frac{f(p)}{2} < f(x) < \frac{3f(p)}{2}. \qquad f(p) - \varepsilon < f(x) < f(p) + \varepsilon$$

This proves the lemma. ∎

See Exercises 2.29 and 2.30 for generalizations of this lemma.

Definition 2.3.1 *We say that f has the* Intermediate Value Property *on an interval I if and only if for all a and b in I with $a < b$ and for all k strictly between $f(a)$ and $f(b)$, there exists $c \in (a, b)$ such that $f(c) = k$.*

Theorem 2.3.1 (Intermediate Value Theorem) *Suppose f is continuous on the interval I. Then f has the Intermediate Value Property on I.*

Proof: We will suppose a and b are in I with $a < b$. As a first case we assume $f(a) < k < f(b)$: For the opposite case see Exercise 2.31. We will let $a_1 = a$ and $b_1 = b$ and we will use the method of interval-halving. Let $m = \frac{a_1 + b_1}{2}$.

1. If $f(m) > k$, then let $b_2 = m$ and $a_2 = a_1$.

2. Otherwise, let $a_2 = m$ and $b_2 = b_1$.

Then halve the interval $[a_2, b_2]$ and proceed as above to construct a decreasing nest of closed finite intervals $[a_n, b_n]$. Note that $|a_n - b_n| \to 0$ as $n \to \infty$. By the *Nested Intervals Theorem* (Theorem 1.6.1) there exists a unique

$$c \in \bigcap_{n \in \mathbb{N}} [a_n, b_n]$$

such that $a_n \to c$ and $b_n \to c$ as $n \to \infty$. Since $c \in [a, b]$, f is continuous at c. Thus

$$f(c) = \lim_{k \to \infty} f(a_k) \leq k \quad \text{and} \quad f(c) = \lim_{k \to \infty} f(b_k) \geq k.$$

These two inequalities can be satisfied only if $f(c) = k$. ∎

■ **EXAMPLE 2.5**

Now we can prove that $\mathbb{R}$ has an $\sqrt[n]{p}$ for all $n \in \mathbb{N}$ and for all $p > 0$. Let $f(x) = x^n$ and note that $f \in C(\mathbb{R})$. Also, $f(0) = 0 < p$ and

$$f(1 + p) = (1 + p)^n$$
$$= 1^n + n1^{n-1}p^1 + \cdots + n1^1 p^{n-1} + p^n$$
$$> np \geq p.$$

Thus $f(0) < p < f(1 + p)$ and so there exists $c \in (0, 1 + p)$ such that $f(c) = c^n = p$, and $c = \sqrt[n]{p}$.

Let us comment a bit more upon this important example. In section 1.3 we considered a sequence of successive *decimal approximations* to $\sqrt{2}$. Thus, a_k was the largest k-decimal place rational number such that $a_k^2 < 2$. Thus

$$\left(a_k + \frac{1}{10^k} \right)^2 > 2 \quad \text{and} \quad \left| a_k - \sqrt{2} \right| < 10^{-k} \to 0$$

as $k \to \infty$. Hence $a_k \to \sqrt{2}$. We remark that the Intermediate Value Theorem, like the method of interval-halving itself, is very useful for the task of solving equations that involve continuous functions. See Exercises 2.32 and 2.33.

If $f \in C[a, b]$, for a closed finite interval [a,b], then we can conclude more than just the continuity of f at each point $c \in [a, b]$. The next theorem shows that f will be *uniformly continuous* on [a,b]. We present a definition first.

Definition 2.3.2 *We say f is* uniformly continuous *on a domain D if and only if for all $\epsilon > 0$ there exists $\delta > 0$ such that for all x_1 and x_2 in D such that $|x_2 - x_1| < \delta$ we have $|f(x_2) - f(x_1)| < \epsilon$.*

That is, f is uniformly continuous on D provided there exists $\delta > 0$ that satisfies the requirements for continuity at x_1 uniformly, meaning at *all* $x_1 \in D$. Intuitively, uniform continuity of f on D imposes a restriction on how fast f can change its values on the entire domain D.

Theorem 2.3.2 *If $f \in C[a, b]$, then f is uniformly continuous on $[a, b]$.*

Proof: We suppose the theorem were false and we will deduce a contradiction. Thus we suppose f is *not* uniformly continuous on [a,b]. Hence there exists $\epsilon > 0$ such that no $\delta > 0$ will suffice the meet the uniform continuity condition. Hence, if $\delta_n = \frac{1}{n}$, there exists $x_n, y_n \in [a, b]$ such that $|x_n - y_n| < \frac{1}{n}$ but

$$|f(x_n) - f(y_n)| \geq \epsilon.$$

Since x_n is a bounded sequence, the Bolzano–Weierstrass Theorem guarantees the existence of a convergent subsequence $x_{n_k} \to p \in [a, b]$ as $k \to \infty$. Now,

$$|y_{n_k} - p| \leq |y_{n_k} - x_{n_k}| + |x_{n_k} - p| \to 0 + 0 = 0.$$

Thus $y_{n_k} \to p$ as well. Therefore, since f is continuous at p,

$$f(p) = \lim_{k \to \infty} f(x_{n_k}) = \lim_{k \to \infty} f(y_{n_k}).$$

This means that

$$|f(x_{n_k}) - f(y_{n_k})| \leq |f(x_{n_k}) - f(p)| + |f(p) - f(y_{n_k})| \to 0 + 0 = 0.$$

But $|f(x_{n_k}) - f(y_{n_k})| \geq \epsilon$ for all k and so we have a contradiction. ∎

EXERCISES

2.29   † Suppose f is continuous at p and $f(p) > c$. Prove: There exists $\delta > 0$ such that $x \in D_f \cap (p - \delta, p + \delta)$ implies $f(x) > c$. (Hint: Consider $g(x) = f(x) - c$.)

2.30   † Suppose f is continuous at p and $f(p) < c$. Prove: There exists $\delta > 0$ such that $x \in D_f \cap (p - \delta, p + \delta)$ implies $f(x) < c$. (Hint: Consider $g(x) = -f(x)$.)

2.31   † Suppose $f \in \mathcal{C}[a, b]$ and $f(a) > k > f(b)$. Prove: There exists $c \in (a, b)$ such that $f(c) = k$. This will complete the proof of Theorem 2.3.1. (Hint: Consider $g(x) = -f(x)$.)

2.32   Let $p(x) = x^3 + 3x^2 - 2x - 1$. Prove that the polynomial equation $p(x) = 0$ has a *root* somewhere in the interval $(0, 1)$.

2.33   Let $p(x) = x^4 + x^3 - 2x^2 + x + 1$. Prove that the polynomial equation $p(x) = 0$ has a root in $(-1, 0)$.

2.34   Let $f(x) = \cos x - x - \sin x$. Prove: There exists a solution to the equation $f(x) = 0$ on the interval $[0, \pi/6]$. (You may assume that $\sin x$ and $\cos x$ are continuous on $\mathbb{R}$.)

2.35   Let $p(x) = a_{2n+1} x^{2n+1} + \cdots + a_1 x + a_0$ be any polynomial of *odd degree*. Prove: The equation $p(x) = 0$ has at least one real root. (Hint: Prove $p(x) \to \pm\infty$ depending on whether $x \to \pm\infty$ and consider then that p must have both positive and negative values.)

2.36   A function $f : (-a, a) \to \mathbb{R}$ is called an *odd function* provided $f(-x) = -f(x)$ for all $x \in (-a, a)$. A function $f : (-a, a) \to \mathbb{R}$ is called an *even function* provided $f(-x) = f(x)$ for all $x \in (-a, a)$. *Prove or give a counterexample* for each of the following statements.

    a) If f is an odd continuous function on $(-a, a)$, then there exists a solution to the equation $f(x) = 0$.

    b) The composition of two odd functions must be odd.

    c) Every function of an even function is even.

    d) An even function of an odd function is even.

    e) An even function of any function is even.

    f) Every function $f : (-a, a) \to \mathbb{R}$ is the sum of an even function with an odd function.

2.37   † Prove the following *fixed point theorem*: Suppose $f \in \mathcal{C}[0, 1]$ and suppose $0 \le f(x) \le 1$ for all $x \in [0, 1]$. Then there exists $c \in [0, 1]$ such that $f(c) = c$. (The point c is then called a *fixed point* for the function f. Hint: Consider $g(x) = f(x) - x$.)

2.38   Prove the following theorem: Suppose $f \in \mathcal{C}[0, 1]$ and suppose $0 \le f(x) \le 1$ for all $x \in [0, 1]$. Let $n \in \mathbb{N}$. Then there exists $c \in [0, 1]$ such that $f(c) = c^n$. (Hint: Compare this problem with Exercise 2.37.)

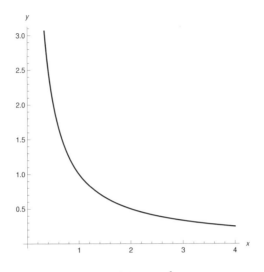

Figure 2.1 $y = \frac{1}{x}$.

2.39   Suppose f and g are in $\mathcal{C}[a, b]$ and suppose $[f(a) - g(a)][f(b) - g(b)] \leq 0$. Prove that there exists $c \in [a, b]$ such that $f(c) = g(c)$. (Hint: Compare this problem with Exercise 2.37.)

2.40   Suppose f is uniformly continuous on $(a, m]$ and also on $[m, b)$. Prove: f is uniformly continuous on (a, b).

2.41   Let $f(x) = \frac{1}{x}$, for all $x \in (0, 1)$. Is f continuous on $(0, 1)$? Is f uniformly continuous on $(0, 1)$? Justify your conclusions. (See Fig. 2.1.)

2.42   Let $f(x) = \frac{1}{x}$, for all $x \in (1, \infty)$. Is f continuous on $(1, \infty)$? Is f uniformly continuous on $(1, \infty)$? Is f uniformly continuous on $(0, \infty)$? Justify your conclusions. (See Fig. 2.1.)

2.43   Let $f(x) = x^2$, for all $x \in \mathbb{R}$. Is $f \in \mathcal{C}(\mathbb{R})$? Is f uniformly continuous on $\mathbb{R}$? Justify your conclusions.

2.44   Suppose f is uniformly continuous on D and suppose $E \subset D$. Prove: f is uniformly continuous on E as well.

2.45   Is $f(x) = x^2$ uniformly continuous on $(0,1)$? Why or why not?

2.46   Let $Q(x) = \sqrt{x}$. Prove that the function Q is uniformly continuous on $[0, \infty)$. (Hint: See Exercise 2.21.)

2.47   Let

$$f(x) = \begin{cases} \sin \frac{1}{x} & \text{if } 0 < x \leq 1, \\ 0 & \text{if } x = 0. \end{cases}$$

You may assume that the function $\sin x$ is continuous. (See Fig. 2.2.)

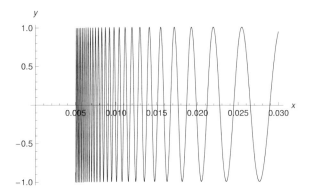

Figure 2.2 $y = \sin \frac{1}{x}$.

a) Show that f does have the *intermediate value property* (Definition 2.3.1) on $[0, 1]$, but $f \notin C[0, 1]$.

b) Show $f \in C(0, 1]$ but f is *not uniformly* continuous on $(0, 1]$.

2.4 EXTREME VALUE THEOREM AND ITS CONSEQUENCES

Definition 2.4.1 *We say a function f is bounded on a domain D, provided that there exist m and M in $\mathbb{R}$ such that*

$$m \leq f(x) \leq M$$

for all $x \in D$.

Observe that $f(x) = x$ is continuous on $\mathbb{R}$ but f is not bounded on $\mathbb{R}$. And $g(x) = \frac{1}{x}$ is continuous on $(0,1)$ but is not bounded, since it lacks an upper bound, though there is a lower bound, 0. We have, however, the following important theorem pertaining to continuous functions on *closed, finite* intervals.

Theorem 2.4.1 *If $f \in C[a, b]$, then f is bounded on $[a, b]$.*

Proof: By Theorem 2.3.2, f is uniformly continuous on $[a, b]$. Thus, for example, if we pick the positive number $\epsilon = 1$, there exists $\delta > 0$ such that for all $x_1, x \in [a, b]$ such that $|x_1 - x| < \delta$ we must have $|f(x_1) - f(x)| < 1$. Now consider the sequence of evenly spaced points defined as follows:

$$x_1 = a, x_2 = a + \delta, x_3 = a + 2\delta, \ldots, x_k = a + (k - 1)\delta, \ldots.$$

By the Archimedean Property of $\mathbb{R}$, there exists $N \in \mathbb{N}$ such that $a + N\delta > b$, whereas $a + (N - 1)\delta \leq b$. In other words, although δ may be a small positive number, if we take enough steps of size δ we must eventually pass b. Now consider the following inequalities.

$$x \in [x_1, x_2) \quad \Longrightarrow \quad f(x_1) - 1 < f(x) < f(x_1) + 1$$
$$x \in [x_2, x_3) \quad \Longrightarrow \quad f(x_2) - 1 < f(x) < f(x_2) + 1$$
$$x \in [x_3, x_4) \quad \Longrightarrow \quad f(x_3) - 1 < f(x) < f(x_3) + 1$$
$$\vdots$$
$$x \in [x_N, b] \quad \Longrightarrow \quad f(x_N) - 1 < f(x) < f(x_N) + 1.$$

If we let

$$m = \min\{f(x_1) - 1, f(x_2) - 1, \dots, f(x_N) - 1\}$$

and if we let

$$M = \max\{f(x_1) + 1, f(x_2) + 1, \dots, f(x_N) + 1\},$$

then we have for all $x \in [a, b]$, $m < f(x) < M$.   ∎

Observe that the function $f(x) = x$ is bounded on $D = (0, 1)$. However, f has *no largest value and no smallest value*, since for all $x \in (0, 1)$, there exists $x' < x < x''$ with $x', x'' \in (0, 1)$.

Definition 2.4.2 *We say that f has a* maximum *value M on a domain D, provided that there exists $x_M \in D$ such that*

$$f(x) \le f(x_M) = M$$

for all $x \in D$. Similarly, we say f has a minimum *value m on D, provided that there exists $x_m \in D$ such that*

$$f(x) \ge f(x_m) = m$$

for all $x \in D$.

The next theorem establishes that on a *closed, finite* interval $[a, b]$, every continuous function must have both a maximum and a minimum value.

Theorem 2.4.2 (Extreme Value Theorem) *If $f \in \mathcal{C}[a, b]$, then f has both a maximum and a minimum value on $[a, b]$.*

Proof: By Theorem 2.4.1, we know that $\{f(x) \mid x \in [a, b]\} \ne \emptyset$ is bounded both above and below. Let $M = \sup\{f(x) \mid x \in [a, b]\}$. Thus M is the least upper bound of the *range* of f. Hence for all $k \in \mathbb{N}$, $M - \frac{1}{k}$ is too small to be an upper bound, so there exists $x_k \in [a, b]$ such that

$$M - \frac{1}{k} < f(x_k) \le M.$$

Since x_k is a bounded sequence, the Bolzano–Weierstrass Theorem guarantees that there exists a *convergent* subsequence x_{n_k}. Thus there exists x_M such that $x_{n_k} \to x_M$ as $k \to \infty$. Since $[a, b]$ is closed, $x_M \in [a, b]$. But

$$M - \frac{1}{n_k} < f(x_{n_k}) \leq M$$

and n_k is an increasing sequence of natural numbers. Thus

$$n_1 \geq 1, n_2 \geq 2, \ldots, n_k \geq k, \ldots.$$

Hence $M - \frac{1}{k} < f(x_{n_k}) \leq M$, for all $k \in \mathbb{N}$. Since $M - \frac{1}{k} \to M$ as $k \to \infty$, and because f is continuous at x_M, $f(x_M) = \lim_{k \to \infty} f(x_{n_k}) = M$. For the minimum point, see Exercise 2.48. ∎

■ EXAMPLE 2.6

Let $p(x) = x^2 - x + 1$. We claim that p must have a minimum value on $\mathbb{R}$. The problem is that although $p \in \mathcal{C}(\mathbb{R})$, $\mathbb{R}$ is not a finite closed interval so the Extreme Value Theorem does not apply directly. Consider

$$p(x) = x^2 \left(1 - \frac{1}{x} + \frac{1}{x^2} \right).$$

The second factor approaches 1 as $x \to \pm\infty$. Thus, there exists M such that $|x| > M$ implies

$$\left(1 - \frac{1}{x} + \frac{1}{x^2} \right) > \frac{1}{2}.$$

Observe that $p(0) = 1$ and if $|x| > \sqrt{2}$ then $x^2 > 2$. Hence if we set $a = \max\{\sqrt{2}, M\}$, then $|x| > a$ implies $p(x) > 1 = p(0)$. Now, $p \in \mathcal{C}[-a, a]$ so the Extreme Value Theorem guarantees that there exists $x_m \in [-a, a]$ such that $p(x_m) \leq p(x)$ for all $x \in [-a, a]$. But this implies that

$$p(x_m) \leq p(0) = 1 \leq f(x)$$

for all values of x since $p(x)$ exceeds 1 outside $[-a, a]$. Hence $p(x_m) \leq p(x)$, for all $x \in \mathbb{R}$.

Theorem 2.4.3 *Let $f \in \mathcal{C}[a, b]$. Then the* range *of f is a closed, finite interval.*

Proof: Recall that the *range* of f is

$$f([a, b]) = \{f(x) \mid x \in [a, b]\}.$$

Since f achieves a maximum value M and a minimum value m, the Intermediate Value Theorem assures us that f achieves every value between M and m. Hence $f([a, b]) = [m, M]$. ∎

Definition 2.4.3 *If f is any function whatever (not necessarily continuous) on a domain of definition D, we define the* sup-norm *of f to be*

$$\|f\|_{\sup} = \sup\{|f(x)| \mid x \in D\}.$$

← $\|f\|_{sup}$ is just the bound M, where
$-M \le f(x) \le M$ i.e $|f(x)|$

Thus $\|f\|_{\sup}$ may be either positive real-valued or else ∞. However, if $f \in \mathcal{C}[a, b]$, then $|f| \in \mathcal{C}[a, b]$ as well, since the absolute value function is continuous and compositions of continuous functions must be continuous. Thus, for all $f \in \mathcal{C}[a, b]$, $\|f\|_{\sup} < \infty$, since f is *bounded*. For any function f the reader will show in Exercise 2.50 that f *is bounded if and only if* $\|f\|_{\sup} < \infty$.

Theorem 2.4.4 *Let f and g be any bounded functions on $[a, b]$ and $\alpha \in \mathbb{R}$. Then*

(*i.*) $\|f\|_{\sup} = 0$ *if and only if $f \equiv 0$.*

(*ii.*) $\|\alpha f\|_{\sup} = |\alpha| \, \|f\|_{\sup}$.

(*iii.*) $\|f + g\|_{\sup} \le \|f\|_{\sup} + \|g\|_{\sup}$.

Item 3 is called the *triangle inequality*.

Proof: Items (i) and (ii) are immediate from the definitions. For 3, consider that for all $x \in [a, b]$ we must have

$$|(f + g)(x)| \le |f(x)| + |g(x)| \le \|f\|_{\sup} + \|g\|_{\sup}.$$

But $\|f + g\|_{\sup}$ is the *least* upper bound of the set of values $|(f + g)(x)|$, and the right side above is *an upper bound* of the same set of values. ∎

In Theorem 2.2.2 we learned that if f and g are in $\mathcal{C}[a, b]$, then $f \pm g \in \mathcal{C}[a, b]$ too, as is αf for all $\alpha \in \mathbb{R}$. Thus continuous functions on [a,b] can be added and subtracted giving other continuous functions. The constant function 0 is an additive identity: $f + 0 = f$. And addition and subtraction of functions in $\mathcal{C}[a, b]$ as well as multiplication by constants enjoy all the usual properties of shared by any so-called *vector space*. We list the axioms of a general (real or complex) vector space in Table 2.1. (See [10] for the concept of a vector space.) In linear algebra, the student will have studied vectors in $\mathbb{R}^n$ primarily, and also the concept of an abstract vector space. So in effect we have learned here that $\mathcal{C}[a, b]$ is a vector space.

In a vector space, it is convenient to have defined a type of positive real-valued function on vectors called a *norm*. (In physics, the norm of a vector is called its *magnitude*.)

Definition 2.4.4 *A normed vector space V, over a field $\mathbb{F}$ (which is either $\mathbb{R}$ or $\mathbb{C}$) is a vector space equipped with a real-valued function called a* norm, *denoted by $\| \cdot \|$, provided that for all $\mathbf{v}$ and $\mathbf{w}$ in V and for all $\alpha \in \mathbb{F}$ we have*

i. $\|\mathbf{v}\| \ge 0$, *and* $\|\mathbf{v}\| = 0$ *if and only if* $\mathbf{v} \equiv \mathbf{0}$ (Positive Definite).

ii. $\|\alpha \mathbf{v}\| = |\alpha| \|\mathbf{v}\|$ (Homogeneous).

iii. $\|\mathbf{v} + \mathbf{w}\| \le \|\mathbf{v}\| + \|\mathbf{w}\|$ *(Triangle Inequality).*

Here $|\alpha|$ *refers to the* absolute value *of* α *if* $\alpha \in \mathbb{R}$ *and to the* modulus *of* α *if* $\alpha \in \mathbb{C}$.

Thus the sup-norm is an example of a norm, and $\mathcal{C}[a, b]$ is a *normed vector space,* or *normed linear space,* as it is also called. The reason a norm is convenient is that it gives us a concept analogous to that of the *length* of a vector, and then the distance between the vectors, or continuous functions f and g, would be understood to be $\|f - g\|_{\sup}$. In the next section we will study convergence of sequences of continuous functions on $[a, b]$ in terms of the sup-norm concept of *distance.*

In a linear algebra class the student will have learned that in a finite-dimensional vector space V there is a *minimal* natural number n, called the dimension, such that any set of more than n vectors must be linearly dependent. That is, if $v_1, \ldots, v_{n+1}$ are in V then there exist $\alpha_1, \ldots, \alpha_{n+1}$ in $\mathbb{R}$, not all 0, such that

$$\sum_{k=1}^{n+1} \alpha_k v_k = 0.$$

Table 2.1 Axioms of a Vector Space

A **vector space** (or *linear space*) over a field $\mathbb{F}$, which may be either the field $\mathbb{R}$ of real numbers or the field $\mathbb{C}$ of complex numbers, is a set V of elements called *vectors* that is equipped with two operations: *addition of vectors* and *multiplication of vectors by scalars* (also called *numbers*) from $\mathbb{F}$. The following axioms must be satisfied for V to be a vector space.

1. If $\mathbf{u}$ and $\mathbf{v}$ in V and $\alpha \in \mathbb{F}$, then $\mathbf{u} + \mathbf{v} \in V$ and $\alpha\mathbf{v} \in V$. (Closure under addition of vectors and scalar multiplication)

2. If $\mathbf{u}$ and $\mathbf{v}$ in V, then $\mathbf{u} + \mathbf{v} = \mathbf{v} + \mathbf{u}$. (Commutativity of vector addition)

3. If $\mathbf{u}, \mathbf{v}$ and $\mathbf{w}$ in V, then $\mathbf{u} + (\mathbf{v} + \mathbf{w}) = (\mathbf{u} + \mathbf{v}) + \mathbf{w}$. (Associativity of vector addition)

4. There exists a unique vector $\mathbf{0} \in V$ such that $\mathbf{u} + \mathbf{0} = \mathbf{u}$ for all $\mathbf{u} \in V$. (Existence of additive identity)

5. For each $\mathbf{u} \in V$ there exists a unique vector $-\mathbf{u} \in V$ such that $-\mathbf{u} + \mathbf{u} = \mathbf{0}$. (Existence of additive inverse)

6. For all $\mathbf{u} \in V$, $1\mathbf{u} = \mathbf{u}$. (The *scalar* $1 \in \mathbb{F}$ is a multiplicative identity.)

7. For all a and b in $\mathbb{F}$ and for all $\mathbf{u} \in V$, $(ab)\mathbf{u} = a(b\mathbf{u})$. (Associativity of scalar multiplication)

8. For all a and b in $\mathbb{F}$ and all $\mathbf{u}$ and $\mathbf{v}$ in
   $V, a(\mathbf{u}+\mathbf{v}) = a\mathbf{u}+a\mathbf{v}$ and $(a+b)\mathbf{u} = a\mathbf{u}+b\mathbf{u}$. (Distributivity of scalar multiplication)

The field $\mathbb{F}$ *will almost always be the field* $\mathbb{R}$ *of real numbers in this book. The one exception is that in* Chapter 6 *(Fourier series)* $\mathbb{F}$ *can be the field* $\mathbb{C}$ *of complex numbers.*

(See, for example, [10].) In Exercise 2.57 below, the reader will show that the vector space $C[a, b]$ must be *infinite-dimensional* because there is no $n \in \mathbb{N}$ that can serve as the dimension of this vector space.

EXERCISES

2.48 † Complete the proof of Theorem 2.4.2 by proving for all $f \in C[a, b]$ the existence of a minimum point x_m, such that $f(x_m) \leq f(x)$ for all $x \in [a, b]$.

2.49 Let $p(x) = a_{2n}x^{2n} + \cdots + a_1 x + a_0$ be any polynomial of *even degree*. Prove: If $a_{2n} > 0$, then p has a *minimum* value on $\mathbb{R}$.

2.50 Prove: A function f is *bounded* if and only if $\|f\|_{\sup} < \infty$. (Hint: See Definition 2.4.1.)

2.51 Give an alternative proof for Theorem 2.4.1: If $f \in C[a, b]$, then f is bounded. (Hint: Suppose false. Then $|f|$ is not bounded above. By Exercise 2.50, for all $n \in \mathbb{N}$ there exists $x_n \in [a, b]$ such that

$$|f(x_n)| > n.$$

Now apply the Bolzano–Weierstrass Theorem to x_n and deduce a contradiction of the fact that $f \in C[a, b]$.)

2.52 Prove that $C[a, b]$ is a vector space.

2.53 Let $\mathcal{B}[a, b]$ denote the set of all bounded functions on $[a, b]$. Is $\mathcal{B}[a, b]$ a vector space? Prove your conclusion.

2.54 For each $n \in \mathbb{N}$, let X_n be the set of polynomials of degree equal to n, and let $P_n = \bigcup_{k=0}^{n} X_k$. Prove or give a counterexample:
   a) The set X_n is a vector space.
   b) The set P_n a vector space.

2.55 Let P denote the set of all functions f on $[0, 1]$ such that $f(x) \geq 0$ for all $x \in [0, 1]$. Is P a vector space? Justify your conclusion.

2.56 Let $\tilde{P}$ denote the set of all functions f on $[0, 1]$ such that there exists at least one point $x \in [0, 1]$, perhaps depending on f, for which $f(x) > 0$. Is P a vector space? Justify your conclusion.

2.57 † Let $S = \{1, x, x^2, x^3, \ldots, x^n, \ldots\}$, a list of infinitely many continuous functions on [a,b]. Show that if $n \in \mathbb{N}$ and if $\alpha_0, \alpha_1, \ldots, \alpha_n$ are not all 0, then it is not possible for $\sum_{k=0}^{n} \alpha_k x^k = 0$, the zero function. (Remark: This exercise shows that $C[a, b]$ is an infinite-dimensional vector space, because it cannot have finite dimension n for any $n \in \mathbb{N}$.)

2.58 Given an example of $f \in C(a, b)$ for which the range of f is
   a) a finite open interval.
   b) an infinite open interval.

2.59 Find $\|f\|_{\sup}$ if

a) $f(x) = x$ on $\left(-1, \frac{1}{2}\right)$.

b) $f(x) = \begin{cases} 0 & \text{if } x \in \mathbb{Q}, \\ -x^2 & \text{if } x \notin \mathbb{Q}. \end{cases}$

2.5 THE BANACH SPACE $\mathcal{C}[a, b]$

We begin with a definition that is applicable whenever a sequence of functions f_n is defined on a common domain, D. We do not require that the functions be continuous in the following definition.

Definition 2.5.1 *If f and f_n are all defined on a domain D, we say that $f_n \to f$ uniformly on D provided*

$$\| f_n - f \|_{\sup} \to 0$$ $ie, \ \sup\left\{ |f_n(x) - f(x)| : x \in D \right\}$

as $n \to \infty$. (Here the sup-norm is taken over the domain D.)

A helpful way to visualize the concept of uniform convergence is to picture, for each given $\epsilon > 0$, the curves $f(x) + \epsilon$ and $f(x) - \epsilon$. Then $f_n \to f$ uniformly on D if and only if for each $\epsilon > 0$ there exists $N \in \mathbb{N}$ such that $n \geq N$ implies that the graph of $y = f_n(x)$ is sandwiched between the graph of $f(x) + \epsilon$ and that of $f(x) - \epsilon$. The following theorem shows that uniform convergence behaves very well with respect to the continuity of functions.

Theorem 2.5.1 *Suppose $f_n \in \mathcal{C}(D)$ for all n, and suppose also that $f_n \to f$ uniformly on D. Then $f \in \mathcal{C}(D)$.*

In words, this theorem says that a uniform limit of continuous functions must be continuous.

Proof: Let $x \in D$ be arbitrary. We must show f is continuous at x. Let $\epsilon > 0$. We need to show there exists $\delta > 0$ such that $x' \in D \cap (x - \delta, x + \delta)$ implies $|f(x') - f(x)| < \epsilon$. Since $f_n \to f$ uniformly on D, there exists $N \in \mathbb{N}$ such that $k \geq N$ implies $\| f_k - f \|_{\sup} < \frac{\epsilon}{3}$. In particular, this means

$$\| f_N - f \|_{\sup} < \frac{\epsilon}{3}.$$

Since f_N is continuous at x, there exists $\delta > 0$ such that $x' \in D \cap (x - \delta, x + \delta)$ implies

$$|f_N(x') - f_N(x)| < \frac{\epsilon}{3}.$$

We claim that this δ works for the function f as well. In fact, if $x' \in D \cap (x - \delta, x + \delta)$ then

$$|f(x') - f(x)| = \Big|[f(x') - f_N(x')] + [f_N(x') - f_N(x)] + [f_N(x) - f(x)]\Big|$$
$$\leq |f(x') - f_N(x')| + |f_N(x') - f_N(x)| + |f_N(x) - f(x)|$$
$$< \frac{\epsilon}{3} + \frac{\epsilon}{3} + \frac{\epsilon}{3}$$
$$= \epsilon.$$

■

Definition 2.5.2 *We say that the sequence of functions $f_n \to f$ pointwise on D if and only if, for all $x \in D$, the sequence of real numbers $f_n(x)$ converges to the real number $f(x)$.*

The following example will illustrate the difference between uniform and pointwise convergence.

■ **EXAMPLE 2.7**

Let $f_n(x) = x^n$, for all $x \in [0, 1]$. We claim f_n does not converge uniformly on $[0, 1]$. If $0 \leq x < 1$, then we know $x^n \to 0$. But if $x = 1$, $x^n = 1^n \to 1$. Thus $f_n \in C[0, 1]$ for all $n \in \mathbb{N}$, and $f_n \to f$ *pointwise* on D, where

$$f(x) = \begin{cases} 0 & \text{if } 0 \leq x < 1, \\ 1 & \text{if } x = 1. \end{cases}$$

So the continuous functions $f_n \to f$ and yet $f \notin C[0, 1]$. The function f could not have failed to be continuous on [0,1] if the convergence had been uniform.

■ **EXAMPLE 2.8**

Now consider $f_n(x) = x^n$, but on the domain $x \in [0, \frac{1}{2}]$. Here we see that

$$\|f_n - 0\|_{\sup} = f_n\left(\frac{1}{2}\right) = \frac{1}{2^n} \to 0$$

so $f_n \to 0$ *uniformly* on $D = [0, \frac{1}{2}]$. This illustrates how the question of uniform convergence is affected by the choice of domain as well as by the sequence of functions f_n. Note that on $[0, \frac{1}{2}]$ the limit function 0 is indeed continuous.

Theorem 2.5.2 *If $f_n \to f$ uniformly on D, then $f_n \to f$ pointwise on D, but not conversely.*

Proof: Observe that for each $x \in D$, we must have

$$|f_n(x) - f(x)| \leq \|f_n - f\|_{\sup}.$$

Thus if $\|f_n - f\|_{\sup} \to 0$, it follows that $|f_n(x) - f(x)| \to 0$ for each $x \in D$. That the converse is false is shown by Example 2.7.   ∎

■ EXAMPLE 2.9

Let

$$f_n(x) = x^{1 + \frac{1}{n}}.$$

It is not hard to see that $f_n(x) \to f(x) = x$ pointwise for all $x \in [0, 1]$. We claim that this convergence is actually uniform. For this purpose it can be convenient to make use of the derivative, even though we have not yet reached our chapter on this subject, but using what we recall from elementary calculus about derivatives. Let $g_n(x) = f(x) - f_n(x)$ for all x, so that $g_n(0) = 0 = g_n(1)$ and $g_n(x) \geq 0$ for all $x \in [0, 1]$. The reader can set $g_n'(x) = 0$ and check that $g_n(x)$ achieves its maximum value at

$$x_n = \frac{1}{\left(1 + \frac{1}{n}\right)^n}$$

and that

$$\|f - f_n\|_{\sup} = g_n(x_n) = \frac{1}{\left(1 + \frac{1}{n}\right)^n} \left(1 - \frac{1}{1 + \frac{1}{n}}\right) \to \frac{1}{e}(1 - 1) = 0.$$

Thus $f_n \to f$ uniformly on $[0, 1]$. The graph showing this sequence of functions converging uniformly to $f(x) = x$ is shown by the right-hand half of the graph in Fig. 4.5.

This theorem assures us that the only possible uniform limit f for a sequence of functions f_n would be the pointwise limit f. If the sequence fails to be pointwise convergent, then it cannot be uniformly convergent. However, a sequence of functions can be pointwise convergent on a domain D without being uniformly convergent on D.

Definition 2.5.3 *In a vector space V equipped with a norm, we say that $\mathbf{v}_n \to \mathbf{v}$ in the norm if and only if $\|\mathbf{v}_n - \mathbf{v}\| \to 0$ as $n \to \infty$. We call a sequence $\mathbf{v}_n$ a* Cauchy sequence *if and only if for all $\epsilon > 0$ there exists $N \in \mathbb{N}$ such that $n \geq m \geq N$ implies $\|\mathbf{v}_n - \mathbf{v}_m\| < \epsilon$. A normed vector space will be called* complete *if and only if every Cauchy sequence converges. A complete normed vector space is known also as a (real)* Banach Space, *after the discoverer, Stefan Banach.*

Theorem 2.5.3 *In any normed vector space V, if a sequence $v_n \in V$ converges, then it is a Cauchy sequence.*

Proof: First, suppose $v_n \to v$, which we understand to mean uniform convergence on $[a, b]$. We need to show that the sequence v_n is Cauchy. Let $\epsilon > 0$. Then there exists $N \in \mathbb{N}$, *corresponding to* ϵ, such that $n \geq N$ implies $\|v_n - v\| < \frac{\epsilon}{2}$. Now, if

n and $m \geq N$,

$$\|v_n - v_m\| = \|[v_n - v] + [v - v_m]\|$$
$$\leq \|v_n - v\| + \|v - v_m\|$$
$$< \frac{\epsilon}{2} + \frac{\epsilon}{2} = \epsilon.$$

Thus the sequence v_n is Cauchy. ∎

■ **EXAMPLE 2.10**

In the vector space $\mathcal{C}[a, b]$, equipped with the sup-norm, we say that $f_n \to f$ in *the norm* if and only if $\|f_n - f\|_{\sup} \to 0$ as $n \to \infty$. This is also called *uniform convergence*. We call the sequence f_n a *Cauchy sequence in the sup-norm* (or a *uniformly Cauchy sequence*) if and only if for all $\epsilon > 0$ there exists $N \in \mathbb{N}$ such that $n \geq m \geq N$ implies $\|f_n - f_m\|_{\sup} < \epsilon$.

Theorem 2.5.4 *The vector space $\mathcal{C}[a, b]$ equipped with the sup-norm is a complete normed vector space.*

Proof: We suppose that f_n is a Cauchy sequence (in the sup-norm sense defined above), and we must prove there exists $f \in \mathcal{C}[a, b]$ such $f_n \to f$. Let $\epsilon > 0$. Then there exists $N \in \mathbb{N}$ such that $n \geq m \geq N$ implies $\|f_n - f_m\|_{\sup} < \frac{\epsilon}{2}$. For each fixed $x \in [a, b]$,

$$|f_n(x) - f_m(x)| \leq \|f_n - f_m\|_{\sup} < \frac{\epsilon}{2}.$$

Thus, for all fixed $x \in [a, b]$, the sequence $f_n(x)$ is a Cauchy sequence of real numbers. By the Completeness Axiom, $f_n(x)$ must converge. So we can define $f(x) = \lim_{n \to \infty} f_n(x)$, for all $x \in [a, b]$. Now, for all n and $m \geq N$,

$$|f_n(x) - f_m(x)| < \frac{\epsilon}{2}.$$

Holding x temporarily fixed in $[a, b]$, let $m \to \infty$, and we find that $|f_n(x) - f(x)| \leq \frac{\epsilon}{2}$. Since this works for each such x, $\|f_n - f\|_{\sup} \leq \frac{\epsilon}{2} < \epsilon$, so $f_n \to f$ uniformly on $[a, b]$, and $f \in \mathcal{C}[a, b]$. ∎

Because uniform limits of continuous functions must be continuous, it is natural to ask the following question. Suppose $f_n \in \mathcal{C}[a, b]$ for all n, and suppose $f_n \to f$ pointwise. *If* the function $f \in \mathcal{C}[a, b]$, would this guarantee that actually $f_n \to f$ uniformly as well? The answer is *no*, as the following example shows.

■ **EXAMPLE 2.11**

We define a sequence $f_n \in \mathcal{C}[a, b]$, for all $n \geq 2$, as follows. Let

$$f_n(x) = \begin{cases} nx & \text{if } 0 \leq x \leq \frac{1}{n}, \\ 2 - nx & \text{if } \frac{1}{n} < x \leq \frac{2}{n}, \\ 0 & \text{if } \frac{2}{n} < x \leq 1. \end{cases}$$

The student should draw the graph of $f_n(x)$ to observe its properties clearly. One sees that $f_n(0) \equiv 0 \to 0$. But if $0 < x \le 1$, then there exists $N \in \mathbb{N}$ such that $n \ge N$ implies $\frac{2}{n} < x$, which implies $f_n(x) = 0$. Thus $f_n \to 0$ pointwise on $[0, 1]$, and indeed the constant function 0 is continuous on $[0, 1]$ as well. Nevertheless, the student sees from the graph that $\|f_n - 0\|_{\sup} \equiv 1 \not\to 0$, so f_n does *not* converge uniformly on $[0, 1]$.

Although the question posed just above Example 2.11 is *no*, there is what may be called a *partial* converse to Theorem 2.5.1.

Theorem 2.5.5 (Dini) *Suppose* $f_n \in C[a, b]$, *for all* $n \in \mathbb{N}$, *and suppose*

$$f_n(x) \to f(x)$$

at least pointwise on $[a, b]$, *where* $f \in C[a, b]$ *as well. If for each fixed* $x \in [a, b]$ *the sequence of numbers* $f_n(x)$ *is* monotone decreasing, *then* $f_n \to f$ *uniformly on* $[a, b]$. *(There is also an* increasing *version of this theorem–see Exercise 2.70.)*

Proof: Let us suppose for all $x \in [a, b]$ the sequence of numbers $f_n(x)$ is a *decreasing* sequence converging to the limit $f(x)$. We wish to show that

$$\|f_n - f\|_{\sup} \to 0$$

as $n \to \infty$. For convenience, denote $h_n = f_n - f$, so $h_n \in C[a, b]$ and for all $x \in [a, b]$, $h_n(x)$ is a decreasing sequence of positive numbers approaching 0. Let $\epsilon > 0$. For each fixed $x \in [a, b]$, there exists $N_x \in \mathbb{N}$ such that $n \ge N_x$ implies $0 \le h_n(x) < \frac{\epsilon}{2}$. Since $h_{N_x} \in C[a, b]$, there exists $r_x > 0$ such that $|x' - x| < r_x$ implies

$$|h_{N_x}(x') - h_{N_x}(x)| < \frac{\epsilon}{2},$$

which implies $0 \le h_{N_x}(x') < \epsilon$ and for all $n \ge N_x$ and for all $x' \in (x - r_x, x + r_x)$, we have $0 \le h_n(x') < \epsilon$. Next observe that

$$[a, b] \subset \bigcup_{x \in [a, b]} (x - r_x, x + r_x),$$

an open cover of $[a, b]$. By the Heine–Borel Theorem, there exists a finite subcover:

$$[a, b] \subset \bigcup_{k=1}^{n} (x_k - r_{x_k}, x + r_{x_k}).$$

Now let

$$N = \max\{N_{x_k} \mid k = 1, 2, \ldots, n\}.$$

If $n \ge N$ and $x \in [a, b]$, then there exists $k \in \{1, \ldots, n\}$ such that

$$x \in (x_k - r_{x_k}, x_k + r_{x_k})$$

and $N \ge N_{x_k}$ as well. Thus $0 \le h_n(x) < \epsilon$; and since $h_n \in C[a, b]$, we have $\|h_n - 0\|_{\sup} < \epsilon$ as well. Thus $h_n \to 0$ uniformly on $[a, b]$. ∎

EXERCISES

In some of these exercises it will be convenient to use the derivative to aid in finding a sup-norm.

2.60   Prove that $f_n(x) = \frac{x}{n} \to 0$ pointwise on $\mathbb{R}$ but *not* uniformly on $\mathbb{R}$. However, prove the convergence *is uniform* on $[0, 1]$.

2.61   Let $f_n(x) = \frac{1}{x^2 + n}$. Prove: $f_n \to 0$ *uniformly* on $\mathbb{R}$.

2.62   Show that $f_n(x) = \frac{x}{x+n}$ converges uniformly on $[0, 1]$ but not on $[0, \infty)$.

2.63   Let $f_n(x) = xe^{-nx}$ for all $x \in [0, \infty)$.
    a) Find $\|f_n\|_{\sup}$ for all n.
    b) Prove that $f_n \to 0$ uniformly on $[0, \infty)$.

2.64   Let $f_n(x) = nxe^{-nx}$ for all $x \in [0, \infty)$.
    a) Find $\|f_n\|_{\sup}$ for all n.
    b) Determine whether or not f_n converges uniformly on $[0, \infty)$, and prove your claim.

2.65   Let $f_n(x) = xe^{-nx^2}$ for all $x \in \mathbb{R}$.
    a) Find $\|f_n\|_{\sup}$ for all n.
    b) Determine whether or not f_n converges uniformly on $\mathbb{R}$ and prove your claim.

2.66   Let $f_n(x) = \frac{x}{1+nx^2}$ for all $x \in \mathbb{R}$.
    a) Find the pointwise limit of f_n.
    b) Does f_n converge uniformly on $\mathbb{R}$? Prove your conclusion.

2.67   Let $f_n(x) = \sin^n x$ and let $\delta > 0$.
    a) Prove: f_n converges uniformly on $[0, \frac{\pi}{2} - \delta]$, but not uniformly on $[0, \frac{\pi}{2}]$.
    b) In which sense does f_n converge on $[0, \pi/2)$? Prove your claim.

2.68   Suppose $f_n(x) = 1 - x^n$. Decide whether or not f_n converges uniformly on each interval, and prove your conclusion.

    a) $[0, 1]$ **b)** $[0, \frac{1}{2}]$
    c) $[0, b]$, where $0 \le b < 1$ **d)** $[0, 1)$

2.69   Let

$$f_n(x) = \begin{cases} n & \text{if } 0 < x < \frac{1}{n}, \\ 0 & \text{if } x \in [0, 1] \setminus \left(0, \frac{1}{n}\right), \end{cases}$$

for all $n \in \mathbb{N}$. For each $x \in [0, 1]$, find $\lim_{n \to \infty} f_n(x)$. Is the convergence uniform on $[0, 1]$? Prove your conclusion.

2.70   † Complete the proof of Theorem 2.5.5 by treating the case in which, for all $x \in [a, b]$, the sequence $f_n(x)$ is an increasing sequence of numbers converging to the limit $f(x)$. If f_n and f are all continuous, prove the convergence is uniform. (Hint: Let $g_n = -f_n$ and use the case proven already.)

2.6 TEST YOURSELF

EXERCISES

2.71   Is the set $\{f_{a,b}(x) = ax + b \mid a \in \mathbb{Q} \text{ and } b \in \mathbb{Q}\}$ of all linear functions with rational coefficients *countable* or *uncountable*?

2.72   True or Give a Counterexample: Every *open* subset of $\mathbb{R}$ is the union of *countably* many *open intervals*.

2.73   True or Give a Counterexample: If $f : [0,1] \to \mathbb{R}$ is *bounded* on $[0,1)$ but *not continuous* at 0, then the $f(x)$ converges to some $L \neq f(0)$ as $x \to 0+$.

2.74   Find all the *cluster points* of the set $\mathbb{Q}$ of all rational numbers.

2.75   True or Give a Counterexample: If $f : [0,1] \to [0,1]$ is a continuous function then there exists $c \in [0,1]$ such that $f(c) = \sqrt{c}$.

2.76   Give an example of a sequence of *unbounded* functions $f_n(x)$ that *converges uniformly* to $f(x) = \log x$ on $\mathbb{R}$.

2.77   Give an example of a *bounded, continuous* function on $(0,1)$ that is *not uniformly continuous* on $(0,1)$.

2.78   Is the given set a *vector space* or *not a vector space*?
   a) The set $\mathcal{P}_{2n-1}$ of all polynomials of odd degree.
   b) The set of all bounded functions on $\mathbb{R}$.

2.79   Find $\|f\|_{\sup}$ if $f(x) = x^3$ on $(-1, 0.5)$.

2.80 $f_n(x) = \frac{e^x}{n}$. Answer True or False:
   a) f_n converges pointwise on $\mathbb{R}$.
   b) f_n converges uniformly on $\mathbb{R}$.

2.81   Let $f_n(x) = xe^{-nx}$ for all $x \in [0, \infty)$. Find $\|f_n\|_{\sup}$.

2.82   Suppose $f_n(x) = 1 - x^n$. True or False: f_n converges uniformly on the given interval.
   a) $[0, b]$, where $0 \leq b < 1$
   b) $[0, 1]$
   c) $[0, 1)$

2.83   Give an example of a sequence $x_n \to 0+$ for which $\sin \frac{\pi}{x_n} = 1$ for all $n \in \mathbb{N}$.

2.84   Let $f(x) = x \sin \frac{\pi}{x}$ for all $x \in (0,1)$. Let $\epsilon > 0$. Find a value of $\delta > 0$ such that if x and x' are in $(0, \delta)$, then $|f(x) - f(x')| < \epsilon$.

2.85   True or False: The function $f(x) = x \sin \frac{\pi}{x}$ is uniformly continuous on $(0,1)$.

2.86   Suppose f is a monotone increasing function on $\mathbb{R}$. True or Give a Counterexample: $\lim_{x \to 0+} f(x) = f(0)$.

2.87   Let $\epsilon > 0$. Find a value of $\delta > 0$ for which whenever x and a are both in $[0, \infty)$ with $|x - a| < \delta$, this implies $|\sqrt{x} - \sqrt{a}| < \epsilon$.

2.88   Let $f_n(x) = 1 + \sin^n \frac{\pi x}{2}$ for all $x \in [0, 1]$. True or False: The sequence f_n is *uniformly* convergent on

   a) $\left[0, \frac{1}{2}\right]$
   b) $[0, b]$ for each $b \in [0, 1)$.
   c) $[0, 1)$

CHAPTER 3

RIEMANN INTEGRAL

3.1 DEFINITION AND BASIC PROPERTIES

The Riemann integral, denoted by $\int_a^b f(x)\,dx$, is an especially useful concept for both pure and applied mathematics. But it is much more difficult to define than the other, simpler limits studied earlier in this book. The integral is defined to meet the following objectives. If f is a positive-valued function defined on an interval $[a, b]$, then the integral should be the *area* of the region of the xy-plane above the interval $[a, b]$ on the x-axis and below the graph of $y = f(x)$. If f is not strictly positive, then the integral is intended to be a *signed area*, by which we mean the area between the x-axis and the positive part of the graph of f, *minus* the area between the x-axis and the negative part of the graph of $y = f(x)$. To these ends, the integral is defined as a limit of so-called Riemann sums, which are determined by f, by a so-called *partition* P of $[a, b]$, and by a choice of so-called *evaluation points* $\bar{x}_i$. Now we define all these terms carefully.

Definition 3.1.1 *A partition P is an ordered list of finitely many points starting with a and ending with $b > a$. Thus $P = \{x_0, x_1, \ldots, x_n\}$, where*

$$a = x_0 < x_1 < \cdots < x_n = b.$$

Advanced Calculus: An Introduction to Linear Analysis. By Leonard F. Richardson
Copyright © 2008 John Wiley & Sons, Inc.

These points are regarded as partitioning $[a, b]$ *into* n *contiguous subintervals,* $[x_{i-1}, x_i], i = 1, \ldots, n.$ *The length of the ith subinterval is given by* $\Delta x_i = x_i - x_{i-1}.$ *The* mesh *of the partition is denoted and defined by*

$$\|P\| = \max\{\Delta x_i \mid i = 1, 2, \ldots n\}.$$

In each of the n *subintervals, we select an arbitrary* evaluation *point* $\bar{x}_i \in [x_{i-1}, x_i].$ *We define the* Riemann sum

$$P(f, \{\bar{x}_i\}) = \sum_{i=1}^{n} f(\bar{x}_i) \Delta x_i.$$

Each summand of the Riemann sum is understood to represent the area of a rectangle of base $[x_{i-1}, x_i]$ and height $f(\bar{x}_i)$, if $f(\bar{x}_i) \geq 0$, or minus such an area if $f(\bar{x}_i) < 0$.

Definition 3.1.2 *The* Riemann integral *on a closed finite interval* $[a, b]$ *of a bounded function* f *is said to exist and have the value* $L \in \mathbb{R}$, *provided that for each* $\epsilon > 0$ *there exists* $\delta > 0$ *such that* $\|P\| < \delta$ *implies*

$$|P(f, \{\bar{x}_i\}) - L| < \epsilon \tag{3.1}$$

independent of the choice *of the partition and the evaluation points. In this case we write*

$$\int_a^b f(x) \, dx = L$$

and we write also that

$$\lim_{\|P\| \to 0} P(f, \{\bar{x}_i\}) = \int_a^b f(x) \, dx.$$

But it must be understood that this limit is in the very intricate sense of Definition 3.1.2. The family of all so-called Riemann integrable *functions* f *is denoted by* $\mathcal{R}[a, b]$. *Thus,* $f \in \mathcal{R}[a, b]$ *if and only if* $\int_a^b f(x) \, dx$ *exists.*

It must be emphasized that the required inequality 3.1 must be satisfied *independent* of the choice of P and *independent* of the choice of $\{\bar{x}_i\}$, just so long as $\|P\| < \delta$.

■ **EXAMPLE 3.1**

Let $f(x) = c$, a constant, for all $x \in [a, b]$. Then $f \in \mathcal{R}[a, b]$ and

$$\int_a^b f(x) \, dx = c(b - a).$$

The proof is simple: Just observe that for all partitions P of $[a, b]$, and for all $\{\bar{x}_i\}$, chosen evaluation points, we have

$$P(f, \{\bar{x}_i\}) = \sum_{i=1}^{n} c \Delta x_i = c \sum_{i=1}^{n} \Delta x_i = c(b - a).$$

■ **EXAMPLE 3.2**

Let

$$f(x) = \begin{cases} 1 & \text{if } x \in \mathbb{Q} \cap [0,1], \\ 0 & \text{if } x \in [0,1] \setminus \mathbb{Q}. \end{cases}$$

Then $f \notin \mathcal{R}[a,b]$. This can be understood as follows. No matter how small we make $\|P\|$, each subinterval determined by the partition must contain both rational and irrational numbers (see Section 1.6). We are free to pick all $\bar{x}_i \in \mathbb{Q}$, in which case

$$P(f, \{\bar{x}_i\}) = \sum_{i=1}^{n} 1 \Delta x_i = 1 - 0 = 1.$$

But we are free also to select all $\bar{x}_i \notin \mathbb{Q}$, in which case $P(f, \{\bar{x}_i\}) = 0$. We cannot bring both values of the Riemann sums within ϵ of the same number L if $0 < \epsilon \leq \frac{1}{2}$.

Theorem 3.1.1 *If $f \in \mathcal{R}[a,b]$, then f is* bounded *on $[a,b]$.*

Proof: We will suppose the theorem were false and deduce a contradiction. Denote by L the value of the $\int_a^b f(x)\,dx$. Then choose $\epsilon = 1$. There exists $\delta > 0$ such that $\|P\| < \delta$ implies $|P(f, \{\bar{x}_k\}) - L| < 1$, so that

$$|P(f, \{\bar{x}_k\})| < |L| + 1. \tag{3.2}$$

Let us fix P of such mesh, so as to satisfy this inequality. But if f is unbounded on $[a,b]$, there exists $i \in \{1, \ldots, n\}$ such that f is also unbounded on $[x_{i-1}, x_i]$. However,

$$|P(f, \{\bar{x}_k\})| = \left| \sum_{k=1}^{n} f(\bar{x}_k) \Delta x_k \right| \geq \left| |f(\bar{x}_i)| \Delta x_i - \left| \sum_{k \neq i} f(\bar{x}_k) \Delta x_k \right| \right|$$

since $|\alpha + \beta| \geq ||\alpha| - |\beta||$ (see Exercise 3.1). But we can hold $\bar{x}_k$ fixed for all $k \neq i$, and vary $\bar{x}_i \in [x_{i-1}, x_i]$ at will so as to make $|f(\bar{x}_i)|$ as large as we wish, violating the bound of inequality (3.2). ∎

Remark 3.1.1 We remind the reader again of the guidance for the proving of theorems in the Introduction on page xxiii. The reader should write out a full analysis of the proof of Theorem 3.1.1 as explained earlier, and just as should be done with regard to every theorem. Note how the indirect proof begins with the supposition that a Riemann integrable function exists that is unbounded. Then we appeal to the definition of the Riemann integral and show that there must be a partition with a subinterval on which the function is unbounded. This introduces an impossible variability in the contribution of that particular integral to the value of the Riemann sum, on account of the total freedom of choice of evaluation point in each interval of the partition.

■ **EXAMPLE 3.3**

Let

$$f(x) = \begin{cases} \frac{1}{x} & \text{if } 0 < x \le 1, \\ 0 & \text{if } x = 0. \end{cases}$$

Then $f \notin \mathcal{R}[0,1]$, since f is unbounded on $[0,1]$.

Theorem 3.1.2 *Let f and g be in $\mathcal{R}[a,b]$ and $c \in \mathbb{R}$. Then*

i. $f + g \in \mathcal{R}[a,b]$ *and*

$$\int_a^b (f+g)(x)\,dx = \int_a^b f(x)\,dx + \int_a^b g(x)\,dx.$$

ii. $cf \in \mathcal{R}[a,b]$ *and*

$$\int_a^b cf(x)\,dx = c\int_a^b f(x)\,dx.$$

Remark 3.1.2 Observe that Theorem 3.1.2 says $\mathcal{R}[a,b]$ is a *vector space*, since sums (and differences) of Riemann integrable functions are again Riemann integrable, and the same is true for constant multiples of Riemann integrable functions. Moreover, if we define a mapping $T : \mathcal{R}[a,b] \to \mathbb{R}$ by

$$T(f) = \int_a^b f(x)\,dx$$

then Theorem 3.1.2 says that T is a *linear* function, since

$$T(cf+g) = cT(f) + T(g).$$

A linear function from a (real) vector space to $\mathbb{R}$ is called a *linear functional*.

Proof:

i. Let $\epsilon > 0$, let

$$L = \int_a^b f(x)\,dx \ \text{ and } \ M = \int_a^b g(x)\,dx.$$

We know there exists $\delta_1 > 0$ such that $\|P\| < \delta_1$ implies

$$|P(f,\{\bar{x}_i\}) - L| < \frac{\epsilon}{2}.$$

Similarly, we know there exists $\delta_2 > 0$ such that $\|P\| < \delta_2$ implies

$$|P(g,\{\bar{x}_i\}) - M| < \frac{\epsilon}{2}.$$

Now let $\delta = \min\{\delta_1, \delta_2\} > 0$. Then $\|P\| < \delta$ implies

$$\begin{aligned}
|P(f + g, \{\bar{x}_i\}) - (L + M)| &= \left| [P(f, \{\bar{x}_i\}) - L] + [P(g, \{\bar{x}_i\}) - M] \right| \\
&\leq \left| [P(f, \{\bar{x}_i\}) - L] \right| + \left| [P(g, \{\bar{x}_i\}) - M] \right| \\
&< \frac{\epsilon}{2} + \frac{\epsilon}{2} = \epsilon.
\end{aligned}$$

Thus $f + g \in \mathcal{R}[a, b]$ and $\int_a^b (f + g)(x)\,dx = L + M$.

ii. This part is Exercise 3.2.

■

Theorem 3.1.3 *Suppose $a \leq b \leq c$. If $f \in \mathcal{R}[a, b]$ and $f \in \mathcal{R}[b, c]$, then $f \in \mathcal{R}[a, c]$ and*

$$\int_a^c f(x)\,dx = \int_a^b f(x)\,dx + \int_b^c f(x)\,dx.$$

Proof: Let $\epsilon > 0$, $L = \int_a^b f(x)\,dx$ and $M = \int_b^c f(x)\,dx$. By hypothesis, there exist $\delta_1 > 0$ and $\delta_2 > 0$ such that if P_1 is a partition of $[a, b]$ with $\|P_1\| < \delta_1$ and if P_2 is a partition of $[b, c]$ with $\|P_2\| < \delta_2$, then we have

$$|P_1(f, \{\bar{x}_i\}) - L| < \frac{\epsilon}{3} \quad \text{and} \quad |P_2(f, \{\bar{x}_i\}) - M| < \frac{\epsilon}{3}.$$

Let B denote $\|f\|_{\text{sup}}$, where the sup-norm refers to the interval $[a, c]$ and is finite. Let

$$\delta = \min\left\{\delta_1, \delta_2, \frac{\epsilon}{9B}\right\}.$$

Suppose now that P is any partition of $[a, c]$ with $\|P\| < \delta$. Unfortunately, P needn't be the union of a partition of $[a, b]$ with a partition of $[b, c]$, since we might not have $b \in P$. So we let $P' = P \cup \{b\} = P_1 \cup P_2$, so $\|P'\| < \delta$ too, and the same is true for P_1 and P_2, which are partitions of $[a, b]$ and $[b, c]$, respectively. Then

$$\begin{aligned}
|P(f, \{\bar{x}_i\}) - (L + M)| &\leq |P(f, \{\bar{x}_i\}) - P'(f, \{\bar{x}_i\})| \\
&\quad + |P'(f, \{\bar{x}_i\}) - (L + M)| \\
&\leq |P(f, \{\bar{x}_i\}) - P'(f, \{\bar{x}_i\})| \\
&\quad + |P_1(f, \{\bar{x}_i\}) - L| \\
&\quad + |P_2(f, \{\bar{x}_i\}) - M| \\
&< \epsilon,
\end{aligned}$$

which we justify as follows.

The sum of the final two summands on the right is less than $\frac{2\epsilon}{3}$ by the choice of δ. For the first summand we observe that if b was in P from the outset, then $P = P'$ and the first summand is then zero. But if $b \notin P$, we see the only non-0 contributions come from the one interval of P that contains b, and from the two intervals of P' that have b as an endpoint, since the contributions of the other intervals cancel. Thus the first summand is less than $3\delta \cdot B = \epsilon/3$.   ■

Remark 3.1.3 If $a \le b$, we *define* $\int_b^a f(x)\,dx = -\int_a^b f(x)\,dx$, which we understand to mean that if $a = b$, then the integral is zero. This enables us to extend Theorem 3.1.3 as follows. First,

$$\int_a^b f(x)\,dx + \int_b^a f(x)\,dx = 0 = \int_a^a f(x)\,dx$$

because of our definition. And if $a \le c \le b$ we have

$$\int_a^c f(x)\,dx + \int_c^b f(x)\,dx = \int_a^b f(x)\,dx$$

by Theorem 3.1.3 above, so that

$$\int_a^c f(x)\,dx = \int_a^b f(x)\,dx - \int_c^b f(x)\,dx$$
$$= \int_a^b f(x)\,dx + \int_b^c f(x)\,dx.$$

If $c < a$, then a similar argument could be given.

EXERCISES

3.1   Prove: For all α and β in $\mathbb{R}$, $|\alpha + \beta| \ge \big|\,|\alpha| - |\beta|\,\big|$. Hint:

$$|\alpha + \beta| = |\alpha - (-\beta)|.$$

3.2   † Prove part (b) of Theorem 3.1.2.

3.3   † Let $p \in [a, b]$ and define the *indicator function*

$$1_{\{p\}}(x) = \begin{cases} 1 & \text{if } x = p, \\ 0 & \text{if } x \ne p. \end{cases}$$

Prove that $1_{\{p\}} \in \mathcal{R}[a, b]$ and

$$\int_a^b 1_{\{p\}}(x)\,dx = 0.$$

(Hint: Find an upper bound on the value of $\mathcal{P}(1_{\{p\}}, \{\bar{x}_i\})$.)

3.4   † Suppose $h(x) = 0$ except at finitely many points $x_1, x_2, \ldots, x_k$ in $[a, b]$. Prove: $h \in \mathcal{R}[a, b]$ and

$$\int_a^b h(x)\,dx = 0.$$

(Hint: Use Exercise 3.3 above and Theorem 3.1.2.)

3.5 Suppose $f \in \mathcal{R}[a, b]$ and $g(x) = f(x)$ except at finitely many points in [a,b]. Prove: $g \in \mathcal{R}[a, b]$ and

$$\int_a^b g(x)\, dx = \int_a^b f(x)\, dx.$$

(Hint: Write $h(x) = g(x) - f(x)$ and apply Exercise 3.4.)

3.6 † Suppose $f \in \mathcal{R}[a, b]$ and $f(x) \geq 0$ for all $x \in [a, b]$. Prove:

$$\int_a^b f(x)\, dx \geq 0.$$

(Hint: Find a lower bound for all Riemann sums $P(f, \{\bar{x}_i\})$.)

3.7 Suppose f and g are in $\mathcal{R}[a, b]$ and $f(x) \leq g(x)$ for all $x \in [a, b]$. Prove:

$$\int_a^b f(x)\, dx \leq \int_a^b g(x)\, dx.$$

(Hint: Let $h(x) = g(x) - f(x)$. Use Theorem 3.1.2 and Exercise 3.6 above.)

3.8 †Define a *step function* on $[a, b]$ as follows. We call σ a step function if there exists a partition

$$a = x_0 < x_1 < \cdots < x_n = b$$

such that $\sigma(x) = c_i$ for all $x \in (x_{i-1}, x_i)$, $i = 1, \ldots, n$. Thus σ is constant on each open interval (x_{i-1}, x_i). The values of σ at the points of the set $\{x_0, x_1, \ldots, x_n\}$ are arbitrary. Prove: The function $\sigma \in \mathcal{R}[a, b]$ and

$$\int_a^b \sigma(x)\, dx = \sum_{i=1}^n c_i(x_i - x_{i-1}).$$

3.9 † Suppose $f \in \mathcal{R}[a, b]$ and $m \leq f(x) \leq M$, for all $x \in [a, b]$. Prove:

$$m(b - a) \leq \int_a^b f(x)\, dx \leq M(b - a).$$

3.10 † (*Mean Value Theorem for Integrals*) Let $f \in \mathcal{C}[a, b]$. You may use the result of Theorem 3.2.2 stating that $\mathcal{C}[a, b] \subset \mathcal{R}[a, b]$. Prove: There exists $\bar{x} \in [a, b]$ such that

$$\int_a^b f(x)\, dx = f(\bar{x})(b - a).$$

(Hint: Let $m = \inf\{f(x) \mid x \in [a, b]\}$ and $M = \sup\{f(x) \mid x \in [a, b]\}$. Use Exercise 3.9 above and the Intermediate Value Theorem for continuous functions.)

3.11   Suppose $f \in \mathcal{R}[a, b]$. Prove that

$$\lim_{n \to \infty} \frac{b-a}{n} \sum_{k=1}^{n} f\left(a + k\frac{b-a}{n}\right) = \int_a^b f(x)\,dx.$$

3.12   Express

$$\lim_{n \to \infty} \frac{2}{n} \sum_{k=1}^{n} \cos\left(1 + \frac{2k}{n}\right)$$

as an integral. (Be sure to include the lower and upper limits of integration.)

3.13   † Let $f \in C[a, b]$. Show that there exists a sequence of step functions σ_n (Exercise 3.8) such that $\sigma_n \to f$ uniformly on $[a, b]$. (Hint: The function f is uniformly continuous on $[a, b]$.)

3.14   ◇ Let

$$f(x) = \begin{cases} 1 & \text{if } x \in \left\{ \frac{1}{n} \mid n \in \mathbb{N} \right\}, \\ 0 & \text{if } x \in [0, 1] \setminus \left\{ \frac{1}{n} \mid n \in \mathbb{N} \right\}. \end{cases}$$

Prove: $f \in \mathcal{R}[0, 1]$ and

$$\int_0^1 f(x)\,dx = 0.$$

Hint: Let $\epsilon > 0$ and show there exists $\delta > 0$ such that $\|P\| < \delta$ implies

$$P(f, \{\bar{x}_i\}) < \epsilon.$$

3.15   Let f be defined as in Exercise 3.14. If σ is any step function (Exercise 3.8) on $[0, 1]$, prove that $\|\sigma - f\|_{\sup} \geq \frac{1}{2}$.

3.2 THE DARBOUX INTEGRABILITY CRITERION

Here we will state and prove an alternative but equivalent criterion for the Riemann integrability of a function $f : [a, b] \to \mathbb{R}$. The new criterion will be very useful for proving important theorems about the Riemann integral. In particular, the Riemann sum of a function f depends on the choice of a partition P and arbitrary evaluation points $\{\bar{x}_i\}$. It is helpful to give a condition for Riemann integrability of f that does *not* refer to evaluation points. Since every $f \in \mathcal{R}[a, b]$ must be *bounded* on $[a, b]$, we can define *Upper Sums* and *Lower Sums* for f as follows.

Definition 3.2.1 *Let f be any bounded function on $[a, b]$ and let P be any partition of $[a, b]$. Let*

$$M_i = \sup\{f(x) \mid x \in [x_{i-1}, x_i]\} \text{ and } m_i = \inf\{f(x) \mid x \in [x_{i-1}, x_i]\}.$$

Note that M_i and m_i are in $\mathbb{R}$. Define the upper sum and the lower sum by

$$U(f, P) = \sum_{i=1}^{n} M_i \Delta x_i \quad and \quad L(f, P) = \sum_{i=1}^{n} m_i \Delta x_i.$$

Observe that

$$L(f, P) \leq P(f, \{\bar{x}_i\}) \leq U(f, P)$$

for all P and for all $\{\bar{x}_i\}$, since $m_i \leq f(\bar{x}_i) \leq M_i$ for all i. We can say more.

Lemma 3.2.1 *If P and P' are any two partitions of $[a, b]$, then*

$$L(f, P) \leq U(f, P').$$

Proof: We begin with the special case in which P and P' differ by a single point. Let $P = \{x_0, \ldots, x_n\}$ be the first partition, and let $z \in (x_{k-1}, x_k)$, for one specific index k between 1 and n. Let $P' = P \cup \{z\}$. Thus P' differs from P only in that the kth interval of P has been subdivided into two smaller intervals by P', and those smaller intervals will be the kth and $(k+1)$th determined by P'. Let m'_k, m'_{k+1}, M'_k, and M'_{k+1} denote the inf and sup of f for the two smaller intervals. Now $m'_k \geq m_k$ and $m'_{k+1} \geq m_k$, whereas $M'_k \leq M_k$ and $M'_{k+1} \leq M_k$. Thus

$$m_k \Delta x_k \leq m'_k(z - x_{k-1}) + m'_{k+1}(x_k - z)$$
$$\leq M'_k(z - x_{k-1}) + M'_{k+1}(x_k - z) \leq M_k \Delta x_k.$$

Hence

$$L(f, P) \leq L(f, P') \leq U(f, P') \leq U(f, P).$$

So far, we have shown that adding one point to P lowers the upper sum and raises the lower sum. If we add a finite number of points to P to form P', we use this argument repeated finitely often to show the same inequality. Now, suppose P and P' are *any* two partitions of $[a, b]$. Then consider their so-called *mutual refinement* $P \cup P' = P''$. Then we have

$$L(f, P) \leq L(f, P'') \leq U(f, P'') \leq U(f, P').$$

∎

Definition 3.2.2 *Define the* upper integral *and the* lower integral *of f on $[a, b]$ by*

$$\overline{\int_a^b} f = \inf\{U(f, P) \mid P \text{ is a partition of } [a, b]\},$$

and

$$\underline{\int_a^b} f = \sup\{L(f, P) \mid P \text{ is a partition of } [a, b]\}$$

respectively.

Since every lower sum is less than or equal to every upper sum, it follows from Exercise 1.32 that $\underline{\int_a^b} f \leq \overline{\int_a^b} f$.

Theorem 3.2.1 (Darboux Integrability Criterion) *Let f be bounded on $[a, b]$. Then $f \in \mathcal{R}[a, b]$ if and only if*

$$\lim_{\|P\| \to 0} [U(f, P) - L(f, P)] = 0.$$

When this condition is satisfied, $\int_a^b f(x)\, dx = \underline{\int_a^b} f = \overline{\int_a^b} f$.

Proof: For the right-to-left implication, we suppose that

$$\lim_{\|P\| \to 0} [U(f, P) - L(f, P)] = 0.$$

Since

$$L(f, P) \leq \underline{\int_a^b} f \leq \overline{\int_a^b} f \leq U(f, P)$$

for all P, we have

$$0 \leq \overline{\int_a^b} f - \underline{\int_a^b} f \leq U(f, P) - L(f, P) \to 0$$

as $\|P\| \to 0$. It follows from the squeeze theorem that

$$\underline{\int_a^b} f = \overline{\int_a^b} f,$$

which we denote by L. But then we know that if $\epsilon > 0$ there exists $\delta > 0$ such that for all P with mesh less than δ we must have *both* L and $P(f, \{\bar{x}_i\})$ between $L(f, P)$ and $U(f, P)$, and thus less than ϵ apart, independent of the choice of evaluation points $\bar{x}_i$. Thus letting $\|P\| \to 0$, we see that

$$\lim_{\|P\| \to 0} P(f, \{\bar{x}_i\}) = L$$

so that $f \in \mathcal{R}[a, b]$ and $\int_a^b f(x)\, dx = L$.

For the left-to-right implication we suppose $f \in \mathcal{R}[a, b]$, and write

$$L = \int_a^b f(x)\, dx.$$

Let $\epsilon > 0$. There exists $\delta > 0$ such that $\|P\| < \delta$ implies

$$|P(f, \{\bar{x}_i\}) - L| < \frac{\epsilon}{4},$$

independent of the choice of $\{\bar{x}_i\}$. We claim it is possible to choose $\{\bar{x}_i\}$ such that

$$|P(f, \{\bar{x}_i\}) - U(f, P)| < \frac{\epsilon}{4},$$

and we could choose $\{\bar{x}_i'\}$ such that $|P(f, \{\bar{x}_i'\}) - L(f, P)| < \frac{\epsilon}{4}$. To prove the first claim, denote $P = \{a = x_0 < \cdots < x_n = b\}$ and select $\bar{x}_i \in [x_{i-1}, x_i]$ such that

$$M_i - \frac{\epsilon}{4(b-a)} < f(\bar{x}_i) \le M_i$$

for all i. Then

$$0 \le U(f, P) - P(f, \{\bar{x}_i\}) = \sum_{i=1}^{n} (M_i - f(\bar{x}_i))\Delta x_i$$

$$< \frac{\epsilon}{4(b-a)} \sum_{i=1}^{n} \Delta x_i$$

$$= \frac{\epsilon}{4}.$$

A similar argument can be given for the second claim. (See Exercise 3.16.) Now, if $\|P\| < \delta$ we have

$$|U(f, P) - L(f, P)| \le |U(f, P) - P(f, \{\bar{x}_i\})| + |P(f, \{\bar{x}_i\}) - L|$$
$$+ |L - P(f, \{\bar{x}_i'\})| + |P(f, \{\bar{x}_i'\}) - L(f, P)|$$
$$< 4\left(\frac{\epsilon}{4}\right) = \epsilon.$$

∎

Theorem 3.2.2 *The space $C[a, b] \subset \mathcal{R}[a, b]$. (That is, every continuous function on a closed finite interval is Riemann integrable.)*

Proof: Let $f \in C[a, b]$. By Theorem 2.3.2, f is also *uniformly* continuous on $[a, b]$. If $\epsilon > 0$, there exists $\delta > 0$ such that $|x_1 - x_2| < \delta$ implies

$$|f(x_1) - f(x_2)| < \frac{\epsilon}{b-a}$$

for all x_1 and x_2 in $[a, b]$. Now let $P = \{a = x_0 < \cdots < x_n = b\}$ such that $\|P\| < \delta$. In each $[x_{i-1}, x_i]$ we have $M_i - m_i < \frac{\epsilon}{b-a}$ since M_i and m_i are actual values of f achieved at points of this interval, less than δ apart. Thus

$$U(f, P) - L(f, P) = \sum_{i=1}^{n} (M_i - m_i)\Delta x_i < \epsilon.$$

Hence

$$\lim_{\|P\| \to 0} [U(f, P) - L(f, P)] = 0,$$

so $f \in \mathcal{R}[a, b]$. ∎

Theorem 3.2.3 *If f is monotone on $[a, b]$, then $f \in \mathcal{R}[a, b]$.*

Proof: Suppose first that f is increasing. Then, for each partition P of $[a, b]$ we have

$$U(f, P) - L(f, P) = \sum_{k=1}^{n} [f(x_i) - f(x_{i-1})] \Delta x_i$$

$$\leq \|P\| \sum_{i=1}^{n} [f(x_i) - f(x_{i-1})]$$

$$= \|P\| [f(b) - f(a)] \to 0$$

as $\|P\| \to 0$. Thus $f \in \mathcal{R}[a, b]$. Finally if f is decreasing, then $-f$ is integrable by the first part of this proof, and so $f = -(-f) \in \mathcal{R}[a, b]$ as well. ∎

Here is a variant of the Darboux Integrability Criterion which is often useful. (See Exercises 3.26 and 3.27 for applications of this variant.)

Theorem 3.2.4 *Let f be bounded on $[a, b]$. If for all $\epsilon > 0$ there exists a partition P of $[a, b]$ such that*

$$U(f, P) - L(f, P) < \epsilon,$$

then $f \in \mathcal{R}[a, b]$.

Proof: Let $M = \|f\|_{\sup} \in \mathbb{R}$, and let $\epsilon > 0$. By hypothesis there exists

$$P' = \{x_0, \ldots, x_N\}$$

such that $U(f, P') - L(f, P') < \frac{\epsilon}{2}$. Let

$$\delta = \frac{\epsilon}{12MN}$$

and suppose $\|P\| < \delta$. Our goal is to show that $U(f, P) - L(f, P) < \epsilon$, establishing that $f \in R[a, b]$ by the Darboux Integrability Criterion. Let $P'' = P \cup P'$, so that $U(f, P'') - L(f, P'') < \frac{\epsilon}{2}$ because $P'' \supseteq P'$. It will suffice to show

$$|U(f, P) - U(f, P'')| < \frac{\epsilon}{4} \text{ and } |L(f, P) - L(f, P'')| < \frac{\epsilon}{4}.$$

By the triangle inequality we see that

$$|U(f, P) - L(f, P)| \leq |U(f, P) - U(f, P'')| + |U(f, P'') - L(f, P'')|$$
$$+ |L(f, P'') - L(f, P)|.$$

But $P'' \supseteq P$ and has at most N points not already in P. Thus at most N subintervals from P and at most $2N$ intervals from P'' are not common to both upper sums. Hence

$$|U(f, P) - U(f, P'')| < NM\delta + 2NM\delta < \frac{\epsilon}{4}.$$

Similar reasoning applies to the lower sums. ∎

EXERCISES

3.16   † Complete the proof of Theorem 3.2.1 by showing that we could choose $\bar{x}_i' \in [x_{i-1}, x_i]$ for all i such that $|P(f, \{\bar{x}_i'\}) - L(f, P)| < \frac{\epsilon}{4}$.

3.17   Let

$$f(x) = \begin{cases} 1 & \text{if } x = 1, \\ 0 & \text{if } x \in [0, 2] \setminus \{1\}. \end{cases}$$

Find a partition P of $[0, 2]$ for which $U(f, P) - L(f, P) < \frac{1}{8}$.

3.18   Suppose

$$f(x) = \begin{cases} 0 & \text{if } x = 0, \\ \frac{1}{n} & \text{if } x \in \left(\frac{1}{n+1}, \frac{1}{n}\right], \end{cases}$$

where $n \in \mathbb{N}$. Prove: $f \in \mathcal{R}[0, 1]$, even though f has infinitely many discontinuities. (Hint: Consider Theorem 3.2.3.)

3.19   Suppose

$$f(x) = \begin{cases} 0 & \text{if } x = 0, \\ \frac{(-1)^n}{n} & \text{if } x \in \left(\frac{1}{n+1}, \frac{1}{n}\right], \end{cases}$$

where $n \in \mathbb{N}$. Prove: $f \in \mathcal{R}[0, 1]$, even though f has infinitely many discontinuities. (Hint: Use the Darboux Criterion.)

3.20   † Let f be any real-valued function on a domain $D \subseteq \mathbb{R}$. Define

$$f^+(x) = \begin{cases} f(x) & \text{if } f(x) \geq 0, \\ 0 & \text{if } f(x) < 0. \end{cases}$$

and let

$$f^-(x) = \begin{cases} -f(x) & \text{if } f(x) < 0, \\ 0 & \text{if } f(x) \geq 0 \end{cases}$$

for all $x \in D$. Prove that

$$f(x) = f^+(x) - f^-(x) \text{ and } |f(x)| = f^+(x) + f^-(x)$$

for all $x \in D$. (Hint: Just check the cases based on the sign of $f(x)$.)

3.21   † Suppose $f \in \mathcal{R}[a, b]$. Prove: f^+ and f^- are in $\mathcal{R}[a, b]$. Hint: Show that

$$U(f^+, P) - L(f^+, P) \leq U(f, P) - L(f, P).$$

3.22   If $f \in \mathcal{R}[a, b]$, prove $|f| \in \mathcal{R}[a, b]$. (Hint: Use Exercises 3.20 and 3.21 above.)

3.23   Prove or give a counterexample: The function $|f| \in \mathcal{R}[a, b] \Leftrightarrow f \in \mathcal{R}[a, b]$.

3.24 † If $f \in \mathcal{R}[a, b]$, prove:

$$\left| \int_a^b f(x)\, dx \right| \leq \int_a^b |f(x)|\, dx.$$

(Hint: Write the left side by expressing $f = f^+ - f^-$ and use the fact that f^+ and f^- are both nonnegative functions.)

3.25 Suppose $f \in \mathcal{C}[a, b]$ is everywhere nonnegative, and suppose there exists $p \in [a, b]$ such that $f(p) > 0$. Prove: $\int_a^b f(x)\, dx > 0$.

3.26 † Suppose $f \in \mathcal{C}(a, b)$ and also that f is bounded on $[a, b]$. Prove: $f \in \mathcal{R}[a, b]$. (Hint: Use Theorem 3.2.4.)

3.27 † Let

$$f(x) = \begin{cases} \sin \frac{\pi}{x} & \text{if } 0 < x \leq 1, \\ 0 & \text{if } x = 0. \end{cases}$$

Prove: $f \in \mathcal{R}[0, 1]$. (Hint: You may use Exercise 3.26.)

3.28 If $f \in \mathcal{R}[a, b]$ and if $[c, d] \subset [a, b]$, prove: $f \in \mathcal{R}[c, d]$. Hint: If P is any partition of $[c, d]$, P can be extended to a partition P^* of $[a, b]$ with $\|P^*\| \leq \|P\|$. Show that

$$U(f, P) - L(f, P) \leq U(f, P^*) - L(f, P^*).$$

3.29 † [9]Let $f \in \mathcal{R}[a, b]$ and let $\epsilon > 0$. Prove there exist step functions (Exercise 3.8) satisfying $\sigma(x) \leq f(x) \leq \sigma'(x)$ for all $x \in [a, b]$, so that

$$\int_a^b \sigma(x)\, dx \leq \int_a^b f(x)\, dx \leq \int_a^b \sigma'(x)\, dx$$

such that

$$\left| \int_a^b \sigma'(x)\, dx - \int_a^b \sigma(x)\, dx \right| < \epsilon$$

and implying also that

$$\int_a^b |f(x) - \sigma(x)|\, dx < \epsilon.$$

(Hint: Use upper and lower Darboux sums.)

3.30 Let f be a bounded function on $[a, b]$.

  a) Prove that $f \in \mathcal{R}[a, b] \Leftrightarrow \overline{\int_a^b} f = \underline{\int_a^b} f$.

  b) Let $f = 1_{\mathbb{Q}}$. Find the numerical values of $\overline{\int_a^b} f$ and $\underline{\int_a^b} f$. Explain why f is not Riemann integrable on $[a, b]$ if $a < b$.

[9]This exercise is used in the proof of Theorem 6.5.1, which establishes the convergence of Fourier series in the L^2-norm.

3.31   †◇ [10] Let f and g be bounded functions on $[a, b]$. Prove that

a) $\overline{\int_a^b}(f + g)(x)\,dx \leq \overline{\int_a^b}f(x)\,dx + \overline{\int_a^b}g(x)\,dx$.

b) $\underline{\int_a^b}(f + g)(x)\,dx \geq \underline{\int_a^b}f(x)\,dx + \underline{\int_a^b}g(x)\,dx$.

3.3 INTEGRALS OF UNIFORM LIMITS

The following example shows that *pointwise* limits of Riemann integrable functions need not be Riemann integrable.

■ **EXAMPLE 3.4**

Since the set $\mathbb{Q}$ of rational numbers is countable, the same is true of the set S of all rational numbers in [0,1]. So write $S = \{q_n \mid n \in \mathbb{N}\}$. Now define a function

$$f_n(x) = \begin{cases} 1 & \text{if } x \in \{q_1, \ldots, q_n\}, \\ 0 & \text{all other } x \in [0, 1]. \end{cases}$$

The reader can observe that for all $x \in [0, 1]$, $f_n(x) \to 1_S(x)$, where

$$1_S(x) = \begin{cases} 1 & \text{if } x \in S, \\ 0 & \text{if } x \in [0, 1] \setminus S, \end{cases}$$

the *indicator function* of the set of rational numbers in $[0, 1]$. Now we know that each $f_n \in \mathcal{R}[0, 1]$, by Exercise 3.4. However, the pointwise limit of the sequence f_n is 1_S, which is not Riemann integrable, as explained in Example 3.2 and also in Exercise 3.30.

The following example shows that if $f_n \in \mathcal{R}[a, b]$ for all n and if $f_n \to f$ *pointwise*, then even if $f \in \mathcal{R}[a, b]$ as well, still it is possible that

$$\int_a^b f_n(x)\,dx \nrightarrow \int_a^b f(x)\,dx$$

as $n \to \infty$.

■ **EXAMPLE 3.5**

Let

$$f_n(x) = \begin{cases} n & \text{if } 0 < x \leq \frac{1}{n}, \\ 0 & \text{if } \frac{1}{n} < x \leq 1, \\ 0 & \text{if } x = 0 \end{cases}$$

for all $n \in \mathbb{N}$.

---

[10]This exercise is used in the proof of Theorem 11.4.1, which is Fubini's theorem.

In Exercise 3.32, the student will prove that $f_n(x) \to f(x) \equiv 0$ pointwise on $[0, 1]$. Also, it is clear that $f_n \in \mathcal{R}[0, 1]$ for all n, and $f \in \mathcal{R}[0, 1]$ as well. Yet

$$\int_0^1 f_n(x)\, dx \equiv 1 \to 1 \neq 0 = \int_0^1 f(x)\, dx.$$

Thus it is *false* in general that

$$\lim_{n \to \infty} \int_a^b f(x)\, dx = \int_a^b \lim_{n \to \infty} f_n(x)\, dx.$$

The following theorem shows, however, that the integral behaves much better with respect to *uniform* convergence.

Theorem 3.3.1 *Suppose $f_n \in \mathcal{R}[a, b]$ for all $n \in \mathbb{N}$, and suppose $f_n \to f$ uniformly on $[a, b]$. Then $f \in \mathcal{R}[a, b]$ and*

$$\int_a^b f_n(x)\, dx \to \int_a^b f(x)\, dx.$$

Proof: First we will use the Darboux Criterion to prove that $f \in \mathcal{R}[a, b]$. Let $\epsilon > 0$. Then there exists $N \in \mathbb{N}$ such that $n \geq N$ implies

$$\|f_n - f\|_{\sup} < \frac{\epsilon}{3(b - a)}.$$

Also, there exists $\delta > 0$ such that $\|P\| < \delta$ implies $U(f_N, P) - L(f_N, P) < \frac{\epsilon}{3}$. Let $P = \{a = x_0 < \cdots < x_m = b\}$ have mesh less than δ, and observe that on each interval $[x_{k-1}, x_k]$ we have

$$m_k^N - \frac{\epsilon}{3(b - a)} \leq f_N(x) - \frac{\epsilon}{3(b - a)}$$
$$< f(x)$$
$$< f_N(x) + \frac{\epsilon}{3(b - a)}$$
$$\leq M_k^N + \frac{\epsilon}{3(b - a)},$$

where M_k^N is the supremum of f_N on $[x_{k-1}, x_k]$ and m_k^N is the corresponding infimum. Thus

$$m_k^N - \frac{\epsilon}{3(b - a)} \leq m_k \leq M_k \leq M_k^N + \frac{\epsilon}{3(b - a)}.$$

Now it follows that

$$|U(f, P) - L(f, P)| = \sum_{k=1}^{n}(M_k - m_k)\Delta x_k$$

$$\leq \sum_{k=1}^{n}\left(M_k^N - m_k^N + \frac{2\epsilon}{3(b-a)}\right)\Delta x_k$$

$$= U(f_N, P) - L(f_N, P) + \frac{2\epsilon}{3}$$

$$< \epsilon.$$

Thus $f \in \mathcal{R}[a, b]$ as claimed.

With this proven, we can now show easily that $\int_a^b f_n(x)\,dx \rightarrow \int_a^b f(x)\,dx$ as $n \rightarrow \infty$. We apply Exercise 3.24 as follows:

$$\left|\int_a^b f_n(x)\,dx - \int_a^b f(x)\,dx\right| = \left|\int_a^b f_n(x) - f(x)\,dx\right|$$

$$\leq \int_a^b |f_n(x) - f(x)|\,dx$$

$$\leq \int_a^b \|f_n - f\|_{\sup}\,dx$$

$$= \|f_n - f\|_{\sup}(b-a) \rightarrow 0$$

as $n \rightarrow \infty$.   Since $\|f_n - f\|_{sup} \to 0$ as $n \to \infty$ ∎

Definition 3.3.1 *If V is a vector space equipped with a* norm $\|\cdot\|$, *we call*

$$T : V \rightarrow \mathbb{R}$$

a linear functional *provided*

$$T(\alpha\mathbf{x} + \mathbf{y}) = \alpha T(\mathbf{x}) + T(\mathbf{y})$$

for all $\mathbf{x}$ and $\mathbf{y}$ in V and for all $\alpha \in \mathbb{R}$. We say that a sequence

$$\mathbf{x}_n \rightarrow \mathbf{x} \in V \Leftrightarrow \|\mathbf{x}_n - \mathbf{x}\| \rightarrow 0$$

as $n \rightarrow \infty$. A linear functional T is called continuous *at $\mathbf{x}$ if and only if for each sequence $\mathbf{x}_n$ in V such that $\mathbf{x}_n \rightarrow \mathbf{x}$ we have $T(\mathbf{x}_n) \rightarrow T(\mathbf{x})$. T is called* continuous *if and only if T is continuous at each $\mathbf{x} \in V$.*

Now recall that $\mathcal{C}[a, b]$ is a complete normed vector space equipped with the sup-norm. If we define $T : \mathcal{C}[a, b] \rightarrow \mathbb{R}$ by

$$T(f) = \int_a^b f(x)\,dx$$

then T is linear by Theorem 3.1.2 and T is continuous by Theorem 3.3.1. Thus T is a continuous linear functional on $\mathcal{C}[a, b]$. The student will meet some more continuous linear functionals on $\mathcal{C}[a, b]$ in the exercises below, and all of the continuous linear functionals on $\mathcal{C}[a, b]$ will be identified by the Riesz Representation Theorem, Theorem 7.4.1.

Lemma 3.3.1 *Let V be any vector space over the real numbers, equipped with a norm $\| \cdot \|$. A linear functional $T : V \to \mathbb{R}$ is continuous $\Leftrightarrow T$ is continuous at $\mathbf{0}$.*

Proof: If T is continuous (at all $\mathbf{x} \in V$) then it must be continuous at $\mathbf{0}$. So we prove the opposite implication. Note that since

$$T(\mathbf{0}) = T(\mathbf{0} + \mathbf{0}) = T(\mathbf{0}) + T(\mathbf{0}),$$

we must have $T(\mathbf{0}) = 0$. Suppose T is continuous at $\mathbf{0}$: That is,

$$\|\mathbf{x}_n - \mathbf{0}\| = \|\mathbf{x}_n\| \to 0$$

implies $T(\mathbf{x}_n) \to 0 = T(\mathbf{0})$. Let $\mathbf{x} \in V$ be arbitrary and suppose $\mathbf{x}_n \to \mathbf{x}$: ie, $\|\mathbf{x}_n - \mathbf{x}\| \to 0$. By hypothesis,

$$T(\mathbf{x}_n - \mathbf{x}) = T(\mathbf{x}_n) - T(\mathbf{x}) \to 0,$$

so $T(\mathbf{x}_n) \to T(\mathbf{x})$.   ∎

Definition 3.3.2 *A linear functional T on a normed vector space V is called* bounded *if and only if there exists $K \in \mathbb{R}$ such that $|T(\mathbf{x})| \leq K \|\mathbf{x}\|$, for all $\mathbf{x} \in V$.*

Theorem 3.3.2 *If T is a linear functional on a normed vector space V, then T is continuous if and only if T is* bounded.

Proof: For the implication from right to left, suppose that T is bounded. It will suffice to prove T is continuous at $\mathbf{0}$. So suppose $\|\mathbf{x}_n\| \to 0$. Then

$$|T(\mathbf{x}_n)| \leq K \|\mathbf{x}_n\| \to 0,$$

and this implies that $T(\mathbf{x}_n) \to \mathbf{0}$.

For the implication from left to right, suppose that T is continuous. We will prove T is bounded by contradiction. So suppose the claim were false. Then for all $n \in \mathbb{N}$ there exists $\mathbf{x}_n \in V$ such that

$$|T(\mathbf{x}_n)\| > n \|\mathbf{x}_n\|.$$

Let $c_n = (\sqrt{n} \|x_n\|)^{-1}$ (since $\mathbf{x}_n \neq \mathbf{0}$) and then

$$|T(c_n \mathbf{x}_n)| > n \|c_n \mathbf{x}_n\| = \sqrt{n}$$

for all $n \in \mathbb{N}$. Denote $\mathbf{y}_n = c_n \mathbf{x}_n$ and observe that $\|\mathbf{y}_n\| = \frac{1}{\sqrt{n}} \to 0$, yet $T(\mathbf{y}_n)$ fails to converge to 0; in fact it is unbounded.   ∎

Because of Theorem 3.3.2, continuous linear functionals on normed linear spaces are often called *bounded linear functionals*. Observe that if $f \in \mathcal{C}[a, b]$ and if we define $T(f) = \int_a^b f(x) \, dx$, then

$$|T(f)| \leq \int_a^b \|f\|_{\sup} \, dx = (b - a)\|f\|_{\sup}.$$

Thus T is bounded, with constant $K = b - a$.

EXERCISES

3.32   † In Example 3.5, show that $f_n(x) \to 0$ pointwise on $[0, 1]$.

3.33   In each example, *find* the pointwise limit and then use Theorem 3.3.1 to show that the convergence is not uniform.

    a) Let f_n be as in Example 3.4.

    b) Let f_n be as in Example 3.5.

3.34   Prove: $\int_1^2 (1 + x^2)^{-n} \, dx \to 0$ as $n \to \infty$. (Hint: Use Theorem 3.3.1, but be sure to prove the necessary hypotheses are satisfied.)

3.35   Find

$$\lim_{n \to \infty} \int_1^2 1 + \frac{1}{(1 + x^2)^n} \, dx$$

and justify your conclusion.

3.36   Let

$$f_n(x) = \begin{cases} n^2 x & \text{if } 0 \leq x \leq \frac{1}{n}, \\ 0 & \text{if } \frac{1}{n} < x \leq 1 \end{cases}$$

for each $n > 1$. Find

    a) $\lim_{n \to \infty} f_n(x)$ for each $x \in [0, 1]$, and

    b) $\lim_{n \to \infty} \int_0^1 f_n(x) \, dx$.

3.37   Prove: $\int_0^{\pi/4} \sin^n(x) \, dx \to 0$ as $n \to \infty$. (Hint: Use Theorem 3.3.1, but be sure to prove the necessary hypotheses are satisfied.)

3.38   In the following exercise, the student may use techniques of integration learned in elementary calculus, including the concept of *improper* integrals. (See page 135.) Find the *mistake* in the following reasoning. We calculate readily that

$$\int_1^n \frac{1}{t \ln n} \, dt = 1$$

for all n. Since $\frac{1}{t \ln n} \to 0$ uniformly on $[1, \infty)$, it follows by taking the limit on n on both sides that

$$\int_1^\infty 0 \, dt = 1.$$

Explain why the latter conclusion is false, and find the error in the reasoning.

3.39   Fix $p \in [a, b]$ and, for all $f \in \mathcal{C}[a.b]$, equipped with the sup-norm, and define $T : \mathcal{C}[a, b] \to \mathbb{R}$ by $T(f) = f(p)$. Prove: T is a continuous linear functional on $\mathcal{C}[a, b]$.

3.40   Let $T : \mathcal{C}[0, 1] \to \mathbb{R}$ be defined by

$$T(f) = \int_0^1 f(x) \left(1 + x^2\right) \, dx.$$

Prove that T is a bounded linear functional. Find a constant K for which

$$|T(f)| \leq K \|f\|_{\sup}$$

for all $f \in \mathcal{C}[0, 1]$.

3.41   Let $T : \mathcal{C}[0, 3] \to \mathbb{R}$ be defined by $T(f) = f(1) - f(2)$. Prove: T is a bounded linear functional on $\mathcal{C}[0, 3]$, equipped with the sup-norm.

3.42   † Consider the vector space $\mathbb{R}^2$ and let $\mathbf{x} = \langle x_1, x_2 \rangle$. Define the Euclidean norm by

$$\|\mathbf{x}\| = \|\langle x_1, x_2 \rangle\| = \sqrt{x_1^2 + x_2^2}$$

and add vectors *componentwise* as usual.

   a) † Prove: If $T : \mathbb{R}^2 \to \mathbb{R}$ is linear, then there exists $\mathbf{a} = \langle a_1, a_2 \rangle \in \mathbb{R}^2$ such that, for all $\mathbf{x}$ we have

$$T(\mathbf{x}) = \mathbf{a} \cdot \mathbf{x} = a_1 x_1 + a_2 x_2.$$

   (Hint: Write $\mathbf{x} = x_1 \langle 1, 0 \rangle + x_2 \langle 0, 1 \rangle$.)

   b) † Prove that for all $\mathbf{a} \in \mathbb{R}^2$, the corresponding linear functional T given in Exercise 3.42.a above is continuous with respect to the Euclidean norm. (Theorem 3.3.2 is helpful.)

   c) Let $C_i : \mathbb{R}^2 \to \mathbb{R}$ be defined by $C_i(\mathbf{x}) = x_i$, $i = 1, 2$. Prove: C_i is a continuous linear functional on $\mathbb{R}^2$.

3.43

   a) Show that $\mathcal{R}[a, b]$, equipped with the sup-norm, is a complete normed vector space.

   b) Suppose that T and T' are both bounded linear functionals defined on $\mathcal{R}[a, b]$ as in (a) above. Suppose $T(\sigma) = T'(\sigma)$ for each step function σ. Prove that $T(f) = T'(f)$ for all $f \in \mathcal{C}[a, b]$. (Hint: Use Exercise 3.13.)

3.44   Let $\mathcal{P}$ denote the vector space of all real polynomial functions (all nonnegative integer degrees allowed) on the interval $[0, 1]$. Endow this vector space with the sup-norm.

   a) Show that the set $\left\{ x^k \mid k \in \mathbb{N} \cup \{0\} \right\}$ is a basis for $\mathcal{P}$.

b) Define $T : \mathcal{P} \to \mathbb{R}$ by letting $T\left(x^k\right) = k$ and show that the domain of T can be extended to all of $\mathcal{P}$ so as to make T a linear functional on $\mathcal{P}$.

c) Prove that T is not bounded, and thus not continuous.

Remark 3.3.1 In Exercise 3.42 above, the reader has shown that every linear functional defined on $\mathbb{R}^2$ is continuous. It is not hard to see that the same argument applies to $\mathbb{R}^n$ for each $n \in \mathbb{N}$ as well. Thus the reader may wonder why theorems are presented for the purpose of proving that a linear functional is continuous, or equivalently, bounded. The reason is that if a normed vector space is infinite-dimensional, then unbounded linear functionals do exist, although this cannot happen in finite-dimensional spaces.

To demonstrate the existence of unbounded linear functionals, we need the concept of a Hamel basis for a vector space V.

Definition 3.3.3 *A Hamel basis B is a (generally uncountable) set $\{e_\alpha | \alpha \in A\}$ having the following two properties.*

i. B *is linearly independent: That is, for each finite set $F \subset A$ we have*

$$\sum_{\alpha \in F} c_\alpha e_\alpha = 0 \implies \text{each coefficient } c_\alpha = 0.$$

ii. *For each vector $\mathbf{v} \in V$ there exists a finite set $F \subset A$ such that*

$$\mathbf{v} = \sum_{\alpha \in F} c_\alpha e_\alpha.$$

It follows easily from the definition that each vector $\mathbf{v} \in V$ can be expressed *uniquely* as a linear combination of finitely many of the basis vectors. The existence of Hamel bases is generally proved in graduate courses in functional analysis [8], as an application of the *Axiom of Choice* from set theory. But if the reader will accept the existence of Hamel bases, then we can demonstrate the existence of unbounded linear functionals on any infinite-dimensional normed vector space.

Let V be any infinite-dimensional normed linear space. Let

$$B = \{e_\alpha \mid \alpha \in A\}$$

be any Hamel basis for V. Define a linear map $T : V \to \mathbb{R}$ as follows. Select a countable set $\{e_i \mid i \in \mathbb{N}\} \subset B$ and define $T(e_i) = i\|e_i\|$ for each $i \in \mathbb{N}$. Then define T to be 0 on each remaining basis vector from B. Extend T in the usual way to a linear transformation of $V \to \mathbb{R}$, and observe that T is unbounded since $\frac{|T(e_i)|}{\|e_i\|}$ is an unbounded sequence. Hence T is an unbounded linear functional. This generalizes Exercise 3.44 above.

Another interesting feature of Exercise 3.42 is that each of the standard *coordinate* functions in $\mathbb{R}^2$ is a bounded linear functional. One way to think of coordinates is that coordinates are bounded linear functionals (real-valued linear functions that are continuous) defined on a normed vector space V having the property that if two vectors $\mathbf{v}_1$ and $\mathbf{v}_2$ are distinct, then there exists a coordinate function C such that $C(\mathbf{v}_1) \neq C(\mathbf{v}_2)$. In a graduate course in functional analysis [8] the reader will study the Hahn–Banach Theorem. One consequence of this theorem is that in every Banach space there are enough bounded linear functionals to serve this same function of distinguishing among the vectors of V.

3.4   THE CAUCHY–SCHWARZ INEQUALITY[11]

If $\mathbf{x} = \langle x_1, x_2 \rangle$ and $\mathbf{y} = \langle y_1, y_2 \rangle \in \mathbb{R}^2$, the student will recall that

$$\mathbf{x} \cdot \mathbf{y} = x_1 y_1 + x_2 y_2 = \|\mathbf{x}\|\|\mathbf{y}\| \cos \theta$$

is the *scalar product* of $\mathbf{x}$ with $\mathbf{y}$, known also as the *dot* product, or the *inner* product, and denoted also by $\langle \mathbf{x}, \mathbf{y} \rangle$. Here θ is the angle between $\mathbf{x}$ and $\mathbf{y}$. Since $|\cos \theta| \leq 1$ for all θ, it follows that

$$|\mathbf{x} \cdot \mathbf{y}| \leq \|\mathbf{x}\|\|\mathbf{y}\|.$$

This is called the Cauchy–Schwarz inequality in $\mathbb{R}^2$. Here

$$\|\mathbf{x}\| = \sqrt{x_1^2 + x_2^2} = \sqrt{\mathbf{x} \cdot \mathbf{x}}$$

and $\|\cdot\|$ has all the properties of a *norm* on $\mathbb{R}^2$. A similar result holds in $\mathbb{R}^n$ as well. It is natural to consider the following generalization of these concepts to $f, g \in \mathcal{R}[a, b]$.

Just as vectors in the plane have two components, we can think of a function $f \in \mathcal{R}[a, b]$ as being a *vector* with infinitely many *components*—namely the numbers $f(x)$ corresponding to the *uncountably* many $x \in [a, b]$. But we cannot possibly add *all* the values of $f(x)g(x)$ for all $x \in [a, b]$. Thus we try defining a *scalar product* of f and g by

$$\langle f, g \rangle = \int_a^b f(x)g(x)\,dx.$$

We need to show however that the product of two Riemann integrable functions is always integrable. This is the next theorem.

Theorem 3.4.1 *If f and g are in $\mathcal{R}[a, b]$, then the product $fg \in \mathcal{R}[a, b]$ as well.*

Proof:

i. We begin with the special case in which $f(x) \geq 0$ and $g(x) \geq 0$ for all $x \in [a, b]$. Let $P = \{x_0, x_1, \ldots, x_n\}$ be any partition. Fix for the moment any one subinterval $[x_{k-1}, x_k]$ and denote by M_f, M_g, and M_{fg} the sup over this interval of f, g, and fg, respectively. Since $f(x)g(x) \leq M_f M_g$ for all

---

[11]This section will be used in Chapter 6 for the study of Fourier series.

$x \in [x_{k-1}, x_k]$, it follows that $M_{fg} \le M_f M_g$. Similarly, $m_{fg} \ge m_f m_g$ on each such subinterval. Hence

$$M_{fg} - m_{fg} \le M_f M_g - m_f m_g$$
$$= (M_f - m_f)M_g + (M_g - m_g)m_f$$
$$\le (M_f - m_f)\|g\|_{\sup} + (M_g - m_g)\|f\|_{\sup}.$$

Summing over $k = 1, 2, \ldots, n$, we find that

$$\lim_{\|P\| \to 0}[U(fg, P) - L(fg, P)] \le \lim_{\|P\| \to 0}[U(f, P) - L(f, P)]\|g\|_{\sup}$$
$$+ \lim_{\|P\| \to 0}[U(g, P) - L(g, P)]\|f\|_{\sup}$$
$$= 0 + 0 = 0.$$

Thus $fg \in \mathcal{R}[a, b]$ by the Darboux Integrability Criterion.

ii. Now suppose f and g in $\mathcal{R}[a, b]$ are arbitrary. Since each such function is bounded, there exists $K \in \mathbb{R}$ such that $f(x) + K \ge 0$ and $g(x) + K \ge 0$ for all $x \in [a, b]$. Since $\mathcal{R}[a, b]$ is a vector space, $f + K$ and $g + K \in \mathcal{R}[a, b]$. By Case (i),

$$(f + K)(g + K) = fg + Kf + Kg + K^2 = F \in \mathcal{R}[a, b]$$

as well. But then $fg = F - Kf - Kg - K^2 \in \mathcal{R}[a, b]$ also. ∎

Definition 3.4.1 *For all f and g in $\mathcal{R}[a, b]$ we define*

$$\langle f, g \rangle = \int_a^b f(x)g(x)\, dx.$$

Define also the L^2-norm[12] of f by

$$\|f\|_2 = \sqrt{\langle f, f \rangle} = \sqrt{\int_a^b |f(x)|^2\, dx}.$$

Note that $\langle f, g \rangle = \langle g, f \rangle$. We observe also that this product is linear in f and g separately: That is,

$$\langle f_1 + cf_2, g \rangle = \langle f_1, g \rangle + c\langle f_2, g \rangle \quad \text{and} \quad \langle f, g_1 + cg_2 \rangle = \langle f, g_1 \rangle + c\langle f, g_2 \rangle.$$

Also, $\langle f, f \rangle \ge 0$ for all $f \in \mathcal{R}[a, b]$. One property of the scalar product of $\mathbb{R}^2$ *fails* to carry over to this case, however: it is possible for $\langle f, f \rangle = 0$ without $f(x) \equiv 0$

---

[12]The letter L in L^2 stands for Lebesgue, and ideally the L^2-norm is studied in the context of the Lebesgue integral. It is still a useful norm to consider however in the context of Riemann integration, as we use the term here.

on $[a, b]$. This means that $\|f\|_2$ is *not* a true norm since it is possible for $\|f\|_2 = 0$ even though $f \neq 0 \in \mathcal{R}[a, b]$. (See Exercises 3.45 and 3.46.)

Remark 3.4.1 It would be possible to remedy the failure of $\| \cdot \|_2$ to be a true norm by forming the vector space of *equivalence classes* of functions in $\mathcal{R}[a, b]$. Here f would be considered *equivalent* to g if and only if $\|f - g\|_2 = 0$. However we will not do this here because of another serious defect: even with such a procedure R[a,b] would still not yield a *complete* normed linear space with respect to the 2-norm $\| \cdot \|_2$. Thus one may wonder, how much would one have to *enlarge* the space $\mathcal{R}[a, b]$ to endow it with a limit for each Cauchy sequence in the 2-norm. The answer is that one must define what is called the *Lebesgue Integral*, a very subtle refinement of the Riemann Integral that can integrate every $f \in \mathcal{R}[a, b]$, with the same results, but which can also integrate many more functions. This subject is left for a graduate course in real analysis, however. Even though the 2-norm is not a true norm on $\mathcal{R}[a, b]$, we can still use our geometrical intuition about norms (or lengths) of vectors to prove the famous Cauchy–Schwarz inequality.

Theorem 3.4.2 (Cauchy–Schwarz) *Suppose f and g are in $\mathcal{R}[a, b]$. Then*

$$\left| \int_a^b f(x)g(x)\, dx \right| \leq \left(\int_a^b f(x)^2\, dx \right)^{\frac{1}{2}} \left(\int_a^b g(x)^2\, dx \right)^{\frac{1}{2}}.$$

Remark 3.4.2 The theorem means that in the notation introduced above

$$|\langle f, g \rangle| \leq \|f\|_2 \|g\|_2,$$

just like the analogous inequality for dot products in $\mathbb{R}^2$.

Proof: For all $t \in \mathbb{R}$, define a polynomial

$$p(t) = \langle tf + g, tf + g \rangle = \int_a^b [tf(x) + g(x)]^2\, dx.$$

Observe that $p(t) \geq 0$ for all t. By linearity of $\langle \cdot, \cdot \rangle$ in each variable (or by elementary algebra with the integral formulation) we see that

$$p(t) = \|f\|_2^2 t^2 + 2\langle f, g \rangle t + \|g\|_2^2 = at^2 + bt + c,$$

where $a = \|f\|_2^2$, $b = 2\langle f, g \rangle$, and $c = \|g\|_2^2$. But a quadratic polynomial $p(t) \geq 0$ for all $t \in \mathbb{R}$ if and only if $b^2 - 4ac \leq 0$, which is equivalent to $b^2 \leq 4ac$. Hence

$$|\langle f, g \rangle|^2 \leq \|f\|_2^2 \|g\|_2^2.$$

∎

EXERCISES

In the following exercises we will assume you know from elementary calculus courses how to evaluate integrals by means of antiderivatives. The necessary Fundamental Theorem of the Calculus will be proven in Chapter 4.

3.45   † Give an example of $f \in \mathcal{R}[a, b]$ such that $\|f\|_2 = 0$ yet $f(x) \not\equiv 0$ on [a,b].

3.46   † Prove that $\| \cdot \|_2$ does satisfy the *triangle inequality:*

$$\|f + g\|_2 \le \|f\|_2 + \|g\|_2.$$

(Hint: Write $\|f + g\|_2^2 = \langle f + g, f + g \rangle$, expand using linearity in each variable, and apply the Cauchy–Schwarz inequality.)

3.47   Suppose $f \in \mathcal{R}[a, b]$.
    a) If there exists $\delta > 0$ such that $f(x) \ge \delta$ for all $x \in [a, b]$, prove that

$$\frac{1}{f} \in \mathcal{R}[a, b].$$

    Hint: On each interval $[x_{k-1}, x_k]$ of any partition, compare

$$M_k^{\frac{1}{f}} = \sup \frac{1}{f(x)} \text{ and } m_k^{\frac{1}{f}} = \inf \frac{1}{f(x)}$$

    with the corresponding numbers M_k^f and m_k^f. Then use the Darboux criterion.
    b) Now suppose only that $|f(x)| \ge \delta > 0$ for all $x \in [a, b]$, for some fixed $\delta > 0$. Show again that

$$\frac{1}{f} \in \mathcal{R}[a, b].$$

    Hint: Apply (a) to $\frac{1}{f^2}$ and then use Theorem 3.4.1.

3.48   Use the Cauchy–Schwarz inequality to show:
    a) $\int_0^\pi \sqrt{x} \sin x \, dx \le \pi$.
    b) $\int_0^{\frac{\pi}{4}} (1 + \tan x) \sqrt{x} \sec x \, dx \le \pi \sqrt{\frac{7}{96}}$.

3.49   Use the triangle inequality for $\| \cdot \|_2$ to show

$$\left[\int_0^{\frac{\pi}{2}} (\sqrt{\cos x} + x)^2 \, dx \right]^{\frac{1}{2}} \le 1 + \sqrt{\frac{\pi^3}{24}}.$$

3.50   Prove: If $f \in \mathcal{R}[0, 1]$, then

$$\int_0^1 x f(x) \, dx \le \frac{1}{\sqrt{3}} \left[\int_0^1 [f(x)]^2 \, dx \right]^{\frac{1}{2}}.$$

3.51   Prove: If $f \in \mathcal{R}[0, 1]$, then

$$\left[\int_0^1 (\sqrt{\cos x} + f(x))^2 \, dx \right]^{\frac{1}{2}} \leq \sqrt{\sin 1} + \left[\int_0^1 (f(x))^2 \, dx \right]^{\frac{1}{2}}.$$

3.52   † If f and g are in $\mathcal{R}[a, b]$, we say f is *orthogonal* to g, denoted by $f \perp g$, if and only if $\langle f, g \rangle = 0$. Prove that

$$f \perp g \Leftrightarrow \|f + g\|_2^2 = \|f\|_2^2 + \|g\|_2^2.$$

(This is a modern analogue of the *Pythagorean Theorem*.)

3.53   Let f and g be in $\mathcal{R}[a, b]$, with $\|g\|_2 > 0$.
   a) Find a constant $c \in \mathbb{R}$ such that $(f - cg) \perp g$. (Hint: Use the definition of orthogonality in Exercise 3.52.)
   b) Now let $f(x) = x$ and $g(x) = 1$ on $[0, 1]$. Find the value of c for which $(f - cg) \perp g$. Using this value of c, would $(f - cg) \perp g$ on the different interval $[-1, 1]$?

3.54   Let $f(x) = \sin x$ and $g(x) = \cos x$ on $[0, \pi]$. Prove $f \perp g$.

3.55   Suppose $\|f\|_2 \|g\|_2 > 0$, where f and g are in $\mathcal{R}[a, b]$.
   a) Prove: $|\langle f, g \rangle| = \|f\|_2 \|g\|_2$ if and only if there exists $t \in \mathbb{R}$ such that

$$\int_a^b [g(x) + tf(x)]^2 \, dx = 0.$$

   (Hint: Let $p(t) = \langle tf + g, tf + g \rangle \geq 0$, as in the proof of the Cauchy–Schwarz inequality.)
   b) The preceding part says that we get *equality* in the Cauchy–Schwarz inequality if and only if g is very close to being a constant multiple of f. What happens to the Cauchy–Schwarz inequality if $\|f\|_2 = 0$ or if $\|g\|_2 = 0$?

3.56   Let $f(x) = \sin x$ on $[0, 1]$. Give an example of g, with $\|g\|_2 > 0$, for which $|\langle f, g \rangle| = \|f\|_2 \|g\|_2$.

3.57   Let $f, g \in \mathcal{R}[a, b]$. Show that $\|f + g\|_2 = \|f\|_2 + \|g\|_2$ if and only if $\langle f, g \rangle = \|f\|_2 \|g\|_2$.

3.58   Let $g \in \mathcal{R}[a, b]$, and define $T_g : \mathcal{C}[a, b] \to \mathbb{R}$ by

$$T_g(f) = \int_a^b f(x) g(x) \, dx$$

for all $f \in \mathcal{C}[a, b]$. Prove: T_g is a bounded linear functional on the Banach space $\mathcal{C}[a, b]$, equipped with the sup-norm.

Remark 3.4.3   On a cover of the Notices of the American Mathematical Society [17] the reader can see a photograph of a clay tablet from the Yale Babylonian Collection.

The sketch on this tablet indicates that Babylonian mathematicians were aware of the geometrical reasoning that justifies the Pythagorean Theorem of plane geometry for the case of a right isosceles triangle. This was approximately one thousand years before the life of Pythagoras - about three thousand years before the present time. It is considered the earliest known record indicating a geometrical proof. We mention this because it is connected to an interesting circle of ideas that we have encountered.

We have used the quadratic formula to prove the Cauchy–Schwarz inequality for the vector space $\mathcal{R}[a, b]$. The reader could easily look ahead to Theorem 8.1.1 to see that essentially the same method of proof establishes the Cauchy–Schwarz inequality in every vector space equipped with a *scalar product* $\langle \mathbf{x}, \mathbf{y} \rangle$, which includes also the Euclidean plane and n-dimensional Euclidean space. The Cauchy–Schwarz inequality yields as a consequence the *triangle inequality* of plane geometry, but in the general context of a vector space with an inner product–a so-called *inner product space*. Although the concept of an inner product space appears at first abstract, the geometry that is employed hinges on the quadratic formula, which embodies the method of *completing the square* learned in high school algebra. This method has been used in school exercises for at least four thousand years. Again, the earliest records appear on clay tablets that were school-room exercise tablets from Babylon [15]. And in Exercise 3.52 above, the reader has proven a version of the Pythagorean Theorem for the vector space $\mathcal{R}[a, b]$. The same proof works in the Euclidean plane, in n-dimensional Euclidean space, or in any inner product space. It is a simple application of the inner product of two vectors.

In other words, we have learned some elements of the geometry of the *function-space* $\mathcal{R}[a, b]$ by applying the quadratic formula from four thousand years ago. Study of this abstract and modern subject from *functional analysis* has shed light on Euclidean Geometry. The tool on which this rests is four thousand years old, and preceded Euclidean Geometry by two millennia. So here is a question that is not part of Mathematics: Is this light shed upon Euclidean Geometry a new light, or an old one?

3.5 TEST YOURSELF

EXERCISES

3.59 Let $f_n(x) = \frac{x}{1+nx^2}$ for all $x \in \mathbb{R}$. Find:
   a) The pointwise limit $f(x) = \lim_{n \to \infty} f_n(x)$ for all $x \in \mathbb{R}$.
   b) $\|f_n\|_{\sup}$ for each $n \in \mathbb{N}$.
   c) Is f_n uniformly convergent on $\mathbb{R}$? (Yes or No)

3.60 Does $f_n(x) = x^n$ converge pointwise on
   a) $(-1, 0]$?
   b) $[-1, 0]$?

3.61   Let

$$f(x) = \begin{cases} 1 & \text{if } 0 \le x < 1, \\ 2 & \text{if } 1 \le x \le 2. \end{cases}$$

If $\epsilon > 0$, find a $\delta > 0$ such that

$$\|P\| < \delta \Rightarrow \left| P(f, \{\bar{x}_i\}) - \int_0^2 f(x)\, dx \right| < \epsilon$$

independent of the choice of evaluation points $\bar{x}_i$.

3.62   Evaluate

$$\lim_{n \to \infty} \frac{2}{n} \sum_{k=1}^n \cos\left(1 + \frac{2k}{n}\right)$$

by expressing it as a suitable definite integral and evaluating this by means of the Fundamental theorem.

3.63   Let

$$f(x) = \begin{cases} 1 & \text{if } x = 1, \\ 0 & \text{if } x \in [0, 2] \setminus \{1\}. \end{cases}$$

Find a partition P of $[0, 2]$ for which $U(f, P) - L(f, P) < \frac{1}{49}$.

3.64   True or Give a Counterexample: If $|f| \in \mathcal{R}[0, 1]$ and if f is *bounded*, then $f \in \mathcal{R}[0, 1]$.

3.65   Suppose the linear map $T : \mathcal{C}[0, 1] \to \mathbb{R}$ is given by

$$T(f) = \int_0^1 f(x) \sin(x)\, dx.$$

Find a *bound* K such that $|T(f)| \le K\|f\|_{\sup}$ for all $f \in \mathcal{C}[0, 1]$.

3.66   Let $T : \mathcal{R}\left[\frac{1}{e}, e\right] \to \mathbb{R}$ be the linear functional defined by

$$T(f) = \int_{\frac{1}{e}}^e \frac{f(x)}{x}\, dx.$$

Find a constant K for which $|T(f)| \le K\|f\|_{\sup}$ for all $f \in \mathcal{R}\left[\frac{1}{e}, e\right]$.

3.67   Let

$$f(x) = \begin{cases} x^2 & \text{if } x \in \mathbb{Q}, \\ -x^2 & \text{if } x \in \mathbb{R} \setminus \mathbb{Q}. \end{cases}$$

a) Find $\overline{\int_0^1} f(x)\, dx$ and $\underline{\int_0^1} f(x)\, dx$.

b) True or False: The function $f \in \mathcal{R}[0, 1]$.

3.68   Suppose $f + g$ and $f - g$ are both Riemann integrable on $[a, b]$. True or False: *Both f and g are in $\mathcal{R}[a, b]$.*

3.69   Decide whether or not each set is a vector space (over the real numbers):
   a) $C[a, b]$, the set of continuous functions on $[a, b]$.
   b) $\mathcal{P}_{10}(\mathbb{R})$, the set of all polynomials of degree exactly equal to 10.

3.70   Let $f_n(x) = xe^{-nx}$ for all $x \in [0, \infty)$. Find $\|f_n\|_{\sup}$ for all n.

3.71   Find $\lim_{n \to \infty} \int_{\pi/4}^{\pi/2} \cos^n x \, dx$.

3.72   Let $T : \mathcal{P}[0, 1] \to \mathbb{R}$ be defined by $Tp = \deg p$ for each polynomial p. True or False: T is a linear map.

3.73   Let
$$f(x) = \begin{cases} x^2 & \text{if } x \in \mathbb{Q}, \\ -x^2 & \text{if } x \in \mathbb{R} \setminus \mathbb{Q}. \end{cases}$$

Find $\overline{\int_0^1} f(x) \, dx$ and $\underline{\int_0^1} f(x) \, dx$.

CHAPTER 4

THE DERIVATIVE

The first limit that is used intensively by every first year calculus student is the derivative. Here we will define the derivative carefully and prove the essential theorems about the derivative. We will discuss carefully the concept of the differential, because this will highlight the very important fact that a function is *differentiable* if and only if its increments can be approximated locally by a *linear* function.

4.1 DERIVATIVES AND DIFFERENTIALS

Definition 4.1.1 *If $x_0 \in D_f$ is a cluster point of the domain D_f of a function f, we define the derivative of f at x_0 to be*

$$f'(x_0) = \lim_{h \to 0} \frac{f(x_0 + h) - f(x_0)}{h},$$

provided that this limit exists. (Note that in accordance with the concept of limit, $h \to 0$ through values such that $x_0 + h \in D_f$. It is also common to denote $x = x_0 + h$ and to write the limit criterion as

$$f'(x_0) = \lim_{x \to x_0} \frac{f(x) - f(x_0)}{x - x_0}$$

Advanced Calculus: An Introduction to Linear Analysis. By Leonard F. Richardson
Copyright © 2008 John Wiley & Sons, Inc.

if the limit exists.)

This limit is often interpreted geometrically as representing the slope of a *tangent line* to the graph of $y = f(x)$ at the point $(x_0, f(x_0))$. The *difference quotient* of which $f'(x_0)$ is the limit is the slope of a chord joining $(x_0, f(x_0))$ to the point $(x_0 + h, f(x_0 + h))$. In effect, if the limit defining $f'(x_0)$ exists, we take the tangent line to be defined by the equation $y - f(x_0) = f'(x_0)(x - x_0)$, which describes the unique line through the base point $(x_0, f(x_0))$ with the slope $f'(x_0)$.

■ **EXAMPLE 4.1**

We present several examples of derivatives.

i. Let $f(x) = |x|$. Then we claim $f'(0)$ does not exist. In fact,

$$\lim_{h \to 0} \frac{|0 + h| - |0|}{h} = \lim_{h \to 0} \frac{|h|}{h}$$

does not exist, since if $h_n \to 0+$ (i.e., from the *right* side of 0), then $\frac{|h_n|}{h_n} \to 1$, but if $h_n \to 0-$ (i.e., from the *left*), then $\frac{|h_n|}{h_n} \to -1$. Since these one-sided limits are different, the limit of the function does not exist.

ii. Again we take $f(x) = |x|$, but we show that if $x_0 > 0$, $f'(x_0) = 1$. In fact, we can use h such that $|h| < x_0$, and then

$$\lim_{h \to 0} \frac{|x_0 + h| - |x_0|}{h} = \lim_{h \to 0} \frac{h}{h} = 1.$$

iii. Now let $f(x) = x^n$, $n \in \mathbb{N}$. Then

$$f'(x) = \lim_{h \to 0} \frac{(x + h)^n - x^n}{h}$$
$$= \lim_{h \to 0} \frac{nx^{n-1}h + \frac{n(n-1)}{2}x^{n-2}h^2 + \cdots + h^n}{h}$$
$$= nx^{n-1}.$$

iv. If $f(x) = c$, a constant, then $f'(x) \equiv 0$. (See Exercise 4.3.)

There is another way to understand the concept of a tangent line that is also very useful, although the definition may seem a bit more technical at the beginning. Imagine now a straight line through $(x_0, f(x_0))$ with slope m. Thus the *rise* of this line corresponding to an increment h in the x-variable is given by a *linear* function $L(h) = mh$. On the other hand, the rise of the graph of $y = f(x)$ will be denoted by $\Delta f = f(x_0 + h) - f(x_0)$. The intuitive idea is that the graph $y = f(x)$ has a tangent line at $(x_0, f(x_0))$, provided that there exists a linear function $L(h) = mh$ that approximates Δf accurately for all sufficiently small values of h. Then the tangent line corresponds to the linear function $L(h)$ that gives this approximation. We make the following formal definition based on this idea.

Definition 4.1.2 *The function f is called* differentiable *at a cluster point $x_0 \in D_f$, provided that there exists a linear function $L(h) = mh$ such that the function given by*

$$\epsilon(h) = \frac{\Delta f - L(h)}{h} \to 0$$

ie, $\dfrac{f(x_0 + h) - f(x_0) - mh}{h}$

as $h \to 0$.

$= \dfrac{f(x_0 + h) - f(x_0)}{h} - m \to 0$

The intuitive idea is that if $\epsilon(h) \to 0$ as $h \to 0$ then $\Delta f - L(h) \to 0$ much faster than $h \to 0$.

Theorem 4.1.1 *A function f is differentiable at a cluster point $x_0 \in D_f$ if and only if $f'(x_0)$ exists. If f is differentiable at x_0, then $L(h) = f'(x_0)h$ is denoted by $df_{x_0}(h)$ and*

$> f'(x_0) \cdot h$

$$\Delta f = f(x_0 + h) - f(x_0) = df_{x_0}(h) + \epsilon(h)h, \tag{4.1}$$

where $\epsilon(h) \to 0$ as $h \to 0$ (in such way that $x_0 + h \in D_f$).

Proof: For the implication from left to right, we suppose that f is differentiable at x_0. Then there exists $m \in \mathbb{R}$ such that

$$\epsilon(h) = \frac{\Delta f - mh}{h} \to 0$$

as $h \to 0$. Thus

$$\frac{\Delta f}{h} = \frac{f(x_0 + h) - f(x_0)}{h} = m + \epsilon(h) \to m$$

as $h \to 0$, so $f'(x_0)$ exists and equals m.
For the opposite direction of implication, we suppose $f'(x_0)$ exists. Let $m = f'(x_0)$. If we define $L(h) = mh$ and let

$$\epsilon(h) = \frac{\Delta f - L(h)}{h} = \frac{f(x_0 + h) - f(x_0)}{h} - f'(x_0),$$

then the hypothesis implies $\epsilon(h) \to 0$ as $h \to 0$. ∎

Remark 4.1.1 We remind the student here to review the Introduction regarding Learning to Write Proofs on page xxiii. The reader should write out a careful analysis of what makes each proof in this course work. We will do this together for the first proof of the chapter.

We are trying to prove that a function f is differentiable if and only if its *increments* Δf can be approximated well locally by the differential, df, in the technical sense described in the statement of the theorem. For the implication from left to right, we suppose f is differentiable, so that Δf can be approximated by df using a suitable error-function ϵ. This enables us to express the difference quotient for the derivative in terms of df and ϵ, and the assumption that $\epsilon \to 0$ as $h \to 0$ implies that the limit that yields the derivative exists.

For the opposite implication we assume $f'(x)$ exists and we *define* ϵ by means of Equation (4.1). All that remains is to prove that $\epsilon \to 0$ as $h \to 0$. By writing ϵ in terms of the difference quotient and the derivative, the existence of the derivative implies that $\epsilon \to 0$ as $h \to 0$.

Corollary 4.1.1 *If f is differentiable at x_0 then f is continuous at x_0.*

Proof: We need to prove $\lim_{x \to x_0} f(x) = f(x_0)$. This is equivalent to proving $\lim_{x \to x_0}(f(x) - f(x_0)) = 0$. Denote $h = x - x_0$, and we have

$$\lim_{h \to 0}(f(x_0 + h) - f(x_0)) = \lim_{h \to 0}[df_{x_0}(h) + \epsilon(h)h]$$
$$= \lim_{h \to 0}[f'(x_0) + \epsilon(h)]h = 0.$$

■

Theorem 4.1.2 *Suppose $f'(x)$ and $g'(x)$ both exist. Then*

   i. $(f \pm g)'(x)$ exists and $(f \pm g)'(x) = f'(x) \pm g'(x)$.

   ii. $(fg)'(x)$ exists and $(fg)'(x) = f'(x)g(x) + f(x)g'(x)$.

   iii. $\left(\frac{f}{g}\right)'(x)$ exists and $\left(\frac{f}{g}\right)'(x) = \frac{g(x)f'(x) - f(x)g'(x)}{g(x)^2}$, if $x \in D_{\frac{f}{g}}$ is a cluster point of $D_{\frac{f}{g}}$.

Proof: Let us prove case (i) for $f + g$.

$$\lim_{h \to 0} \frac{(f + g)(x + h) - (f + g)(x)}{h}$$
$$= \lim_{h \to 0}\left(\frac{f(x + h) - f(x)}{h} + \frac{g(x + h) - g(x)}{h}\right)$$
$$= f'(x) + g'(x)$$

The proof for $f - g$ is very similar. For (ii) and (iii) see Exercises 4.4 and 4.5. ■

Theorem 4.1.3 *(The Chain Rule) Suppose g is differentiable at x_0 and f is differentiable at $g(x_0)$. Then $(f \circ g)'(x_0)$ exists and*

$$(f \circ g)'(x_0) = f'(g(x_0))g'(x_0).$$

Proof: To show that the derivative of the composition $f \circ g$ exists at x_0, we denote $k = g(x_0 + h) - g(x_0) \to 0$ as $h \to 0$ since g is continuous at x_0 by Corollary 4.1.1.

Also, $k = g'(x_0)h + \tilde{\epsilon}(h)h$, where $\tilde{\epsilon}(h) \to 0$ as $h \to 0$. Next we form the following difference quotient noting that $\epsilon(k) \to 0$ as $k \to 0$:

$$\frac{f(g(x_0 + h)) - f(g(x_0))}{h}$$

$$= \frac{f'(g(x_0))k + \epsilon(k)k}{h}$$

$$= \frac{f'(g(x_0))[g'(x_0) + \tilde{\epsilon}(h)]h + \epsilon(k)[g'(x_0) + \tilde{\epsilon}(h)]h}{h}$$

$$\to f'(g(x_0))g'(x_0)$$

as $h \to 0$.   ∎

EXERCISES

4.1   If $f(x) = |x|$ and if $x_0 < 0$, show that $f'(x_0) = -1$. (See Example 4.1(ii).)

4.2   Prove directly from Definition 4.1.1 that if $f'(a)$ exists, then f must be continuous at a. (Hint: Suppose $x_n \to a$ with each $X_n \in D_f \setminus a$. Prove that $f(x_n) - f(a) \to 0$ as $x_n \to a$.)

4.3   † Prove the conclusion of Example 4.1(iv).

4.4   † Prove conclusion (ii) of Theorem 4.1.2. (Hint: You may use the conclusion of Corollary 4.1.1.)

4.5   †

    a) Prove: if $g'(x)$ exists and $g(x) \neq 0$, then $\left(\frac{1}{g}\right)'(x)$ exists and is equal to $\frac{-g'(x)}{g(x)^2}$. (Hint: You may use the conclusion of Corollary 4.1.1.)

    b) Now use part (a) above together with Theorem 4.1.2(ii) to prove conclusion (iii) of Theorem 4.1.2.

4.6   Let

$$f(x) = \begin{cases} x \sin\left(\frac{\pi}{x}\right) & \text{if } x \in (0, 1], \\ 0 & \text{if } x = 0, \end{cases}$$

as in Fig. 4.1.

    a) Prove: $f \in C[0, 1]$. (Be sure to consider $x = 0$. You can use the *squeeze* theorem for functions.)

    b) Prove: $f'(x)$ exists for all $x \in (0, 1]$, but $f'(0)$ does *not* exist. (We assume you know about the derivative of $\sin x$ from elementary calculus.)

4.7   † Let

$$f(x) = \begin{cases} x^2 \sin\left(\frac{\pi}{x}\right) & \text{if } x \neq 0, \\ 0 & \text{if } x = 0. \end{cases}$$

See Fig. 7.1, p. 217.

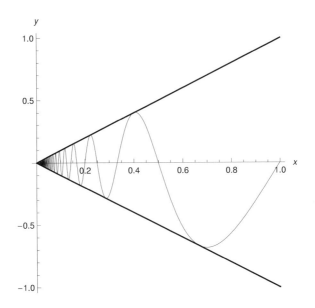

Figure 4.1 $f(x) = x \sin\left(\frac{\pi}{x}\right)$, with envelope $u(x) = x$, $l(x) = -x$

a) Prove: $f'(x)$ exists for all $x \in \mathbb{R}$. (Be sure to consider $x = 0$.)
b) Prove: $f \in \mathcal{C}(\mathbb{R})$.
c) Prove: f' is *not* continuous at $x = 0$.

4.8   Let

$$f(x) = \begin{cases} x^2 & \text{if } x \in \mathbb{Q}, \\ 0 & \text{if } x \notin \mathbb{Q}. \end{cases}$$

Prove that f is continuous at one and only one point $x = a$. Find the value of a and prove that $f'(a)$ exists.

4.9   Suppose f is one-to-one (also called *injective*) and differentiable on D_f and suppose g is a *differentiable* inverse: ie, $f(g(x)) \equiv x$. Suppose $x \in D_g$ is a cluster point of D_g.
a) Prove:

$$g'(x) = \frac{1}{f'(g(x))}.$$

b) Use the result of (a) to derive the formula for $g'(x)$ if $f(x) = \sin x$ so that $g(x) = \sin^{-1} x$, the inverse sine of x.

4.2   THE MEAN VALUE THEOREM

The theorems in this section play a vital role in such diverse applications as extreme value problems, proofs of inequalities, and as we shall see in Section 4.3, even the proof of the Fundamental Theorem of Calculus.

Definition 4.2.1 *The* local maximum *and* local minimum *points of* f, *defined below, are called* local extreme points.

  i. *A function* f *is said to have a* local maximum point *at* $a \in D_f$ *if and only if there exists* $\delta > 0$ *such that* $x \in D_f \cap (a - \delta, a + \delta)$ *implies* $f(x) \leq f(a)$.

  ii. *The point* $a \in D_f$ *is called a* local minimum point *if and only if there exists* $\delta > 0$ *such that* $x \in D_f \cap (a - \delta, a + \delta)$ *implies* $f(x) \geq f(a)$.

The idea behind the preceding definition is that a local extreme point need not be either the highest or lowest point on the entire graph of the function, but will be a local high point or a local low point–meaning just within its own vicinity. To study local extreme points with the derivative, we need the concept of an *interior point* of a domain.

Definition 4.2.2 *A point* a *is called an* interior point *of* D_f *if and only if there exists* $\delta > 0$ *such that* $(a - \delta, a + \delta) \subseteq D_f$. *The set of all interior points of* D_f *is denoted by* D_f^o, *the* interior *of* D_f.

In other words, an interior point a of a domain D is a point which has no points from the complement of D within some small specified radius $\delta > 0$ of a.

Theorem 4.2.1 *If* f *is differentiable at a local extreme point* $\mu \in D_f^o$, *then* $f'(\mu) = 0$.

Proof:

  i. Consider first the case in which f has a local maximum point at μ. Note that there exists $\delta > 0$ such that $0 < h < \delta$ implies

$$\frac{f(\mu + h) - f(\mu)}{h} \leq 0,$$

whereas $-\delta < h < 0$ implies

$$\frac{f(\mu + h) - f(\mu)}{h} \geq 0.$$

Thus

$$\lim_{h \to 0+} \frac{f(\mu + h) - f(\mu)}{h} \leq 0,$$

whereas

$$\lim_{h \to 0-} \frac{f(\mu + h) - f(\mu)}{h} \geq 0.$$

Because $f'(\mu)$ exists, it follows that

$$f'(\mu) = \lim_{h \to 0} \frac{f(\mu + h) - f(\mu)}{h} = 0$$

since it is *both* nonnegative and nonpositive.

ii. If f has a local minimum at μ, consider $g = -f$, which has a local maximum at μ, to reach the desired conclusion.

∎

Theorem 4.2.2 (Rolle's Theorem) *If $f \in C[a, b]$ is such that*

$$f(a) = 0 = f(b)$$

and such that $f'(x)$ exists at least for all $x \in (a, b)$, then there exists $\mu \in (a, b)$ such that $f'(\mu) = 0$.

Remark 4.2.1 Rolle's Theorem says that if the chord joining the two endpoints $(a, f(a))$ and $(b, f(b)$ lies on the x-axis, then there exists a tangent line at some point of the graph that is *parallel* to the chord.

Proof: If $f \equiv 0$ then the claim of the theorem is trivial. If there exists $x_0 \in [a, b]$ such that $f(x_0) > 0$ then the maximum point of f on $[a, b]$ must occur at an *interior* point $\mu \in (a, b)$. By Theorem 4.2.1, $f'(\mu) = 0$. The remaining case, in which there exists $x_0 \in [a, b]$ such that $f(x_0) < 0$, is similar.

∎

Theorem 4.2.3 (Mean Value Theorem) *Suppose $f \in C[a, b]$ and differentiable at least on (a, b). Then there exists $\mu \in (a, b)$ such that*

$$f'(\mu) = \frac{f(b) - f(a)}{b - a}.$$

Remark 4.2.2 The Mean Value Theorem states that under the given hypotheses there must be a tangent to the graph that is parallel to the chord joining the endpoints. (See Fig. 4.2.) Observe that this is a generalization of Rolle's Theorem.

Proof: Observe that the straight line that is the chord joining the endpoints of the graph of f is described by the equation

$$y = f(a) + \frac{f(b) - f(a)}{b - a}(x - a).$$

Now, for all $x \in [a, b]$, let

$$h(x) = f(x) - y = f(x) - f(a) - \frac{f(b) - f(a)}{b - a}(x - a),$$

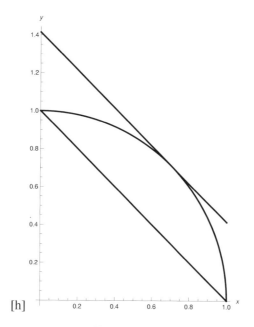

[h]

Figure 4.2 $f(x) = \sqrt{1 - x^2}$, chord with parallel tangent.

the difference in height between the graph of f and the graph of the chord. Observe that $h \in C[a, b]$ and $h(a) = 0 = h(b)$. Also, h' exists at least on (a, b). Thus Rolle's Theorem implies there exists $\mu \in (a, b)$ such that $h'(\mu) = 0$. This implies that

$$f'(\mu) = \frac{f(b) - f(a)}{b - a}.$$

■

Corollary 4.2.1 *If $f' \equiv 0$ on an interval I, then f is a constant function on I.*

Proof: Fix any $a \in I$ and let $x \in I$ be arbitrary. Then the Mean Value Theorem implies there exists $\mu \in I$ such that

$$f(x) - f(a) = f'(\mu)(x - a) = 0(x - a) = 0$$

so that $f(x) \equiv f(a)$, a constant function. ■

Corollary 4.2.2 *If $f' \equiv g'$ on an interval I, then there exists $c \in \mathbb{R}$ such that $f(x) \equiv g(x) + c$.*

Proof: Let $h(x) = f(x) - g(x)$ and observe that $h' \equiv 0$ on I. Now use Corollary 4.2.1. ■

Corollary 4.2.3 *Suppose $f \in C[a, b]$ is differentiable at least on (a, b). Then*

i. *If $f'(x) \geq 0$ for all $x \in (a, b)$, then f is increasing (denoted by $f \nearrow$) on $[a, b]$.*

ii. *If $f'(x) > 0$ for all $x \in (a, b)$, then f is* strictly *increasing (denoted by $f \uparrow$) on $[a, b]$.*

iii. *If $f'(x) \leq 0$ for all $x \in (a, b)$, then f is decreasing (denoted by $f \searrow$) on $[a, b]$.*

iv. *If $f'(x) < 0$ for all $x \in (a, b)$, then f is* strictly *decreasing (denoted by $f \downarrow$) on $[a, b]$.*

Proof:

i. Suppose that $f'(x) \geq 0$ for all $x \in (a, b)$. Then if $x_1 < x_2$ are points of $[a, b]$, the Mean Value Theorem implies there exists $\mu \in (a, b)$ such that

$$f(x_2) - f(x_1) = f'(\mu)(x_2 - x_1) \geq 0$$

so that $f(x_2) \geq f(x_1)$, and f is increasing on $[a, b]$.

For the remaining cases, see Exercises 4.10–4.12. ∎

■ **EXAMPLE 4.2**

If $x > 0$, then $\sin x < x$.

Proof: Let $f(x) = x - \sin x$. Then $f(0) = 0$ and $f'(x) = 1 - \cos x > 0$ on $(0, 2\pi)$. By Corollary 4.2.3, f is increasing strictly on $[0, 2\pi]$, so that $f(x) > 0$ for all $x \in (0, 2\pi]$. Hence $\sin x < x$ if $0 < x \leq 2\pi$. If $x > 2\pi$, however, it follows that $\sin x \leq 1 < x$ as well. ∎

Definition 4.2.3 *A function $f : I \to \mathbb{R}$, where I is an interval, is called* monotone *on I if and only if f is either increasing throughout I or else decreasing throughout I. The function f is called* strictly monotone *provided it is either strictly increasing or else strictly decreasing.*

Definition 4.2.4 *A function $f : D \to \mathbb{R}$, where $D \subseteq \mathbb{R}$, is called* injective *if and only if f is one-to-one. That is, f is injective if and only if*

$$f(x_1) = f(x_2) \Leftrightarrow x_1 = x_2$$

for all x_1 and x_2 in D.

■ **EXAMPLE 4.3**

A function f that is strictly monotone on an interval I must be injective on I.

EXERCISES

4.10   † Prove Case ii. of Corollary 4.2.3.

4.11   † Prove Case iii. of Corollary 4.2.3.

4.12   † Prove Case iv of Corollary 4.2.3.

4.13   Give an example of a differentiable function f for which $f' \in C[a, b]$ such that f increases strictly on $[a, b]$ yet it is *false* that $f'(x) > 0$ for all $x \in [a, b]$.

4.14   Give an example of f increasing strictly on $[a, b]$ yet $f \notin C[a, b]$.

4.15   † Let $n \in \mathbb{N}$. Prove:

$$\left(1 - x^2\right)^n \geq 1 - nx^2$$

for all $x \in [0, 1]$. Hint: Let $f(x) = \left(1 - x^2\right)^n - \left(1 - nx^2\right)$.

4.16   Prove: $\log(1 + x) < x$ for all $x > 0$.

4.17   Prove: $1 - \frac{x^2}{2} < \cos x$ for all $x \neq 0$.

4.18   Prove:

$$\frac{x}{1 + x^2} < \tan^{-1} x < x$$

for all $x > 0$.

4.19   Let p be a polynomial of degree n and let

$$E = \left\{ x \in \left(0, \frac{\pi}{2}\right) \;\middle|\; \sin x = p(x) \right\}.$$

Prove that the number of elements in the set E, denoted by $|E|$, satisfies the inequality $|E| \leq n + 1$. (Hint: Apply Rolle's Theorem.)

4.20   Suppose $|f'(x)| \leq M \in \mathbb{R}$ for all $x \in I$, an interval.
   a) Prove: f is *uniformly continuous* on I.
   b) If I is a finite interval (a, b), prove that $\lim_{x \to a+} f(x)$ exists. (Hint: It is enough for this problem to know that f is uniformly continuous, by the previous part. Consider first any sequence $x_n \to a+$ and prove that $f(x_n)$ is Cauchy.)
   c) Consider the function $s(x) = \sin \frac{1}{x}$ on $(0, 1)$. Does s satisfy the hypothesis of this exercise? Investigate statements (a) and (b) for the case of the function s.

4.21   Let $q(x) = \sqrt{x}$ for all $x \in [0, 1]$. True or False:
   a) The function q is uniformly continuous on $(0, 1)$.
   b) The function q' is bounded on $(0, 1)$.

4.22   † Suppose $f \in C[a, b]$. Suppose f is differentiable at a and at b, and also that

$$f'(a)f'(b) < 0.$$

Prove f is not injective on $[a, b]$. (Hint: Consider the extreme points of f.)

4.23   If $D \subseteq \mathbb{R}$, then prove D°, the *interior* of D, is the union of all open subsets of D.

4.3   THE FUNDAMENTAL THEOREM OF CALCULUS

Theorem 4.3.1 (Fundamental Theorem, Version 1) *Suppose $f \in \mathcal{R}[a, b]$, and suppose also there exists F such that $F'(x) \equiv f(x)$ on $[a, b]$. Then*

$$\int_a^b f(x)\, dx = F(b) - F(a).$$

Remark 4.3.1 We will see by example below that it is possible for a derivative to exist throughout $[a, b]$ and yet not be integrable, which implies that a derivative can exist without being either continuous or monotone, for example.

Proof: By hypothesis if $\epsilon > 0$ there exists $\delta > 0$ such that $\|P\| < \delta$ implies

$$\left| P(f, \{\bar{x}_i\}) - \int_a^b f(x)\, dx \right| < \epsilon.$$

Let $P = \{x_0, \ldots, x_n\}$ be any partition of $[a, b]$ such that $\|P\| < \delta$. By the Mean Value Theorem in each sub-interval $[x_{i-1}, x_i]$ we are free to pick $\bar{x}_i$ in such a way that $F'(\bar{x}_i)\Delta x_i = F(x_i) - F(x_{i-1})$. Then

$$\begin{aligned}
P(f, \{\bar{x}_i\}) &= \sum_{i=1}^n f(\bar{x}_i)\Delta x_i \\
&= \sum_{i=1}^n F'(\bar{x}_i)\Delta x_i \\
&= \sum_{i=1}^n [F(x_i) - F(x_{i-1})] \\
&= F(b) - F(a).
\end{aligned}$$

Since this implies $\left| F(b) - F(a) - \int_a^b f(x)\, dx \right| < \epsilon$ for all $\epsilon > 0$, we must have $\int_a^b f(x)\, dx = F(b) - F(a)$.   ∎

■ **EXAMPLE 4.4**

Let

$$F(x) = \begin{cases} x^2 \sin \frac{1}{x^2} & \text{if } 0 < |x| \le 1, \\ 0 & \text{if } x = 0. \end{cases}$$

The student can show that $F'(0)$ exists and equals 0, and $F'(x)$ can be calculated using the Chain Rule at any $x \neq 0$. Thus F' exists throughout $[-1, 1]$, yet $F' \notin \mathcal{R}[-1, 1]$ and also $F' \notin \mathcal{C}[-1, 1]$. (See Exercise 4.24.)

Theorem 4.3.2 (Fundamental Theorem, Version 2) *Suppose* $f \in \mathcal{C}[a, b]$, *and define*

$$F(x) = \int_a^x f(t)\, dt$$

for all $x \in [a, b]$. *Then* $F'(x)$ *exists and* $F'(x) \equiv f(x)$.

Proof: The difference quotient

$$
\begin{aligned}
\frac{F(x + h) - F(x)}{h} &= \frac{1}{h} \int_x^{x+h} f(t)\, dt \\
&= \frac{1}{h} f(\bar{x}) h \\
&= f(\bar{x}) \to f(x)
\end{aligned}
$$

as $h \to 0$. Here $\bar{x}$ comes from the Mean Value Theorem for Integrals (Exercise 3.10), and $f(\bar{x}) \to f(x)$ since $\bar{x} \to x$ as $h \to 0$ and $f \in \mathcal{C}[a, b]$. ∎

We have seen in Example 4.4 that a derivative can exist without being continuous. Nevertheless, all derivatives on intervals do share one property in common with continuous functions, as we see below.

Theorem 4.3.3 (Intermediate Value Theorem for Derivatives) *If* $f'(x)$ *exists for all* $x \in I$, *an interval, then* f' *has the* Intermediate Value Property *on* I: *Namely, if* $a, b \in I$ *and if* y *lies strictly between* $f'(a)$ *and* $f'(b)$, *then there exists* $\bar{x}$ *between* a *and* b *such that* $f'(\bar{x}) = y$.

Proof: Suppose $a < b$ and $f'(a) < y < f'(b)$. (For the opposite inequality, just consider $g = -f$.) Define $\phi(x) = f(x) - yx$ so that $\phi \in \mathcal{C}[a, b]$ and is also differentiable on $[a, b]$. It would suffice to show there exists $\bar{x} \in (a, b)$ such that $\phi'(\bar{x}) = 0$. By the Extreme Value Theorem, ϕ must have both a maximum and a minimum point in $[a, b]$. If such an extreme point is also an interior point, we are done. So we will suppose the extreme points occur only at the endpoints a and b, and we will deduce a contradiction. We will present the argument in three cases.

i. Suppose $\phi(a)$ is a maximum point and $\phi(b)$ is a minimum point. Then

$$\phi'(a) = \lim_{h \to 0+} \frac{\phi(a + h) - \phi(a)}{h} \leq 0$$

so that $f'(a) \leq y$. And

$$\phi'(b) = \lim_{h \to 0-} \frac{\phi(b + h) - \phi(b)}{h} \leq 0$$

so that $f'(b) \leq y$ as well. This contradicts the hypothesis.

ii. Suppose $\phi(a)$ is a minimum and $\phi(b)$ is a maximum. Then a similar contradiction can be deduced. (See Exercise 4.30.)

iii. Finally, if one endpoint is both a maximum and a minimum point, then ϕ is a constant function and $\phi'(x) \equiv 0$.

■

EXERCISES

4.24   † Prove that $F'(x)$ exists for all $x \in [-1, 1]$ in Example 4.4, yet F' is *not* bounded, and so $F' \notin \mathcal{R}[-1, 1]$ and also $F' \notin \mathcal{C}[-1, 1]$.

4.25   Let

$$F(x) = \int_0^{x^2} e^{-t^2} dt$$

for all $x \in \mathbb{R}$. Find $F'(x)$. (Hint: Let $G(u) = \int_0^u e^{-t^2} dt$ where $u = U(x) = x^2$. Thus $F = G \circ U$.)

4.26   Suppose $f \in \mathcal{R}[a, b]$ and let

$$F(x) = \int_a^x f(t) \, dt$$

for all $x \in [a, b]$. Prove: $F \in \mathcal{C}[a, b]$. (Hint: f must be bounded.)

4.27   Let

$$f(x) = \begin{cases} \sin \frac{\pi}{x} & \text{if } 0 < x \leq 1, \\ 0 & \text{if } x = 0. \end{cases}$$

Let

$$F(x) = \int_0^x f(t) \, dt$$

for all $x \in [0, 1]$. Prove: $F \in \mathcal{C}[0, 1]$. (Hint: See Exercise 3.27.)

4.28   Suppose $f \in \mathcal{R}[a, b]$ and let $F(x) = \int_a^x f(t) \, dt$ for all $x \in [a, b]$. If $f(x) \geq 0$ for all $x \in [a, b]$, prove F is increasing on $[a, b]$.

4.29   Suppose f' and g' exist and suppose $f', g' \in \mathcal{R}[a, b]$. Prove:

$$\int_a^b f(x)g'(x) \, dx = f(b)g(b) - f(a)g(a) - \int_a^b g(x)f'(x) \, dx.$$

(Hint: Consider $\int_a^b (fg)'(x) \, dx$.)

4.30   † Prove case (ii) of Theorem 4.3.3.

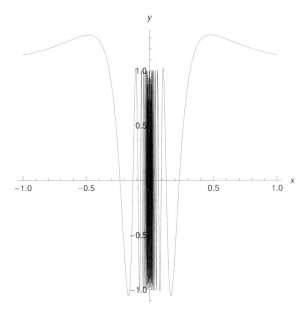

Figure 4.3 $f(x) = 2x \sin\left(\frac{1}{x}\right) - \cos\left(\frac{1}{x}\right)$.

4.31   Suppose

$$f(x) = \begin{cases} 0 & \text{if } 0 \le x < 1, \\ 1 & \text{if } 1 \le x \le 2. \end{cases}$$

a) Does there exist a function F on $[0, 2]$ such that $F'(x) \equiv f(x)$?

b) Let $F(x) = \int_0^x f(t)\, dt$ for all $x \in [0, 2]$. Prove that $F \in C[0, 2]$ but that it is false that $F'(x) \equiv f(x)$ on $[0, 2]$.

4.32   Let

$$F(x) = \begin{cases} x^2 \sin \frac{1}{x} & \text{if } 0 < |x| \le 1, \\ 0 & \text{if } x = 0 \end{cases}$$

and let $f(x) = F'(x)$. (See Fig. 4.3.)

a) Prove that $F'(x)$ exists for all $x \in [-1, 1]$.

b) Find $f(x)$ for all $x \in [-1, 1]$, and prove $f \in \mathcal{R}[-1, 1] \setminus C[-1, 1]$. (Hint: Apply Exercise 3.26.)

c) Find $\int_{-1}^1 f(x)\, dx$.

4.33   Find $\int_0^1 g(x)\, dx$ if

$$g(x) = \begin{cases} 2x \cos \frac{\pi}{x} - \pi \sin \frac{\pi}{x} & \text{if } x \in (0, 1], \\ 0 & \text{if } x = 0. \end{cases}$$

4.34 Suppose $f : (-a, a) \to \mathbb{R}$ is a differentiable function. *Either prove* the following statements *or give counterexamples*:

  a) If f is an *odd* function, then f' is an *even* function.
  b) If f is an *even* function, then f' is an *odd* function.
  c) If f' is odd, then f is even.
  d) If f' is even, then f is odd.

4.4 UNIFORM CONVERGENCE AND THE DERIVATIVE

We have learned that if $f_n \in \mathcal{C}[a, b]$ and if $\|f_n - f\|_{\sup} \to 0$, then $f \in \mathcal{C}[a, b]$ as well. We have seen that if $f_n \in \mathcal{R}[a, b]$ and if $\|f_n - f\|_{\sup} \to 0$, then $f \in \mathcal{R}[a, b]$ and, moreover,

$$\int_a^b f_n(x)\, dx \to \int_a^b f(x)\, dx.$$

It is understandable that the student now anticipates learning a virtually identical result for differentiation. The following examples show, however, that differentiation behaves in a more delicate manner with respect to uniform convergence (meaning *sup-norm* convergence).

■ EXAMPLE 4.5

Let $f_n(x) = \frac{\sin nx}{n}$ for all $x \in \mathbb{R}$, and let $g(x) \equiv 0$. Then

$$\|f_n - g\|_{\sup} = \sup\left\{ \frac{|\sin nx|}{n} \,\middle|\, x \in \mathbb{R} \right\} = \frac{1}{n} \to 0$$

so that $f_n \to g$ uniformly on $\mathbb{R}$. However, $f_n'(x) = \cos nx \nrightarrow g'(x) \equiv 0$ for some values of x.

For example, if $x = 0$, $\cos nx \equiv \cos 0 = 1 \nrightarrow 0$, and this failure of $f_n'(x)$ to converge to $g'(x)$ occurs at many other values of x as well. (See Fig. 4.4.)

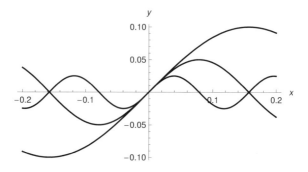

Figure 4.4 $f_n(x) = \frac{1}{n} \sin nx$, $n = 10, 20, 40$.

■ **EXAMPLE 4.6**

Let

$$f_n(x) = |x|^{1 + \frac{1}{n}}$$

for all $x \in [-1, 1]$. Then f_n is increasing as n increases and $f_n(x) \to f(x) = |x|$ uniformly on $[-1, 1]$ by Dini's Theorem. (An alternative approach to proving this uniform convergence is illustrated in Example 2.9.)

If $x > 0$, $f(x) \equiv x$ and $f'(x) = 1$. If $x < 0$, then $f(x) \equiv -x$ and $f'(x) = -1$. If $x = 0$,

$$\lim_{x \to 0+} \frac{f_n(x) - f_n(0)}{x - 0} = \lim_{x \to 0+} \frac{x^{1 + \frac{1}{n}}}{x}$$
$$= \lim_{x \to 0+} x^{\frac{1}{n}} = 0$$

and

$$\lim_{x \to 0-} \frac{f_n(x) - f_n(0)}{x - 0} = \lim_{x \to 0-} \frac{(-x)^{1 + \frac{1}{n}}}{x}$$
$$= \lim_{x \to 0-} -(-x)^{\frac{1}{n}} = 0$$

so that $f_n'(0) \equiv 0 \to 0$ and yet $f'(0)$ does not exist. (See Fig. 4.5.)

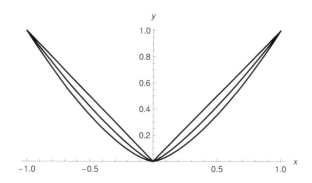

Figure 4.5 $|x|^{1 + \frac{1}{2}}, |x|^{1 + \frac{1}{4}}, \ldots$ increases and approaches $|x|$.

Nevertheless, there is a useful theorem about uniform convergence and the derivative, as we see below.

Theorem 4.4.1 *Suppose f_n is defined on a finite interval I and $f_n' \in C(I)$. Suppose f_n' converges uniformly on I. Suppose moreover that there exists at least one point $a \in I$ such that $f_n(a)$ is a convergent sequence of real numbers. Then there exists a differentiable function f such that $f_n \to f$ uniformly on I, and*

$$f'(x) \equiv \lim_{n \to \infty} f_n'(x)$$

on I.

Remark 4.4.1 Notice that the key unexpected hypothesis is that it is the sequence of *derivatives* f'_n which we must assume to be uniformly convergent.

Proof: Let g denote temporarily the uniform limit of f'_n on I, so $g \in C[a, x]$ for each x such that $[a, x] \subseteq I$. By the Fundamental Theorem (Version 1) we can define $f(x)$ for all $x \in I$ as follows:

$$f_n(x) = f_n(a) + \int_a^x f'_n(t) \, dt$$

$$\to \lim_{n \to \infty} f_n(a) + \int_a^x g(t) \, dt = f(x).$$

By the Fundamental Theorem (Version 2) $f'(x) = g(x)$ for all $x \in I$. Thus $f_n \to f$ at least pointwise. To show $\|f_n - f\|_{\sup} \to 0$, observe that

$$|f_n(x) - f(x)| = \left| f_n(a) + \int_a^x f'_n(t) \, dt - f(a) - \int_a^x g(t) \, dt \right|.$$

Denoting by L the finite length of I, we have

$$\|f_n - f\|_{\sup} \le |f_n(a) - f(a)| + L \, \|f'_n - g\|_{\sup} \to 0 + 0 = 0$$

as $n \to \infty$.   ∎

EXERCISES

4.35   Let

$$f_n(x) = \frac{1}{n} \sin \left(n^2 x \right).$$

Prove: f_n converges uniformly on $\mathbb{R}$ to a differentiable function, yet $f'_n(0)$ *diverges*.

4.36   In Example 4.5, find a value of x for which $f'_n(x)$ *diverges* as $n \to \infty$.

4.37   Give an example of a sequence f_n for which $f'_n \to 0$ uniformly on $\mathbb{R}$, yet $f_n(x)$ *diverges* for all $x \in \mathbb{R}$.

4.38   Let $f_n(x) = \sin^n x$ for all $x \in [0, \pi]$. Prove that f'_n is *not* uniformly convergent on $[0, \pi]$. (Hint: Suppose false and apply Theorem 4.4.1 to deduce a contradiction.)

4.39   Denote by $C^1[a, b]$ the space of functions having continuous derivatives on $[a, b]$. Define

$$\|f\| = \|f\|_{\sup} + \|f'\|_{\sup}$$

for each $f \in C^1[a, b]$.
  a) Prove that $\| \cdot \|$ is a norm on the vector space $C^1[a, b]$.
  b) ◇ Prove that $C^1[a, b]$ is a *complete* normed linear space using the norm $\| \cdot \|$.

4.5 CAUCHY'S GENERALIZED MEAN VALUE THEOREM

Theorem 4.5.1 (Cauchy's Generalized Mean Value Theorem) *Let f and g be in $\mathcal{C}[a,b]$ and suppose that both f and g are differentiable at least on (a,b), with $g'(x) \neq 0$ for any $x \in (a,b)$. Then there exists $\bar{x} \in (a,b)$ such that*

$$\frac{f(b) - f(a)}{g(b) - g(a)} = \frac{f'(\bar{x})}{g'(\bar{x})}.$$

Remark 4.5.1 Observe that the pair of functions $x = g(t)$, $y = f(t)$ with $t \in [a,b]$ defines a *smooth* curve in the xy-plane parametrically. This curve need not be the graph of a function of the form $y = \phi(x)$, since it is possible that more than one y corresponds to x. However, geometrically, the conclusion of the Generalized MVT is essentially the same as that of the ordinary MVT: There is a tangent line to the parametric curve that is parallel to the chord joining the endpoints. (See Fig. 4.6.)

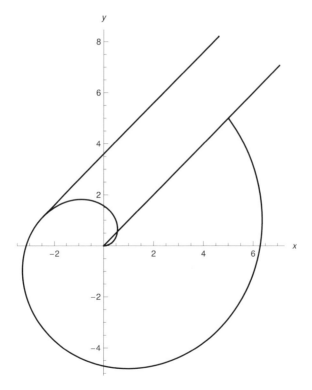

Figure 4.6 $x = t \cos t$, $y = t \sin t$, chord with parallel tangent.

Proof: Using the parametric functions defined in the remark above, we write an equation for the straight line through the endpoints of the parametric curve:

$$y = f(a) + \frac{f(b) - f(a)}{g(b) - g(a)}((x - g(a)).$$

Note that $g(b) - g(a) \neq 0$ because of the ordinary Mean Value Theorem. Now we define a function $h(t)$ to be the difference in height between the parametric curve and the chord at the point with first coordinate $x = g(t)$, $t \in [a, b]$. Thus

$$h(t) = f(t) - \left[f(a) + \frac{f(b) - f(a)}{g(b) - g(a)}(g(t) - g(a)) \right]$$

for all $t \in [a, b]$. We see that $h \in C[a, b]$ and h' exists at least on (a, b). And $h(a) = 0 = h(b)$. By Rolle's Theorem, there exists $\bar{x} \in (a, b)$ such that

$$h'(\bar{x}) = 0 = f'(\bar{x}) - \frac{f(b) - f(a)}{g(b) - g(a)} g'(\bar{x}).$$

■

As an application of the Cauchy Generalized Mean Value Theorem, we prove one of many useful versions of L'Hôpital's Rule.

Theorem 4.5.2 *(L'Hôpital's Rule) Suppose f and g are in $C[a, b]$, both differentiable at least on (a, b), with $g'(x)$ nowhere vanishing. If*

$$\lim_{x \to a} f(x) = 0 = \lim_{x \to a} g(x)$$

and if

$$\lim_{x \to a} \frac{f'(x)}{g'(x)} = L,$$

then $\lim_{x \to a} \frac{f(x)}{g(x)}$ exists and

$$\lim_{x \to a} \frac{f(x)}{g(x)} = L.$$

Proof: Observe that under the hypotheses we must have $f(a) = 0 = g(a)$. By Cauchy's Generalized MVT we have

$$\frac{f(x)}{g(x)} = \frac{f(x) - f(a)}{g(x) - g(a)} = \frac{f'(\bar{x})}{g'(\bar{x})} \to L$$

as $x \to a$ because this forces $\bar{x} \to a$. ■

Remark 4.5.2 We observe that a similar theorem could be proved with the limit as $x \to b$ and that $g(x)$ can't be 0 in this theorem.

■ **EXAMPLE 4.7**

We give some examples of limits that can be computed by means of L'Hôpital's rule.

i. $\lim_{x \to 1} \frac{\log x}{x-1} = \lim_{x \to 1} \frac{1/x}{1} = 1$.

ii. $\lim_{x \to 0} \frac{1-\cos x}{x^2} = \lim_{x \to 0} \frac{\sin x}{2x} = \lim_{x \to 0} \frac{\cos x}{2} = \frac{1}{2}$.

Next, we prove two of many possible variations on L'Hôpital's Rule. (A similar theorem could be proven for $x \to -\infty$.)

Theorem 4.5.3 *Suppose f and g are both differentiable on (b, ∞) with $g'(x)$ nowhere 0 and $b > 0$. Suppose that*

$$\lim_{x \to \infty} f(x) = 0 = \lim_{x \to \infty} g(x)$$

and also that

$$\lim_{x \to \infty} \frac{f'(x)}{g'(x)} = L.$$

Then $\lim_{x \to \infty} \frac{f(x)}{g(x)}$ exists and is equal to L.

Proof: Define

$$F(u) = \begin{cases} f\left(\frac{1}{u}\right) & \text{if } 0 < u < \frac{1}{b}, \\ 0 & \text{if } u = 0 \end{cases}$$

and

$$G(u) = \begin{cases} g\left(\frac{1}{u}\right) & \text{if } 0 < u < \frac{1}{b}, \\ 0 & \text{if } u = 0. \end{cases}$$

We see that $F, G \in \mathcal{C}[0, 1/b)$. Also, $G'(u) = -g'(1/u)/u^2 \neq 0$ on $(0, 1/b)$. Thus

$$\begin{aligned} \lim_{x \to \infty} \frac{f(x)}{g(x)} &= \lim_{u \to 0+} \frac{F(u)}{G(u)} \\ &= \lim_{u \to 0+} \frac{F'(u)}{G'(u)} \\ &= \lim_{u \to 0+} \frac{-f'\left(\frac{1}{u}\right)\frac{1}{u^2}}{-g'\left(\frac{1}{u}\right)\frac{1}{u^2}} \\ &= \lim_{u \to 0+} \frac{f'\left(\frac{1}{u}\right)}{g'\left(\frac{1}{u}\right)} \\ &= \lim_{x \to \infty} \frac{f'(x)}{g'(x)} = L. \end{aligned}$$

■

Theorem 4.5.4 *Suppose for all $x > b$ that f and g are both differentiable with $g'(x)$ nowhere 0. Suppose that*

$$\lim_{x \to \infty} f(x) = \infty = \lim_{x \to \infty} g(x)$$

and that

$$\lim_{x \to \infty} \frac{f'(x)}{g'(x)} = L.$$

Then

$$\lim_{x \to \infty} \frac{f(x)}{g(x)}$$

exists and is equal to L.

Proof: If $x > x_0 > b$, there exists $\bar{x}$ such that

$$\frac{f'(\bar{x})}{g'(\bar{x})} = \frac{f(x) - f(x_0)}{g(x) - g(x_0)}$$

$$= \frac{f(x)}{g(x)} \left[\frac{1 - \frac{f(x_0)}{f(x)}}{1 - \frac{g(x_0)}{g(x)}} \right],$$

where the expression in brackets $\to 1$ as $x \to \infty$ and where $\bar{x}$ depends upon x. We can write this as

$$\frac{f(x)}{g(x)} = \frac{f'(\bar{x})}{g'(\bar{x})} \left[\frac{1 - \frac{g(x_0)}{g(x)}}{1 - \frac{f(x_0)}{f(x)}} \right].$$

If $\epsilon > 0$, there exists x_0 such that $\bar{x} > x_0$ implies

$$L - \frac{\epsilon}{2} < \frac{f'(\bar{x})}{g'(\bar{x})} < L + \frac{\epsilon}{2}.$$

Let $1 > \eta > 0$. There exists $B > x_0$ such that $x \geq B$ implies

$$\left| \frac{1 - \frac{g(x_0)}{g(x)}}{1 - \frac{f(x_0)}{f(x)}} - 1 \right| < \eta.$$

Thus

$$\left(L - \frac{\epsilon}{2} \right)(1 - \eta) < \frac{f(x)}{g(x)} < \left(L + \frac{\epsilon}{2} \right)(1 + \eta)$$

since $1 - \eta > 0$. By picking $\eta > 0$ sufficiently small we can guarantee that

$$\left| \frac{f(x)}{g(x)} - L \right| < \epsilon$$

for all $x \geq B$. Thus $\frac{f(x)}{g(x)} \to L$ as $x \to \infty$. ■

■ **EXAMPLE 4.8**

Here are more examples utilizing l'Hôpital's Rule.

i.

$$\lim_{n \to \infty} \left(1 + \frac{1}{n}\right)^n = \lim_{x \to \infty} \left(1 + \frac{1}{x}\right)^x$$

$$= \lim_{x \to \infty} e^{x \log\left(1 + \frac{1}{x}\right)}$$

$$= e^{\lim_{x \to \infty} \frac{\log\left(1 + \frac{1}{x}\right)}{1/x}}$$

$$= e^{\lim_{x \to \infty} \frac{1}{1 + \frac{1}{x}}} = e.$$

ii. $\lim_{x \to \infty} \frac{x^n}{e^x} = \lim_{x \to \infty} \frac{nx^{n-1}}{e^x} = \cdots = \lim_{x \to \infty} \frac{n!}{e^x} = 0.$ Here $n \in \mathbb{N}$.

EXERCISES

4.40   Let $f(t) = t^2$ and $g(t) = t^3$, for all $t \in [0, 1]$.
   a) Find the value(s) of $\bar{t}_1 \in [0, 1]$ such that

$$f(1) - f(0) = f'(\bar{t}_1)(1 - 0).$$

   b) Find the value(s) of $\bar{t}_2 \in [0, 1]$ such that

$$g(1) - g(0) = g'(\bar{t}_2)(1 - 0).$$

   c) Find the value(s) of $\bar{t} \in [0, 1]$ such that

$$\frac{f(1) - f(0)}{g(1) - g(0)} = \frac{f'(\bar{t})}{g'(\bar{t})}.$$

4.41   Let p be a polynomial of degree n and let $E = \{x \mid e^x = p(x)\}$. Prove that the number of elements in the set E, denoted by $|E|$, satisfies the inequality $|E| \le n + 1$. (Hint: Use Cauchy's Generalized Mean Value Theorem.)

4.42   Find the **error** in the following *attempt* to apply L'Hôpital's Rule:

$$\lim_{x \to 2} \frac{x^2 - x - 2}{x^2 - 2x} = \lim_{x \to 2} \frac{2x - 1}{2x - 2} = \lim_{x \to 2} \frac{2}{2} = 1$$

Show that in fact $\lim_{x \to 2} \frac{x^2 - x - 2}{x^2 - 2x} = \frac{3}{2}$.

4.43   Find

$$\lim_{x \to 0+} x^x.$$

(Hint: See Example 4.8(i).)

4.44   If $n \in \mathbb{N}$ and if $p > 0$, find $\lim_{x \to \infty} \frac{(\log x)^n}{x^p}$.

4.45   If P is any polynomial, find

$$\lim_{h \to 0} \frac{P(x + 3h) + P(x - 3h) - 2P(x)}{h^2}.$$

4.46   Find

$$\lim_{x \to 0} \frac{e^{-\frac{1}{x^2}}}{x^k}$$

where $k \in \mathbb{N}$.

4.6   TAYLOR'S THEOREM

We will see that Taylor's Theorem is another generalization of the Mean Value Theorem. Denote by $\mathcal{C}^n(a - r, a + r)$ the set of all functions f such that at least the first n derivatives of f are continuous: Thus $f^{(n)} \in \mathcal{C}(a - r, a + r)$. We would like to approximate f by means of a polynomial of degree n, expressed in powers of $x - a$:

$$f(x) \approx c_0 + c_1(x - a) + \cdots + c_n(x - a)^n.$$

If we actually had $f(x) = c_0 + c_1(x - a) + \cdots + c_n(x - a)^n$ then it is easy to see by substitution that

$$f(a) = c_0, \ f'(a) = c_1, \ \ldots, f^{(n)}(a) = n!c_n.$$

Thus we could in this case write

$$c_k = \frac{f^{(k)}(a)}{k!}$$

for all $k = 0, \ldots, n$, with the understandings that $0! = 1$ and $f^{(0)}(a) = f(a)$, by definition.

Definition 4.6.1 *We define the nth Taylor Polynomial*

$$P_n(x) = \sum_{k=0}^{n} \frac{f^{(k)}(a)}{k!}(x - a)^k$$

for all $f \in \mathcal{C}^n(a - r, a + r)$, where $r > 0$.

The problem is to describe the difference $f(x) - P_n(x)$, which we call $R_n(x)$, the nth Taylor Remainder term. This remainder is zero only if f is actually equal to the polynomial P_n.

Theorem 4.6.1 (Taylor's Theorem) *If $f^{(n+1)}(x)$ exists for all x in the interval $(a - r, a + r)$, then we have $f(x) = P_n(x) + R_n(x)$, where*

$$R_n(x) = \frac{f^{(n+1)}(\mu)}{(n+1)!}(x - a)^{n+1}$$

for some suitable value of μ between a and x.

Remark 4.6.1 We can regard Taylor's Theorem as a generalization of the Mean Value Theorem in the following sense. The Mean Value Theorem says that if f' exists on $(a - r, a + r)$, then

$$f(x) = f(a) + f'(\mu)(x - a)$$

for all $x \in (a - r, a + r)$ and for some suitable μ strictly between a and x. Note that $f(a) = P_0(x)$, a constant polynomial. And $f'(\mu)(x - a)$ is in the correct form to be $R_0(x)$. Thus the Mean Value Theorem implies the special case of Taylor's Theorem in which $n = 0$. Note that if $f^{(n+1)}(x)$ exists for all $x \in (a - r, a + r)$, then $f \in C^n(a - r, a + r)$.

Proof: Fix x for now. If $x = a$ we see easily that $f(a) = P_n(a)$, so that $R_n(a) = 0$. On the other hand, if $x \neq a$, then

$$\frac{(x - a)^{n+1}}{(n+1)!} \neq 0,$$

so there exists $K \in \mathbb{R}$ such that

$$R_n(x) = \frac{K}{(n+1)!}(x - a)^{n+1}.$$

Our goal is to show there exists μ between a and x such that $K = f^{(n+1)}(\mu)$. The trick is to replace a by a variable α and to define

$$h(\alpha) = f(x) - \left[\sum_{k=0}^{n} \frac{f^{(k)}(\alpha)}{k!}(x - \alpha)^k + \frac{K}{(n+1)!}(x - \alpha)^{n+1} \right].$$

If we set $\alpha = a$, we see that

$$h(a) = f(x) - \left[P_n(x) + \frac{K}{(n+1)!}(x - a)^{n+1} \right] = 0$$

by definition of K. On the other hand, if we set $\alpha = x$, then we see that

$$h(x) = f(x) - [f(x) + 0 + \cdots + 0] = 0.$$

Moreover, as a function of α, h is differentiable for all α between a and x. By Rolle's Theorem, there exists μ between a and x such that $h'(\mu) = 0$. However,

$$h'(\alpha) = 0 - \left\{ (f'(\alpha)) + (-f'(\alpha) + f''(\alpha)(x - \alpha)) \right.$$

$$+ \left(-f''(\alpha)(x - \alpha) + \frac{f^{(3)}(\alpha)}{2!}(x - \alpha)^2 \right)$$

$$+ \cdots + \left(-\frac{f^{(n)}(\alpha)}{(n-1)!}(x - \alpha)^{n-1} \right.$$

$$\left. + \frac{f^{(n+1)}(\alpha)}{n!}(x - \alpha)^n \right) - \frac{K}{n!}(x - \alpha)^n \right\}$$

$$= \frac{f^{(n+1)}(\alpha)}{n!}(x - \alpha)^n - \frac{K}{n!}(x - \alpha)^n,$$

so that $h'(\mu) = 0$ and $K = f^{(n+1)}(\mu)$. ∎

■ **EXAMPLE 4.9**

Let $f(x) = e^x$ and write $f(x) = P_n(x) + R_n(x)$ in powers of $x = x - 0$. We claim that $R_n(x) \to 0$ as $n \to \infty$ for all $x \in \mathbb{R}$, so that

$$e^x = \lim_{n \to \infty} \sum_{k=0}^{n} \frac{1}{k!} x^k$$

for all x $\in \mathbb{R}$.

It suffices to show for all x that

$$R_n(x) = \frac{f^{(n+1)}(\mu)}{(n+1)!} x^{n+1} = \frac{e^\mu}{(n+1)!} x^{n+1} \to 0$$

as $n \to \infty$. However,

$$0 \le |R_n(x)| = \frac{e^\mu}{(n+1)!} |x|^{n+1} \le e^{|x|} \frac{|x|^{n+1}}{(n+1)!}.$$

Thus it suffices to show that for all $x \in \mathbb{R}$,

$$\frac{|x|^{n+1}}{(n+1)!} \to 0$$

as $n \to \infty$. So fix x arbitrarily and observe that there exists $N \in \mathbb{N}$ such that $\frac{|x|}{N} < \frac{1}{2}$. Now, for all $n \ge N$, write $n = N + k$, and

$$\frac{|x|^n}{n!} \le \frac{|x|^N}{N!} \cdot \frac{1}{2^k} \to 0$$

as $k \to \infty$, which is necessary since $n \to \infty$.

EXERCISES

4.47   Suppose f is a polynomial of degree n on $(a - r, a + r)$. Prove that $f(x) \equiv P_n(x)$, the nth Taylor Polynomial defined in Definition 4.6.1. (Hint: Prove that $R_n(x) \equiv 0$.)

4.48   If $x > 0$, prove that

$$1 + x + \frac{x^2}{2} < e^x < 1 + x + \frac{x^2}{2}e^x.$$

(Hint: Use $e^x = P_1(x) + R_1(x)$.)

4.49   Prove that e is irrational. (Hint: Suppose false, so that $e = \frac{p}{q}$, where $p, q \in \mathbb{N}$. Write $e = e^1 = P_n(1) + R_n(1)$, multiply both sides by $n!$, and deduce a contradiction when $n \in \mathbb{N}$ is sufficiently large.)

4.50   Expand the polynomial $p(x) = 3x^3 + 2x^2 - x + 1$ as a polynomial in powers of $(x - 1)$: That is, show that

$$p(x) = \sum_{k=0}^{3} c_k (x - 1)^k$$

and find the values of the constants $c_0, \ldots, c_3$.

4.51   Let $f(x) = \sin x$ and find the nth Taylor Polynomial $P_n(x)$ in powers of $x = x - 0$ (that is, use $a = 0$). Prove that $P_n(x) \to \sin x$ as $n \to \infty$ for all $x \in \mathbb{R}$. (See Fig. 4.7.) Prove also that no polynomial $P(x) \equiv \sin x$ on any interval of positive length.

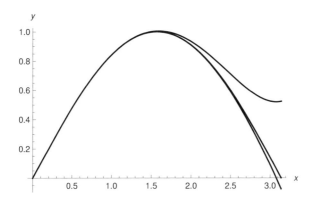

Figure 4.7 $\sin(x)$, $P_5(x)$, and $P_7(x)$ on $[0, \pi]$. Can you see which curve belongs to which function?

4.52   Let $f(x) = \cos x$ and find the nth Taylor Polynomial $P_n(x)$ in powers of $x = x - 0$ (i.e., use $a = 0$). Prove that $P_n(x) \to \cos x$ as $n \to \infty$ for all $x \in \mathbb{R}$.

4.7 TEST YOURSELF

EXERCISES

4.53   True or False: $\ln(1+x) < x$ for all $x > 0$.

4.54   Give an example of a function F that is differentiable at every $x \in [0,1]$ yet $F'(x)$ is *not bounded* on $[0,1]$.

4.55   True or False: If $f \in \mathcal{R}[a,b]$ and $F(x) = \int_a^x f(t)\, dt$ for all $x \in [a,b]$, then $F'(x) = f(x)$ for all $x \in [a,b]$.

4.56   True or False: If $F'(x)$ exists and is Riemann integrable on $[a,b]$, then

$$\int_a^b F'(x)\, dx = F(b) - F(a).$$

4.57   Suppose that the derivative $f'(x)$ exists and is bounded on the closed interval $[a,b]$ and that f' is continuous at on the open interval (a,b) but is *not* continuous at a or at b. True or False: The function f' is Riemann integrable on $[a,b]$.

4.58   Give an example of an *un*bounded function f on $[0,1]$ that is equal *everywhere* to the derivative of another function F.

4.59   Define

$$\|f\| = \|f\|_{\mathrm{sup}} + \|f'\|_{\mathrm{sup}}$$

for each $f \in \mathcal{C}^1[a,b]$. True or False: If $T : \mathcal{C}^1[a,b] \to \mathbb{R}$ is defined by

$$Tf = f'\left(\frac{a+b}{2}\right),$$

then T is continuous with respect to $\| \cdot \|$.

4.60   Let $f(t) = t^2$ and $g(t) = t^3$, for all $t \in [0,1]$. Find the value(s) of $\bar{t} \in [0,1]$ such that $\frac{f(1)-f(0)}{g(1)-g(0)} = \frac{f'(\bar{t})}{g'(\bar{t})}$.

4.61   If P is any polynomial, find $\lim_{h \to 0} \frac{P(x+4h)+P(x-4h)-2P(x)}{h^2}$.

4.62   Find $\lim_{x \to \infty} \frac{(\log x)^{10}}{x^{0.1}}$.

4.63   Let $T(f) = \int_0^1 f(x) \sin \frac{\pi}{x}\, dx$ for each $f \in \mathcal{R}[0,1]$. Is T continuous?

CHAPTER 5

INFINITE SERIES

An infinite series is a sum of infinitely many numbers. Infinite series appear through-out pure and applied mathematics, and they were important even long ago. For example, the student probably recalls that the infinite decimal expansion of the frac-tion $\frac{1}{3}$ is $0.33\underline{3}\ldots$. The underlined $\underline{3}$ connotes endless repetition of the digit 3. Such an endless decimal expansion can be understood as representing

$$\frac{3}{10} + \frac{3}{100} + \frac{3}{1000} + \cdots,$$

where again the 3 dots indicate that the additions continue without end. What does it mean to add infinitely many numbers? Can anyone actually perform an infinite number of additions? These are some of the questions we will address in the present chapter.

5.1 SERIES OF CONSTANTS

If x is a sequence of real numbers, we can think of x as being a *function* defined on the natural numbers $\mathbb{N}$ as follows: for all $k \in \mathbb{N}$, $x(k) = x_k \in \mathbb{R}$. Usually the kth term of the sequence x is denoted x_k rather than $x(k)$, but it is useful to have the

Advanced Calculus: An Introduction to Linear Analysis. By Leonard F. Richardson
Copyright © 2008 John Wiley & Sons, Inc.

symbol x for the sequence as a whole (viewed as a function, for example) so that we can write conveniently about various *global* properties of the sequence, as we will see in the next several sections.

The first thing we wish to understand about a sequence x is whether or not it is possible to define in a useful way the concept of the *sum* of the infinitely many numbers x_k. What is clear is that for all $n \in \mathbb{N}$ we can define the nth *partial sum*

$$s_n = \sum_{k=1}^{n} x_k = x_1 + x_2 + \cdots + x_n$$

by virtue of finitely many repetitions of the closure of $\mathbb{R}$ under addition.

Definition 5.1.1 *We say the infinite series*

$$\sum_{k=1}^{\infty} x_k = x_1 + x_2 + \cdots + x_n + \cdots$$

converges to the sum s provided the sequence *of partial sums*

$$s_n = \sum_{k=1}^{n} x_k \to s$$

as $n \to \infty$. If $\sum_{k=1}^{\infty} x_k$ converges to s, then we call the sequence x summable. A series that does not converge is called divergent.

Theorem 5.1.1 (nth Term Test) *If x is a summable sequence, then $x_n \to 0$ as $n \to \infty$.*

Proof: Denote $s_n = \sum_{k=1}^{n} x_k$. Since x is summable, there exists $L \in \mathbb{R}$ such that $s_n \to L$ as $n \to \infty$. Hence the sequence $s_{n-1} \to L$ as well. Therefore

$$x_n = s_n - s_{n-1} \to L - L = 0.$$

■

Remark 5.1.1 We remind the reader here, for one final time, of the importance of learning how to read mathematical proofs, and how this contributes to learning to prove theorems oneself. The reader should review the Introduction on page xxiii. It is very important for each reader to write out a careful analysis of each proof studied, taking careful note of exactly how it works and how each step is justified.

The proof of the nth term test is very brief. Even so, it offers lessons. The reader will see in Example 5.1 that it is possible for $x_n \to 0$, although $\sum_{1}^{\infty} x_n$ may diverge. The insight that enables us to prove this theorem is that we can express x_n in terms of s_n and s_{n-1}. Then one must recognize that if $s_n \to s$, so also must $s_{n-1} \to s$. In fact, because $s_n \to s$ we know that for each $\epsilon > 0$ there exists $N \in \mathbb{N}$ such that

$n \geq N$ implies that $|s_n - s| < \epsilon$. Hence if $n \geq N + 1$ we must have $|s_{n-1} - s| < \epsilon$ as well, showing that $s_{n-1} \to s$, lagging only one step behind s_n itself.

It is important to understand that the nth term test can be used to prove *divergence* of an infinite series, but it is a one-directional implication and does not prove convergence, as shown by the following important example.

■ **EXAMPLE 5.1**

Let $x_k = \frac{1}{k}$ for all $k \in \mathbb{N}$. We claim that even though $x_k = \frac{1}{k} \to 0$ as $k \to \infty$, the so-called *harmonic series* $\sum_{k=1}^{\infty} x_k$ diverges.

As usual, let s_n denote the Nth partial sum. If s_n were convergent, then s_n would be bounded, and every subsequence, such as s_{2^n}, would be bounded as well. However,

$$
\begin{aligned}
s_{2^n} =& (1) + \left(\frac{1}{2}\right) + \left(\frac{1}{3} + \frac{1}{4}\right) + \left(\frac{1}{5} + \cdots + \frac{1}{8}\right) \\
& + \cdots + \left(\frac{1}{2^{n-1} + 1} + \cdots + \frac{1}{2^n}\right) \\
> & 1 + \frac{n}{2} \to \infty
\end{aligned}
$$

as $n \to \infty$. This is a contradiction. Hence the *harmonic series* diverges. See Exercise 5.1.

■ **EXAMPLE 5.2**

Since the harmonic series diverges, it is interesting to note that the so-called *alternating harmonic series*

$$
\sum_{k=1}^{\infty} \frac{(-1)^{k+1}}{k} = 1 - \frac{1}{2} + \frac{1}{3} - \frac{1}{4} + \frac{1}{5} + \cdots
$$

converges.

This is a consequence of the following theorem.

Theorem 5.1.2 (Alternating Series Test) *Suppose the sequence x_k is a decreasing sequence converging to the limit 0. That is, suppose $x_1 \geq x_2 \geq \cdots$ and $x_k \to 0$ as $k \to \infty$. Then*

$$
\sum_{k=1}^{\infty} (-1)^{k+1} x_k = x_1 - x_2 + x_3 - x_4 + \cdots
$$

converges. Moreover, if s denotes the sum of the infinite series, then we have the following estimate of the difference between the partial sum s_n and the sum s:

$$
|s_n - s| \leq x_{n+1}
$$

for all n.

Proof: Observe that $s_1 = x_1 \geq s_2 = x_1 - x_2$. However, $s_3 = s_2 + x_3 \leq s_1$ because the positive sequence x_k decreases. Extending this reasoning, we observe the following:

$$s_2 \leq s_4 \leq \cdots \leq s_{2n} \leq s_{2n-1} \leq \cdots \leq s_3 \leq s_1.$$

In other words, the subsequence s_{2n} is increasing with every odd partial sum as an upper bound, and s_{2n-1} is decreasing with every even partial sum as a lower bound. Thus

$$s_{2n} \nearrow L,$$

the least upper bound of the even partial sums. Hence L is a lower bound of the odd partial sums. And

$$s_{2n-1} \searrow G \geq L,$$

where G is the greatest lower bound of the odd partial sums. And

$$0 \leq G - L \leq s_{2n-1} - s_{2n} = x_{2n} \searrow 0.$$

Thus $G - L = 0$ and $L = G$. Moreover, since the successive terms of the sequence s_n are on opposite sides of L for all n

$$|s_n - L| \leq |s_n - s_{n+1}| = x_{n+1} \searrow 0.$$

Thus $s_n \to L = s$ and the theorem is proven. ∎

One of the things we learn from the harmonic and the alternating harmonic series is that a series $\sum_{k=1}^{\infty} x_k$ can converge even though $\sum_{k=1}^{\infty} |x_k|$ fails to converge. We see next why the opposite phenomenon cannot occur.

Definition 5.1.2 *A series*

$$\sum_{k=1}^{\infty} x_k$$

is called absolutely convergent *if the series*

$$\sum_{k=1}^{\infty} |x_k|$$

converges. In this case the sequence x is called absolutely summable. A series

$$\sum_{k=1}^{\infty} x_k$$

that converges but fails *to converge absolutely is called* conditionally convergent, *or* conditionally summable.

Thus the alternating harmonic series is conditionally convergent.

Theorem 5.1.3 *Every absolutely summable sequence x is summable.*

Proof: If

$$s_n = \sum_{k=1}^{n} x_k,$$

it will suffice to show that s_n is a Cauchy sequence. What we know by hypothesis is that

$$\sigma_n = \sum_{k=1}^{n} |x_k|$$

is Cauchy. Thus if $\epsilon > 0$, there exists N such that $n > m \geq N$ implies $|\sigma_n - \sigma_m| < \epsilon$. However, if $n > m \geq N$, then

$$|s_n - s_m| = \left| \sum_{k=m+1}^{n} x_k \right| \leq \sum_{k=m+1}^{n} |x_k| = |\sigma_n - \sigma_m| < \epsilon.$$

■

■ **EXAMPLE 5.3**

We present the *Geometric Series Test*. We call the series

$$\sum_{k=0}^{\infty} ar^k = a + ar + ar^2 + \cdots + ar^n + \cdots$$

a *geometric series* with *common ratio r*.

Observe that if the constant $a \neq 0$ and $|r| \geq 1$, then $ar^n \not\to 0$ as $n \to \infty$, so $\sum_{k=0}^{\infty} ar^k$ diverges by the nth term test. However, we claim that if $|r| < 1$ then the corresponding geometric series must be absolutely convergent, and thus convergent as well. To prove

$$\sum_{k=0}^{\infty} |ar|^k$$

converges provided $|r| < 1$, it suffices to prove

$$s_n = \sum_{k=0}^{n} ar^k = a + ar + ar^2 + \cdots + ar^n$$

converges. But $s_n - rs_n = a - ar^{n+1}$. Hence

$$s_n = \frac{a - ar^{n+1}}{1 - r} \to \frac{a}{1 - r}$$

as $n \to \infty$. That is

$$\sum_{k=0}^{\infty} ar^k = \frac{a}{1 - r}, \tag{5.1}$$

provided that $|r| < 1$, so

$$\sum_{k=1}^{\infty} |ar^k|$$

converges as well.

For example, the series

$$\sum_{k=1}^{\infty} \left(-\frac{1}{2}\right)^k$$

converges absolutely.

■ **EXAMPLE 5.4**

The geometric series test enables us to resolve the famous problem known as Zeno's paradox. The ancient Greek philosopher Zeno proposed this scenario. The legendary warrior Achilles was challenged to a foot race against a tortoise. For fairness, the tortoise is given a head start of d yards. The claim is that Achilles can never catch the tortoise. Suppose the tortoise runs at r times the speed of Achilles, where $0 < r < 1$. The idea is that if it takes Achilles t seconds to reach the point where the tortoise was at the beginning of the race, then when he reaches that point the tortoise will have advanced to a position dr yards ahead of Achilles new location. Then it will take tr seconds to reach the new location of the tortoise. However, during that time the tortoise will have advanced to a position dr^2 yards ahead of Achilles. In effect, each time Achilles reaches the previous position of the tortoise, the determined quadruped has moved a bit farther ahead. So Achilles can never reach the tortoise.

The resolution is that the time required for Achilles to reach the tortoise is the sum of an infinite series:

$$T = \sum_{0}^{\infty} tr^n = t \sum_{0}^{\infty} r^n,$$

which is finite since $|r| < 1$, by the geometric series test. Of course, Zeno realized that the tortoise would lose the race to Achilles. Zeno's point was that in his era it was not yet understood how to deal with limits, such as the sum of an infinite series, with logical rigor.

EXERCISES

5.1 † In Example 5.1, try regrouping s_{2^n-1} in such a way as to show $s_{2^n-1} \leq n$, for all n. If $s_N \geq 100$, how big must N be? *Remark:* The reader may be surprised at how many terms a computer would have to add to make the partial sums rise to a total of only 100, even though the harmonic series diverges to ∞. How many years would your PC have to compute in order to add the required number of terms to make $s_N \geq 100$? What would be the round-off error from so many additions?

5.2 Test for absolute convergence, conditional convergence, or divergence:

a) $\sum_{k=1}^{\infty}(-1)^{k+1}$

b) †

$$\sum_{k=1}^{\infty}\frac{1}{k(k+1)}$$

(Hint: Write

$$\frac{1}{k(k+1)} = \frac{1}{k} - \frac{1}{k+1}$$

and calculate s_n. This is an example of what is called a *telescoping series*.)

c) $\sum_{k=1}^{\infty} a_k$, where

$$a_k = \begin{cases} \frac{1}{\sqrt{k}} & \text{if } k \text{ is a perfect square,} \\ 0 & \text{if } k \text{ isn't a perfect square.} \end{cases}$$

d) $\sum_{k=1}^{\infty}\frac{(-1)^{k+1}}{\sqrt{k}}$

e) $\sum_{k=1}^{\infty} a_k$ where

$$a_k = \begin{cases} \frac{1}{j} & \text{if } k = 2j - 1, \\ \frac{-1}{j} & \text{if } k = 2j. \end{cases}$$

f) $\sum_{k=0}^{\infty}\left(\frac{-\pi}{3}\right)^k$ **g)** $\sum_{k=0}^{\infty}\left(\frac{e}{3}\right)^k$

5.3 Using Exercise 5.2.b above as a model, find a formula for a_k such that

$$s_n = \sum_{k=1}^{n} a_k = \sqrt{n}$$

for all n, so that s_n diverges to ∞ although $\frac{s_n}{n} \to 0$ as $n \to \infty$.

5.4 Find a formula for a_k so that $s_n = \sum_{k=1}^{n} a_k = \log n$, for all n.

5.5 Given an example of $x_k \to 0$ for which $\sum_{k=1}^{\infty}(-1)^{k+1}x_k$ *diverges*.

5.6 If x and y are sequences and $c \in \mathbb{R}$, we define $(cx)_k = cx_k$ and $(x+y)_k = x_k + y_k$. Prove: If x and y are summable, then

a) $x + y$ is summable and

$$\sum_{k=1}^{\infty}(x_k + y_k) = \sum_{k=1}^{\infty}x_k + \sum_{k=1}^{\infty}y_k.$$

b) cx is summable and

$$\sum_{k=1}^{\infty}cx_k = c\sum_{k=1}^{\infty}x_k.$$

We note that this exercise shows that the family of summable sequences is a *vector space*.

5.7   A function $f : \mathbb{R} \to \mathbb{R}$ is called a *contraction* of $\mathbb{R}$ if and only if there exists a constant $r \in [0, 1)$ such that for all x and x' in $\mathbb{R}$ we have

$$|f(x) - f(x')| \leq r|x - x'|.$$

Let f be a contraction of $\mathbb{R}$, with corresponding constant r.

a) Prove that f is uniformly continuous on $\mathbb{R}$.

b) Let $x_0 \in \mathbb{R}$ be arbitrary and define a sequence x_n by $x_n = f(x_{n-1})$, for each $n \in \mathbb{N}$. Show that $|x_{n+1} - x_n| \leq r^n |x_1 - x_0|$.

c) Prove that the sequence x_n in 5.7.b is a Cauchy sequence. (Hint: Exercise 1.25 may be helpful.)

d) Let $p = \lim_{n \to \infty} x_n$, and prove that $f(p) = p$. (A point q is called a *fixed point* of f if and only if $f(q) = q$. You have just shown that every *contraction* of $\mathbb{R}$ has a *fixed point*.)

e) If p and q are both fixed points of f, prove that $p = q$.

f) Let $f : \mathbb{R} \to \mathbb{R}$ be a differentiable function such that $\|f'\|_{\sup} = r < 1$. Prove that f is a contraction of $\mathbb{R}$.

5.2   CONVERGENCE TESTS FOR POSITIVE TERM SERIES

Since every absolutely summable sequence is summable, it is desirable to have tests designed to determine whether or not a series of exclusively nonnegative terms converges. We will see that absolutely convergent series play a very important role in the applications of infinite series. Throughout the present section, all terms of infinite series will be nonnegative unless otherwise noted.

We observe first that if $x_k \geq 0$ for all $k \in \mathbb{N}$, then

$$s_n = \sum_{k=1}^{n} x_k$$

is a monotone increasing sequence: s_n is increasing as n increases strictly. Thus we know that

$$s_n \to \sup\{s_m \mid m \in \mathbb{N}\}.$$

It follows that s_n is *convergent* if and only if $\sup\{s_m \mid m \in \mathbb{N}\} < \infty$. This observation leads us to the following useful test.

Theorem 5.2.1 (Comparison Test) *Suppose there exists $K \in \mathbb{N}$ such that*

$$0 \leq x_k \leq y_k$$

for all $k \geq K$. (In words, suppose y_k eventually dominates x_k.) Then we have the following conclusions.

i. *If $\sum_{k=1}^{\infty} y_k$ converges, then $\sum_{k=1}^{\infty} x_k$ converges.*

ii. If $\sum_{k=1}^{\infty} x_k$ diverges, then $\sum_{k=1}^{\infty} y_k$ diverges.

Proof: Let $s_n = \sum_{k=1}^{n} x_k$ and $\sigma_n = \sum_{k=1}^{n} y_k$. Let $B = \sum_{k=1}^{K-1} |x_k - y_k|$.

i. We know that

$$\sigma_n \nearrow \varsigma = \sup\{\sigma_n | n \in \mathbb{N}\} \in \mathbb{R}.$$

But

$$s_n \leq \sigma_n + B \leq \varsigma + B$$

for all $n \in \mathbb{N}$. Since the sequence s_n is bounded above by $\varsigma + B \in \mathbb{R}$, s_n converges to its least upper bound, and so $\sum_{k=1}^{\infty} x_k$ converges.

ii. This part is a consequence of the previous part, being its contrapositive. In particular, if $\sum_{k=1}^{\infty} y_k$ did not diverge then it would converge. Then part (i) would imply that $\sum_{k=1}^{\infty} x_k$ converges, which contradicts the hypothesis.

∎

For the next theorem, recall from elementary calculus that if $f \in \mathcal{R}[1, b]$ for all $b \geq 1$, we define the *improper* integral

$$\int_1^{\infty} f(x)\, dx = \lim_{b \to \infty} \int_1^b f(x)\, dx$$

if this limit exists. The infinite length of the interval on which the integral takes place is responsible for the term *improper integral.*

Theorem 5.2.2 (Integral Test) *Suppose f is a monotone decreasing function approaching 0 on $[1, \infty)$: that is, $f \searrow 0$ on $[1, \infty)$. Let $a_k = f(k)$, for all $k \in \mathbb{N}$. Then $\sum_{k=1}^{\infty} a_k$ converges if and only if*

$$\int_1^{\infty} f(x)\, dx < \infty.$$

Remark 5.2.1 Since $f(x) \geq 0$ for all $x \geq 0$ in this theorem, $\int_1^b f(x)\, dx$ is an *increasing* function of b. Thus the condition $\int_1^{\infty} f(x)\, dx < \infty$ means the same thing as

$$\int_1^{\infty} f(x)\, dx = \lim_{b \to \infty} \int_1^b f(x)\, dx$$

exists in this case.

Proof:

i. First we prove the implication from right to left, which is the *if* part. Because f is decreasing, we have

$$a_2 \leq \int_1^2 f(x)\, dx,$$

$$a_3 \leq \int_2^3 f(x)\, dx,$$

$$\vdots$$

$$a_n \leq \int_{n-1}^n f(x)\, dx.$$

Thus

$$s_n = \sum_{k=1}^n a_k \leq a_1 + \int_1^n f(x)\, dx$$

$$\leq a_1 + \int_1^\infty f(x)\, dx < \infty$$

for all $n \in \mathbb{N}$. Thus s_n converges.

ii. Now we prove the implication from left to right, which is the *only if* part. suppose $\sum_{k=1}^\infty a_k < \infty$. Observe for all k that $a_k \geq \int_k^{k+1} f(x)\, dx$. Thus

$$\int_1^n f(x)\, dx \leq \sum_{k=1}^n a_k \leq \sum_{k=1}^\infty a_k < \infty$$

for all $n \in \mathbb{N}$. Thus $\int_1^b f(x)\, dx$ is an increasing function of b that is bounded above. Hence

$$\lim_{b \to \infty} \int_1^b f(x)\, dx = \int_1^\infty f(x)\, dx$$

exists.

∎

Theorem 5.2.3 (Ratio Test) *Suppose $x_k > 0$ for all k and suppose*

$$\frac{x_{k+1}}{x_k} \to L$$

as $k \to \infty$. Then we have the following conclusions.

i. *If $L > 1$, $\sum_{k=1}^\infty x_k$ diverges.*

ii. *If $L < 1$, $\sum_{k=1}^\infty x_k$ converges.*

iii. *If $L = 1$, the test fails to determine convergence or divergence.*

Proof:

i. If $L > 1$, then there exists $K \in \mathbb{N}$ such that $k \geq K$ implies $\frac{x_{k+1}}{x_k} > 1$, which implies $0 < x_k < x_{k+1}k$, and $x_k \nrightarrow 0$. Thus $\sum_{k=1}^{\infty} x_k$ fails to converge because of the nth term test.

ii. If $L < 1$, fix any number r such that $L < r < 1$. Then there exists $K \in \mathbb{N}$ such that $k \geq K$ implies $\frac{x_{k+1}}{x_k} < r$ which implies $x_{k+1} < x_k r$. It follows that for all $j \in \mathbb{N}$ we have $x_{K+j} < x_K r^j$. Thus the convergent geometric series

$$\sum_{j=1}^{\infty} x_K r^j \text{ dominates } \sum_{j=1}^{\infty} x_{K+j},$$

which must also converge. Hence the partial sums of $\sum_{k=1}^{\infty} x_k$ are all bounded above by

$$\sum_{k=1}^{K} x_k + \sum_{j=1}^{\infty} x_{K+j},$$

so that $\sum_{k=1}^{\infty} x_k$ converges.

iii. See Exercise 5.14.

∎

EXERCISES

5.8 † Prove the *Limit Comparison Test:* Suppose $x_k \geq 0$ and $y_k > 0$ for all $k \in \mathbb{N}$. Suppose $\frac{x_k}{y_k} \to L \in \mathbb{R}$.
   a) If $L > 0$ then x is summable if and only if y is summable. (Hint: Apply the comparison test.)
   b) If $L \geq 0$ and if y is summable, then x is summable as well.

5.9 † Use the result of Exercise 5.8 above to test for convergence:
   a) $\sum_{k=1}^{\infty} \frac{2k}{2k^2-1}$
   b) $\sum_{k=1}^{\infty} \frac{k-1}{k2^k}$

   c) $\sum_{k=1}^{\infty} \frac{1}{2k-1}$ **d)** $\sum_{k=1}^{\infty} \frac{1}{2k}$

5.10 † Now suppose x and y are sequences that need not have entirely nonnegative terms. If x is absolutely summable and if y is *bounded*, prove that xy is absolutely summable, where we define the product xy by $(xy)_k = x_k y_k$ for all $k \in \mathbb{N}$, so that $\sum_{k=1}^{\infty} x_k y_k$ resembles an *inner product* of two vectors.

5.11 † Apply the integral test to prove:
   a) If $p > 1$, then $\sum_{k=1}^{\infty} \frac{1}{k^p}$ is convergent.
   b) If $0 \leq p \leq 1$, then $\sum_{k=1}^{\infty} \frac{1}{k^p}$ is divergent.

5.12 Test for convergence or divergence:

a) $\sum_{k=1}^{\infty} \frac{1}{\sqrt{k}}$

b) $\sum_{k=2}^{\infty} \left(\frac{1}{k \log k} \right)$ c) $\sum_{k=1}^{\infty} \frac{5k^3}{k^5+1}$

5.13 $\sum_{k=1}^{\infty} kr^{k-1}$, where $|r| < 1$.

5.14   Prove the final part of Theorem 5.2.3 by finding two sequences, x_k and y_k such that $\frac{x_{k+1}}{x_k} \to 1$ and $\frac{y_{k+1}}{y_k} \to 1$, but $\sum_{k=1}^{\infty} x_k$ converges whereas $\sum_{k=1}^{\infty} y_k$ diverges. Prove your claims. (Hint: Consider Exercise 5.11.)

5.15   Prove that the series $\sum_{k=0}^{\infty} \frac{x^k}{k!}$ is absolutely convergent for all $x \in \mathbb{R}$. Note that $0! = 1$ by definition.

5.16   Prove the nth *Root Test:* Suppose $x_k \geq 0$ for all $k \in \mathbb{N}$, and suppose $\sqrt[k]{x_k} \to L$ as $k \to \infty$. Then we have the following conclusions.
    a) If $L > 1$, $\sum_{k=1}^{\infty} x_k$ diverges.
    b) If $L < 1$, $\sum_{k=1}^{\infty} x_k$ converges.
    c) If $L = 1$, then the test fails.
        Hint: Try to do for kth roots what was done for ratios in the proof of Theorem 5.2.3.

5.17   † Let $x_k \geq 0$ for all $k \in \mathbb{N}$. Note that $\lim_{k \to \infty} \sqrt[k]{x_k}$ need not exist.
    a) If $\limsup \sqrt[k]{x_k} < 1$, prove that the sequence x is summable.
    b) If $\limsup \sqrt[k]{x_k} > 1$, prove that $\sum_{k=1}^{\infty} x_k = \infty$.

5.18   In Theorem 5.2.3 $\lim \frac{x_{k+1}}{x_k}$ need not exist. Let $S = \limsup \frac{x_{k+1}}{x_k}$.
    a) Prove: If $S < 1$, then the positive-term series $\sum_{k=1}^{\infty} x_k$ converges.
    b) Give an example in which $S > 1$ yet the series $\sum_{k=1}^{\infty} x_k$ converges.
    c) Make up and prove a valid lim inf version of the ratio test.

5.19   Test for convergence or divergence:
    a) $\sum_{k=0}^{\infty} \frac{k!}{k^k}$
    b) $\sum_{k=0}^{\infty} \frac{k}{e^{k^2}}$
    c) $\sum_{k=2}^{\infty} \frac{1}{(\log k)^k}$

5.3 ABSOLUTE CONVERGENCE AND PRODUCTS OF SERIES

In adding lists of finitely many real numbers, we know that the numbers may be added in any order we find convenient: the result will be independent of order. With infinite series, however, this is actually false. The following definition and example show what can go wrong.

Definition 5.3.1 *A series $\sum_{j=1}^{\infty} y_j$ is called a rearrangement of the series $\sum_{k=1}^{\infty} x_k$ provided every term y_j appears among the x_k's and each x_k appears exactly once among the y_j's. Thus the only difference between the sequence y and sequence x is that the terms appear in a possibly different order. Note that for y to be a* *rearrangement of the sequence* x, y *must be a* single *sequence containing each term of* x *exactly one time.*

■ **EXAMPLE 5.5**

Consider the alternating harmonic series

$$\sum_{k=1}^{\infty} \frac{(-1)^{k+1}}{k} = 1 - \frac{1}{2} + \frac{1}{3} - \frac{1}{4} + \frac{1}{5} - \frac{1}{6} + \cdots,$$

which converges by the alternating series test, but is only conditionally convergent since the harmonic series diverges.

In Exercise 5.9 the reader has shown that the series of positive terms is divergent; That is,

$$\sum_{k=1}^{\infty} \frac{1}{2k - 1} = 1 + \frac{1}{3} + \frac{1}{5} + \cdots = \infty$$

and that the series of negative terms diverges as well:

$$\sum_{k=1}^{\infty} \frac{-1}{2k} = -\frac{1}{2} - \frac{1}{4} - \frac{1}{6} - \cdots = -\infty.$$

We will show that we can re-arrange the terms of the alternating harmonic series in such a way that the new series diverges. We will construct the rearranged series as follows. Begin by selecting enough of the positive terms so that their sum exceeds 2. Then follow these terms with the first negative term of the alternating harmonic series–namely, $-\frac{1}{2}$. After this negative term, add several more positive terms using enough to increase the total sum thus far by at least 2. We can do this, since the sum of the positive terms is infinite. Follow this with the second negative term: $-\frac{1}{4}$. We see that after the first negative term is included, the partial sum to this point is > 1. After the second negative term is included, the partial sum to that point is > 2. And so on. The partial sums diverge to infinity.

Because of this example, we regard conditionally convergent series as inherently *unstable*, in the sense that their convergence depends critically on the order in which the terms appear, because of the cancelations that are critical to the convergence occurring. The next theorem shows that absolutely convergent series do not exhibit this instability.

Theorem 5.3.1 (Dirichlet) *Suppose* $\sum_{k=1}^{\infty} x_k$ *is any absolutely convergent series, and suppose* $\sum_{k=1}^{\infty} y_k$ *is any rearrangement of this series. Then* $\sum_{k=1}^{\infty} y_k$ *is also absolutely convergent, and*

$$\sum_{k=1}^{\infty} y_k = \sum_{k=1}^{\infty} x_k.$$

Proof: Define

$$x_k^+ = \begin{cases} x_k & \text{if } x_k \geq 0, \\ 0 & \text{if } x_k < 0 \end{cases}$$

and

$$x_k^- = \begin{cases} |x_k| & \text{if } x_k \leq 0, \\ 0 & \text{if } x_k > 0. \end{cases}$$

Observe that $x_k \equiv x_k^+ - x_k^-$ and $|x_k| \equiv x_k^+ + x_k^-$. Note also that $0 \leq x_k^+ \leq |x_k|$ and $0 \leq x_k^- \leq |x_k|$. Define corresponding terminology for the sequence y. We see that

$$\sum_{k=1}^{n} x_k^+ \leq \sum_{k=1}^{n} |x_k| \leq \sum_{k=1}^{\infty} |x_k| < \infty$$

so that

$$\sum_{k=1}^{\infty} x_k^+ = P < \infty.$$

Similarly,

$$\sum_{k=1}^{\infty} x_k^- = Q < \infty.$$

Now,

$$s_n = \sum_{k=1}^{n} x_k = \sum_{k=1}^{n} (x_k^+ - x_k^-)$$

$$= \sum_{k=1}^{n} x_k^+ - \sum_{k=1}^{n} x_k^- \to P - Q.$$

Thus the absolute summability of x implies $\sum_{k=1}^{\infty} x_k = P - Q$.

Now we consider y_k^+ and y_k^- similarly defined. Since each y_k^+ is equal to some x_k^+, given n there exists N such that

$$\sum_{k=1}^{n} y_k^+ \leq \sum_{k=1}^{N} x_k^+ \leq P.$$

Thus $\sum_{k=1}^{\infty} y_k^+ = P' \leq P < \infty$. Similarly, $\sum_{k=1}^{\infty} y_k^- = Q' \leq Q < \infty$. Now,

$$\sum_{k=1}^{n} |y_k| = \sum_{k=1}^{n} y_k^+ + \sum_{k=1}^{n} y_k^- \leq P' + Q'$$

for all n, so y is absolutely summable, as claimed. Moreover, we see as for x that

$$\sum_{k=1}^{\infty} y_k = P' - Q'.$$

To show that the sum of the y_k's is identical to that for the x_k's it would suffice to show that $P' = P$ and $Q' = Q$. What we know is that because y is a rearrangement of x,

$$P' \leq P \text{ and } Q' \leq Q.$$

However, x is *also a rearrangement* of y. Thus $P \leq P'$ and $Q \leq Q'$, so that $P = P'$ and $Q = Q'$, as desired. ∎

The Cauchy Product

Sometimes, as in Exercise 5.20, Theorem 5.3.1 helps us find the sum of an absolutely convergent series. But the primary importance of this theorem is that there are applications of series that create a need for order-independent sums, as guaranteed by this theorem for absolutely convergent series. Later in this chapter, we will study the expansion of functions as sums of infinite series of functions–for example, power series expansions of the form

$$f(x) = \sum_{k=0}^{\infty} a_k x^k.$$

If f and g are two functions with such expansions, we would like to know how to multiply two such infinite series together in such a way that the product will equal $f(x)g(x)$. In Exercise 5.10 we encountered the so-called inner product of two sequences x and y, for which $(xy)_k = x_k y_k$. But $\sum_{k=1}^{\infty} x_k y_k$ does not usually equal $\left(\sum_{k=1}^{\infty} x_k\right)\left(\sum_{k=1}^{\infty} y_k\right)$.

For example, if

$$x_k = \frac{(-1)^{k+1}}{\sqrt{k}} = y_k,$$

then x and y are conditionally summable, but

$$\sum_{k=1}^{\infty} x_k y_k = \infty,$$

which is divergent. Thus the sum of the inner product could not be the product of the two convergent sums.

Even for absolutely convergent series, this problem is apparent. For example, if

$$x_k = \frac{1}{2^k} = y_k,$$

then

$$\sum_{k=0}^{\infty} x_k = 2 = \sum_{k=0}^{\infty} y_k.$$

In this case

$$\sum_{k=0}^{\infty} x_k y_k = \sum_{k=0}^{\infty} \frac{1}{4^k} = \frac{4}{3},$$

but the product of the sums is 4.

If we seek a method of multiplying infinite series that will have the product of the sums equal to the sum of the product series, then we need to take our cue from the distributive law for finite arithmetic processes. Thus

$$\left(\sum_{j=1}^{\infty} x_j \right) \left(\sum_{k=1}^{\infty} y_k \right)$$

needs to be the sum of *all* products of the form $x_j y_k$ where j and k vary independently in $\mathbb{N}$. These individual products which will be the summands of the new series can be conveniently arranged in an infinite rectangular array as shown below.

$$
\begin{array}{ccccc}
x_1 y_1 & x_1 y_2 & x_1 y_3 & \cdots & x_1 y_k & \cdots \\
x_2 y_1 & x_2 y_2 & x_2 y_3 & \cdots & x_2 y_k & \cdots \\
x_3 y_1 & x_3 y_2 & x_3 y_3 & \cdots & x_3 y_k & \cdots \\
\vdots & \vdots & \vdots & & \vdots & \\
x_k y_1 & x_k y_2 & x_k y_3 & \cdots & x_k y_k & \cdots \\
\vdots & \vdots & \vdots & & \vdots &
\end{array}
$$

In Section 1.8 we saw that such a family as this, arranged in an infinite table, is a countable set and can be organized in many different ways into a sequence. One easy pattern for doing this is to begin with

$$x_1 y_1, x_1 y_2, x_2 y_1, x_1 y_3, x_2 y_2, x_3 y_1, \ldots .$$

Here we are proceeding systematically along a family of diagonals on which we have $x_j y_k$ with $j + k = 2$, then $j + k = 3$, then $j + k = 4$, etc. But there are many other patterns that work, and we want to make sure that the sum of the resulting series is independent of the order in which the terms have been lined up in a sequence. This means we will need to make sure we have an absolutely convergent series in at least one order, which will then mean every rearrangement is also absolutely convergent and has the same sum. If this is established, then we can write

$$\sum_{j,k \in \mathbb{N}} x_j y_k,$$

where this notation carries the meaning that the sum will be independent of the ordering of the terms into a sequence.

Theorem 5.3.2 *If x and y are both absolutely summable sequences, then the count-able family of terms of the form $x_j y_k$ is absolutely summable in every order. Moreover,*

$$\sum_{j,k \in \mathbb{N}} x_j y_k = \left(\sum_{j=1}^{\infty} x_j \right) \left(\sum_{k=1}^{\infty} y_k \right).$$

Proof: Fix temporarily a sequence z that is comprised precisely of the terms of the form $x_j y_k$ to be summed. Let $s_n = \sum_{i=1}^{n} z_i$ and let N denote the largest index j or

k that appears among the terms $x_j y_k$ corresponding to $z_1, z_2, \ldots, z_n$. Clearly,

$$\sum_{i=1}^{n} |z_i| \leq \left(\sum_{j=1}^{N} |x_j| \right) \left(\sum_{k=1}^{N} |y_k| \right)$$

$$\leq \left(\sum_{j=1}^{\infty} |x_j| \right) \left(\sum_{k=1}^{\infty} |y_k| \right) < \infty$$

for all $n \in \mathbb{N}$. Thus z is absolutely summable. By Theorem 5.3.1, the terms $x_j y_k$ are absolutely summable and have the same sum independent of how they are ordered into a sequence. That is, $\sum_{j,k=1}^{\infty} x_j y_k = \lim_{n \to \infty} s_n = s$.

It remains only to show that s has the value claimed in the theorem. Since the reasoning thus far in the present proof was independent of the choice of z, let us pick z now according to convenience as follows. We will follow a pattern of expanding squares in the upper left-hand corner:

$$x_1 y_1, x_2 y_1, x_2 y_2, x_1 y_2, x_3 y_1, x_3 y_2, x_3 y_3, x_2 y_3, x_1 y_3, \ldots,$$

where we have listed explicitly the terms from the squares that are 1 by 1, 2 by 2, and 3 by 3. So $s_n \to s$. Hence $s_{n^2} \to s$ as well, since this is a subsequence of the convergent sequence s. However,

$$s_{n^2} = \left(\sum_{j=1}^{n} x_j \right) \left(\sum_{k=1}^{n} y_k \right) \to \left(\sum_{j=1}^{\infty} x_j \right) \left(\sum_{k=1}^{\infty} y_k \right) = s.$$

Thus s has the value claimed in the theorem.   ∎

■ EXAMPLE 5.6

There is a special form of the product of two sequences, called the *Cauchy Product*, that we will use in the study of power series.

Consider the following ordering for the sequence z used in the proof of Theorem 5.3.2: $x_1 y_1, x_1 y_2, x_2 y_1, x_1 y_3, x_2 y_2, x_3 y_1, \ldots$. Here we are proceeding systematically along the family of diagonals on which we have the terms $x_j y_k$ with $j + k = 2$, $j + k = 3$, $j + k = 4$, and in general $j + k = l$. Thus

$$c_l = \sum_{j+k=l} x_j y_k,$$

the sum of the terms on the lth diagonal. Then the sums $\sum_{l=2}^{n} c_l$ form a subsequence of the sequence s of partial sums of z. Thus

$$\sum_{l=2}^{\infty} c_l = \left(\sum_{j=1}^{\infty} x_j \right) \left(\sum_{k=1}^{\infty} y_k \right).$$

The absolutely summable sequence c is called the *Cauchy Product* of the sequences x and y. Thus the sum of the Cauchy Product sequence is the product of the sum of x with the sum of y. Of course, the same idea works with sequences x and y indexed by integers starting with 0 instead of 1. Then c_l is indexed by $l \geq 0$.

Another important application of absolute convergence appears in the following theorem about so-called *double summations*.

Theorem 5.3.3 *If the countable set*

$$\{a_{j,k} \mid (j,k) \in \mathbb{N}^2\}$$

is absolutely summable and if

$$c_k = \sum_{j=1}^{\infty} a_{j,k},$$

then c_k is also absolutely summable, and

$$\sum_{k=1}^{\infty} c_k = \sum_{j,k} a_{j,k}.$$

Remark 5.3.1 Here c_k denotes the sum of all the entries $a_{j,k}$ in the kth column of the infinite matrix or array indexed by j for the rows and k for the columns. Thus this theorem asserts that if the double sum is absolutely convergent, then the sum of the individual *column sums* is absolutely convergent and has the same sum as that of the original double sum. At first this might appear to be a special case of the rearrangements addressed in Dirichlet's theorem. This is not the case. Each term c_k represents the sum of an infinite series found in one entire column of the infinite square array $a_{j,k}$. Dirichlet's theorem addresses the rearrangement of one infinite series into *one* (differently ordered) infinite series—*not* into a series of infinitely many different *subseries* of the original series.

Proof: Let $S = \sum_{j,k} a_{j,k}$ and $P = \sum_{j,k} |a_{j,k}|$. We observe that for each k we have $|c_k| \leq \sum_{j=1}^{\infty} |a_{j,k}|$ and for each $K \in \mathbb{N}$ we have

$$\sum_{k=1}^{K} |c_k| \leq \sum_{k=1}^{K} \left(\sum_{j=1}^{\infty} |a_{j,k}| \right) \leq P.$$

(See Exercise 5.28.a.) Thus c_k is absolutely summable. It remains to show that $\sum_{k=1}^{\infty} c_k = \sum_{j,k} a_{j,k}$.

Let $S' = \sum_{k=1}^{\infty} c_k$. We must show $S' = S$. Let $\epsilon > 0$ be arbitrary. There exists K_1 such that $K \geq K_1$ implies

$$\left| S' - \sum_{k=1}^{K} c_k \right| < \frac{\epsilon}{2}.$$

Also, there exists K_2 such that $K \geq K_2$ implies

$$\sum_{j \in \mathbb{N},\, k > K_2} |a_{j,k}| < \frac{\epsilon}{2}$$

so that

$$\left| S - \sum_{j \in \mathbb{N},\, k \leq K} a_{j,k} \right| < \frac{\epsilon}{2}.$$

(See Exercises 5.28.b and 5.28.c.) Selecting $K = \max(K_1, K_2)$ we see by the triangle inequality that $|S - S'| < \epsilon$ for all ϵ. Hence $S = S'$. ∎

Law of the Unconscious Statistician

Absolute summability is particularly useful in the study of probability. A fundamental object of study in probability is a *random variable*. A random variable X is called *discrete* if it has at most countably many values, so that the *range* $S_X = \{x_n \mid n \in \mathbb{N}\}$. To each value x_n of X there is assigned a probability $\mathcal{P}(X = x_n) \in [0, 1]$, and $\sum_{n=1}^{\infty} \mathcal{P}(X = x_n) = 1$.

Definition 5.3.2 *If X is a discrete random variable with range S_X, the* expectation $\mathcal{E}(X) = \sum_{x \in S_X} x \mathcal{P}(X = x)$, *provided this series is* absolutely *convergent.*

The requirement of absolute convergence is very important in the definition of expectation, though it is commonly omitted from elementary text books about probability. Consider for example a discrete random variable X for which $S_X = \mathbb{Q}$, the set of all rational numbers. There is no one preferred or most natural way in which to order $\mathbb{Q}$ into a sequence. If we did not require absolute convergence in the definition of expectation, then the value of $\mathcal{E}(X)$ could be any real number, or even $\pm\infty$, depending on how we ordered S_X in the sum. Thus expectation would fail to be well-defined.

An important theorem in elementary probability pertains to a function g defined on the range S_X of a discrete random variable X. If we let $Y = g(X)$, then Y is again a random variable and is necessarily discrete since the image under g of a countable set is at most countable itself. Thus S_Y is necessarily countable at most. It is a fundamental property of the probability function $\mathcal{P}$ that

$$\mathcal{P}(Y = y) = \sum_{x \in g^{-1}(y)} \mathcal{P}(X = x).$$

Here, $g^{-1}(y) = \{x \in S_X \mid g(x) = y\}$. By the definition of expectation we know that

$$\mathcal{E}(Y) = \sum_{y \in S_Y} y \mathcal{P}(Y = y).$$

The following important theorem is sometimes called the *Law of the Unconscious Statistician* because it is frequently not recognized as a theorem that requires proof in elementary texts.

Theorem 5.3.4 (Law of the Unconscious Statistician) *Let X be a discrete random variable and let $Y = g(X)$ be another discrete random variable. If the sum*

$$\sum_{x \in S_X} g(x)\mathcal{P}(X = x)$$

is absolutely *convergent, then $\mathcal{E}(Y)$ exists and equals the latter sum.*

Proof: We will apply Theorem 5.3.3. Let $S_Y = \{y_k \mid k \in \mathbb{N}\}$. We can denote $g^{-1}(y_k) = \{x_{j,k}\}$, which is indexed by j and is a countable set at most. We can write

$$\sum_{x \in S_X} g(x)\mathcal{P}(X = x) = \sum_{(j,k) \in \mathbb{N}^2} a_{j,k},$$

where

$$a_{j,k} = g(x_{j,k})\mathcal{P}(X = x_{j,k})$$

for all (j, k) if $g^{-1}(y_k)$ is countable. If, however, $g^{-1}(y_k)$ is finite, we use the lower values of j to index the actual members of $g^{-1}(y_k)$ and we enter 0 for $a_{j,k}$ for all larger values of j. Thus for each $y_k \in S_Y$ we have the corresponding *column sum*

$$c_k = \sum_{j=1}^{\infty} a_{j,k} = y_k \mathcal{P}(Y = y_k).$$

Since the double sum is absolutely convergent by hypothesis, Theorem 5.3.3 tells us that $\mathcal{E}(Y) = \sum_k c_k$ exists and that it is given by the following absolutely convergent sum:

$$\mathcal{E}(Y) = \sum_{x \in S_X} g(x)\mathcal{P}(X = x).$$

∎

EXERCISES

5.20   † Use the formula for the sum of an absolutely convergent series given in the proof of Theorem 5.3.1 to find the sum of the series

$$\frac{1}{3} - \frac{1}{4} + \frac{1}{9} - \frac{1}{16} + - \cdots + \frac{1}{3^k} - \frac{1}{4^k} + - \cdots .$$

5.21   If c is a summable sequence and if $x \in [0, 1]$, prove that $\sum_{k=0}^{\infty} c_k x^k$ converges.

5.22   Suppose $|f'(x)|$ exists and is less than 1 for all $x \in (0, 1]$.

  a) Prove:

$$\lim_{n \to \infty} f\left(\frac{1}{n}\right)$$

  exists and equals some number $L \in \mathbb{R}$. Hint: Show that the series

$$\sum_{n=1}^{\infty} \left[f\left(\frac{1}{n+1}\right) - f\left(\frac{1}{n}\right) \right]$$

is absolutely summable.

    b) Prove that $\lim_{x\to 0+} f(x)$ exists and equals L.

5.23   † If $\sum_{k=1}^{\infty} x_k$ is conditionally convergent, prove that

$$\sum_{k=1}^{\infty} x_k^+ = \infty = \sum_{k=1}^{\infty} x_k^-.$$

Here x_k^+ and x_k^- are defined as in the proof of Theorem 5.3.1.

5.24   Use the result of Exercise 5.23 above to show that every conditionally convergent series can be rearranged so that it diverges to $-\infty$.

5.25   If x is conditionally summable and y is a sequence that is not identically 0, prove that the countable family of terms $x_j y_k$ cannot be absolutely summable in any order.

5.26   Let c denote the Cauchy Product of the sequence x with the sequence y, and let d be the Cauchy Product of y with x. Show that $c_l = d_l$ for all $l \in \mathbb{N}$.

5.27   Let $x, y \in \mathbb{R}$, and form the sequences

$$e_j = \frac{x^j}{j!} \text{ and } f_k = \frac{y^k}{k!}$$

$j, k = 0, 1, \dots$. Let c be the Cauchy Product of e with f. Use the Binomial Theorem to prove that

$$c_l = \frac{(x+y)^l}{l!}.$$

(The reader who knows that $e^x = \sum_{j=0}^{\infty} \frac{x^j}{j!}$ will see that this exercise gives a direct proof from the Maclaurin series expansion of e^x that $e^x e^y = e^{x+y}$.)

5.28   † Provide missing details as follows in the proof of Theorem 5.3.3.

    a) Prove that

$$\sum_{k=1}^{K} |c_k| \leq \sum_{k=1}^{K} \left(\sum_{j=1}^{\infty} |a_{j,k}| \right) \leq P.$$

    (Hint: Consider $\sum_{k=1}^{K} \left(\sum_{j=1}^{J} a_{j,k} \right)$ and let $J \to \infty$.)

    b) Prove that there exists K_1 as claimed.

    c) Prove that there exists K_2 as claimed. Hint: Consider

$$\sum_{k=1}^{K} \left(\sum_{j=1}^{J} |a_{j,k}| \right).$$

5.29   Demonstrate the importance of the assumption in Theorem 5.3.3 that the sum over all $(j, k) \in \mathbb{N}^2$ be absolutely convergent by giving an example in which each *column sum* $c_k = 0$, making the sequence c_k absolutely summable, and yet

$$\sum_{(j,k) \in \mathbb{N}^2} |a_{j,k}|$$

diverges.

5.4   THE BANACH SPACE l_1 AND ITS DUAL SPACE

Recall that a Banach space is any vector space, equipped with a nonnegative function called a *norm*, having the property of being complete with respect to the norm: ie, such that every sequence that is Cauchy in the norm converges to a point of the space. We have seen that $\mathbb{R}^n$ is a Banach space in the Euclidean norm, and $C[a, b]$ is a Banach space in the sup-norm. Now we will define another space, together with a nonnegative function that we will show makes this space a Banach space.

Definition 5.4.1 *Denote by l_1 the set of all absolutely summable sequences x. If x is any sequence, let*

$$\|x\|_1 = \sum_{k=1}^{\infty} |x_k|.$$

Thus $x \in l_1$ if and only if $\|x\|_1 < \infty$.

Remark 5.4.1 The reader should be sure to use *lower case* l_1 to denote this space, as L^1 is used to denote a different space of absolutely integrable functions in the sense of Lebesgue. Both spaces are named after Henri Lebesgue.

Theorem 5.4.1 *The set l_1 is a vector space, $\| \cdot \|_1$ is a norm, and with this norm l_1 is a Banach space.*

Proof: For all x and y in l_1 and $c \in \mathbb{R}$, $(x + y)_k = x_k + y_k$ and $(cx)_k = cx_k$. To show that l_1 is closed under addition and scalar multiplication, it will suffice to show that

$$\|x + y\|_1 \le \|x\|_1 + \|y\|_1 < \infty$$

and

$$\|cx\|_1 = |c| \|x\|_1 < \infty,$$

which will simultaneously establish two of the properties required of a norm. However,

$$\|x + y\|_1 = \sum_{k=1}^{\infty} |x_k + y_k| \le \sum_{k=1}^{\infty} (|x_k| + |y_k|)$$

$$= \sum_{k=1}^{\infty} |x_k| + \sum_{k=1}^{\infty} |y_k|$$

$$= \|x\|_1 + \|y\|_1 < \infty.$$

The second property is Exercise 5.30. To complete the proof that $\| \cdot \|_1$ is indeed a norm, see Exercise 5.31.

To complete the proof, we need to show l_1 is complete in the so-called l_1-norm. To avoid confusion with the indices, we will denote a Cauchy sequence $\{x^{(n)}\}_{n=1}^{\infty} \subset l_1$. This means that for all $n \in \mathbb{N}$ we have $x^{(n)} \in l_1$, so that $x^{(n)}$ is itself a sequence with kth term denoted by $x_k^{(n)} \in \mathbb{R}$. Since $x^{(n)}$ is a Cauchy sequence in the l_1-norm, for all $\epsilon > 0$ there exists $N \in \mathbb{N}$ such that

$$m, n \geq N \implies \left\| x^{(m)} - x^{(n)} \right\|_1 = \sum_{k=1}^{\infty} \left| x_k^{(m)} - x_k^{(n)} \right| < \frac{\epsilon}{2}. \qquad (5.2)$$

However, this implies for all fixed k that $m, n \geq N$ implies $\left| x_k^{(m)} - x_k^{(n)} \right| < \frac{\epsilon}{2}$ as well. Hence, for all k, $\left\{ x_k^{(n)} \right\}_{n=1}^{\infty}$ is a Cauchy sequence in $\mathbb{R}$. Thus we can define a sequence x by setting $x_k = \lim_{n \to \infty} x_k^{(n)}$, for all $k \in \mathbb{N}$. We need to show that $x \in l_1$ and that $\|x^{(m)} - x\|_1 \to 0$ as $m \to \infty$. Fix arbitrarily $p \in \mathbb{N}$ and $m, n \geq N$, so that Equation (5.2) above implies that

$$\sum_{k=1}^{p} \left| x_k^{(m)} - x_k^{(n)} \right| < \frac{\epsilon}{2}.$$

Letting $n \to \infty$ we see that

$$\sum_{k=1}^{p} \left| x_k^{(m)} - x_k \right| \leq \frac{\epsilon}{2}.$$

Here we have used the finite upper summation limit p so that we could employ p iterations of the theorem about limits of sums of convergent sequences. Since

$$\sum_{k=1}^{p} \left| x_k^{(m)} - x_k \right| \leq \frac{\epsilon}{2}$$

for all $p \in \mathbb{N}$, we have

$$\sum_{k=1}^{\infty} \left| x_k^{(m)} - x_k \right| = \left\| x^{(m)} - x \right\|_1 \leq \frac{\epsilon}{2}.$$

Hence $x^{(m)} - x \in l_1$ and $x^{(m)} \in l_1$ implies $x^{(m)} - [x^{(m)} - x] = x \in l_1$ as well. Since $m \geq N$ implies

$$\left\| x^{(m)} - x \right\|_1 \leq \frac{\epsilon}{2} < \epsilon,$$

we see that $x^{(m)} \to x$ as $m \to \infty$. ∎

For examples illustrating the concepts in the proof of Theorem 5.4.1, see Exercises 5.36–5.38.

In Section 3.3 we identified several different bounded linear functionals on the Banach space $C[a, b]$, and we showed in Exercises 3.42.a and 3.42.b of that section that *all* the bounded linear functionals on $\mathbb{R}^2$ could be identified with the elements of $\mathbb{R}^2$ itself. (The bounded linear functionals on $\mathbb{R}^n$ can be identified with $\mathbb{R}^n$ in a similar way.) Here we are going to identify *all* the bounded linear functionals on l_1. But before we do this, let us observe that in the exercises just cited, we showed that the bounded linear functionals on $\mathbb{R}^2$ correspond to $\mathbb{R}^2$ itself, and $\mathbb{R}^2$ is a Banach space equipped with the Euclidean norm. It is not ordinarily the case that the family of bounded linear functionals on a Banach space can be identified with the space itself. But it is true that this family of bounded linear functionals is always a Banach space. We will prove this first.

Definition 5.4.2 *Let V denote any normed linear space. Let V' be the set of all $T : V \to \mathbb{R}$ such that T is linear and bounded. We call V' the* dual space *of V.*

Theorem 5.4.2 *Let V be any normed linear space. Then V' is a Banach space, with norm defined by*

$$\|T\| = \inf \left\{ K \,\middle|\, |T(v)| \le K\|v\| \; \forall v \in V \right\}.$$

Remark 5.4.2 The symbol $\|T\|$ denotes the *norm* of the bounded linear functional T, that is defined in the theorem. In Exercise 5.32.a, the student will show that $|T(v)| \le \|T\| \cdot \|v\|$ for all $v \in V$.

Proof: The reader will recall from a course in linear algebra that the sum of any two linear maps is linear and that any constant times a linear map is linear, so we will see that V' is a vector space if we can show that the function $\| \cdot \|$ defined on V' is a norm, which will prove also that the sum of two bounded linear functionals is again bounded, and the same for scalar multiples. Observe first that

$$|T(v)| \le \|T\| \cdot \|v\|$$

for all $v \in V$, so that

$$|cT(v)| = |c||T(V)| \le |c| \cdot \|T\| \cdot \|v\|$$

so that

$$\|cT\| \le |c| \cdot \|T\| < \infty.$$

But $T = \frac{1}{c}(cT)$, so that $\|T\| \le \frac{1}{|c|}\|cT\|$. Thus $\|cT\| = |c|\|T\|$. Observe next that

$$|(T_1 + T_2)(v)| \le |T_1(v)| + |T_2(v)| \le (\|T_1\| + \|T_2\|)\|v\|.$$

Thus we see that $\|T_1 + T_2\| \le \|T_1\| + \|T_2\|$. To complete the proof that $\| \cdot \|$ is a norm on V', see Exercise 5.32.b.

It remains to be shown that V' is complete in the given norm. Let T_n be any Cauchy sequence in V'. Let $\epsilon > 0$. Then there exists N such that m and $n \ge N$ implies

$$\|T_m - T_n\| < \frac{\epsilon}{2}.$$

Thus, for all $v \in V$,

$$|T_m(v) - T_n(v)| < \frac{\epsilon}{2}\|v\|.$$

Hence $\{T_n(v)\}_{n=1}^{\infty}$ is a Cauchy sequence in $\mathbb{R}$ and we can define

$$T(v) = \lim_{n \to \infty} T_n(v)$$

for all $v \in V$. The proof that T is linear is Exercise 5.33. Finally, we must show that T is bounded and that $\|T_n - T\| \to 0$. But if m and $n \geq N$, we know that

$$|T_m(v) - T_n(v)| \leq \frac{\epsilon}{2}\|v\|.$$

Letting $n \to \infty$, we see that

$$|T_m(v) - T(v)| \leq \frac{\epsilon}{2}\|v\|$$

for all $v \in V$. Thus

$$\|T_m - T\| \leq \frac{\epsilon}{2} < \epsilon,$$

so $T_m \to T$ in norm. Moreover,

$$\|T\| = \|T_m - (T_m - T)\| \leq \|T_m\| + \|T_m - T\| < \infty,$$

so T is bounded as claimed. Thus V' is a Banach space. ∎

Now we will proceed to identify l_1' as a Banach space.

Definition 5.4.3 *Denote by l_∞ the set of all bounded sequences, and*

$$\|y\|_\infty = \sup\left\{|y_k| \,\middle|\, k \in \mathbb{N}\right\} \ \forall y \in l_\infty.$$

The norm $\|\cdot\|_\infty$ is also called a sup-norm.

Theorem 5.4.3 *For all $y \in l_\infty$ and for all $x \in l_1$ define*

$$T_y(x) = \sum_{k=1}^{\infty} y_k x_k.$$

We claim that l_∞ is a Banach space, the mapping $T : y \to T_y$ from l_∞ is linear, injective (meaning one-to-one), and surjective *(meaning onto) l_1', and*

$$\|y\|_\infty = \|T_y\|,$$

which means that T preserves norms.

Remark 5.4.3 Because of the properties described in Theorem 5.4.3, the mapping T is called an *isomorphism* from the Banach space l_∞ to l_1'.

Proof: For all $y \in l_\infty$

$$T_y : x \to \sum_{k=1}^{\infty} y_k x_k$$

is an absolutely convergent series (Exercise 5.10), and $T_y : l_1 \to \mathbb{R}$ is clearly linear. T_y is bounded since

$$|T_y(x)| = \left| \sum_{k=1}^{\infty} y_k x_k \right| \le \sum_{k=1}^{\infty} |y_k x_k| \le \|y\|_\infty \|x\|_1$$

for all $x \in l_1$. Thus $\|T_y\| \le \|y\|_\infty$. The reader will prove that $\|T_y\| = \|y\|_\infty$ (Exercise 5.34). The mapping $T : y \to T_y$ carries l_∞ linearly into l_1'. Since $\|T_y\| = \|y\|_\infty$, $T_y = 0$ if and only if $y = 0$. That is, the kernel of the map $T : y \to T_y$ is $\{0\}$, so T is one-to-one from l_∞ *into* l_1'.

It remains to be proven that T is *onto* l_1'. So we let $L \in l_1'$ be arbitrary, and we must show there exists $y \in l_\infty$ such that $L \equiv T_y$. For each $j \in \mathbb{N}$, let $e^{(j)} \in l_1$ be defined by

$$e_k^{(j)} = \begin{cases} 1 & \text{if } k = j, \\ 0 & \text{if } k \ne j. \end{cases}$$

Define a sequence y by letting $y_j = L\left(e^{(j)}\right)$ for all $j \in \mathbb{N}$. Then

$$|y_j| \le \|L\| \left\| e^{(j)} \right\|_1 = \|L\|$$

for all j. Hence $y \in l_\infty$. We claim that for all $x \in l_1$, $L(x) = T_y(x)$, so that $L = T_y$. Observe that

$$\left\| \sum_{k=1}^{n} x_k e^{(k)} - x \right\|_1 = \sum_{k=n+1}^{\infty} |x_k| = \|x\|_1 - \sum_{k=1}^{n} |x_k| \to 0$$

as $n \to \infty$. Therefore

$$L\left(\sum_{k=1}^{n} x_k e^{(k)} \right) = \sum_{k=1}^{n} y_k x_k \to L(x)$$

as $n \to \infty$, since every bounded linear functional L must be continuous. That is, $L(x) = \sum_{k=1}^{\infty} y_k x_k = T_y(x)$.

Finally, we explain briefly how the mapping $y \to T_y$ establishes that l_∞ must be a Banach space itself, and why this Banach space is said to be isomorphic to l_1'. We know already that the sup-norm is a norm, but this follows also from the properties of T. For example,

$$\|y + z\|_\infty = \|T_{y+z}\| = \|T_y + T_z\|$$
$$\le \|T_y\| + \|T_z\| = \|y\|_\infty + \|z\|_\infty.$$

The other properties of a norm can be established for the l_∞-norm similarly. To see that l_∞ is complete, suppose $y^{(n)}$ is a sequence of bounded sequences that is Cauchy in the l_∞-norm. Thus $T_{y^{(n)}}$ is a Cauchy sequence in l'_1. Hence there exists $y \in l_\infty$ such that $T_{y^{(n)}} \to T_y$. It follows that $y^{(n)} \to y$ since

$$\left\| y^{(n)} - y \right\|_\infty = \left\| T_{y^{(n)}} - T_y \right\| \to 0.$$

The mapping T is called an *isomorphism of Banach spaces* because it preserves all the vector space operations, it preserves the norm, and it preserves convergence and completeness in that norm. ∎

EXERCISES

5.30   † Prove for all $x \in l_1$ and for all $c \in \mathbb{R}$, $\|cx\|_1 = |c| \|x\|_1 < \infty$.

5.31   † Show for all $x \in l_1$ that $\|x\|_1 \geq 0$ and $\|x\|_1 = 0$ if and only if $x = 0$, the identically 0 sequence (ie, $x_k = 0$ for all k).

5.32   Let V be a normed linear space.
   a) † If $T \in V'$, prove $|T(v)| \leq \|T\| \cdot \|v\|$ for all $v \in V$.
   b) † If $T \in V'$, prove $T = 0$, the zero functional, if and only if $\|T\| = 0$.

5.33   † In the proof of Theorem 5.4.2, we defined a function $T : V \to \mathbb{R}$ by

$$T(v) = \lim_{n \to \infty} T_n(v),$$

where $T_n \in V'$. Prove that T is linear.

5.34   † If $y \in l_\infty$, $x \in l_1$, and $T_y(x) = \sum_{k=1}^\infty y_k x_k$, prove $\|T_y\| = \|y\|_\infty$. (Hint: The direction "$\leq$" was established in the proof of Theorem 5.4.3. Try applying T_y to $e^{(j)}$ from the proof of that theorem.)

5.35   Prove directly that l_∞ with the sup-norm is a Banach space, without appealing to Theorem 5.4.3.

5.36   For all $n \in \mathbb{N}$ define a sequence $x^{(n)} \in l_1$ by letting $x_k^{(n)} = \frac{n+1}{n2^k}$, for all $k \in \mathbb{N}$.
   a) Show that $x^{(n)} \in l_1$ by calculating $\left\| x^{(n)} \right\|_1$.
   b) Define a sequence x by letting $x_k = \lim_{n \to \infty} x_k^{(n)}$, for all $k \in \mathbb{N}$. Show that $x \in l_1$ by calculating $\|x\|_1$.
   c) Prove: $\left\| x^{(n)} - x \right\|_1 \to 0$ as $n \to \infty$.

5.37   For all $n \in \mathbb{N}$ define a sequence $x^{(n)} \in l_1$ by letting

$$x_k^{(n)} = \begin{cases} 1 & \text{if } k \leq n, \\ \frac{1}{k^2} & \text{if } k > n. \end{cases}$$

   a) Show that $x^{(n)} \in l_1$ by showing $\left\| x^{(n)} \right\|_1 < \infty$.

b) Define a sequence x by letting $x_k = \lim_{n\to\infty} x_k^{(n)}$, for all $k \in \mathbb{N}$. Is $x \in l_1$? Prove your answer, yes or no.

c) Is $x^{(n)}$ a Cauchy sequence in l_1? Prove your answer, yes or no.

5.38   For all $n \in \mathbb{N}$ define a sequence $x^{(n)} \in l_1$ by letting

$$x_k^{(n)} = \begin{cases} \frac{1}{k} & \text{if } k \leq n, \\ 0 & \text{if } k > n. \end{cases}$$

a) Show that $x^{(n)} \in l_1$ by showing $\left\|x^{(n)}\right\|_1 < \infty$.

b) Define a sequence x by letting $x_k = \lim_{n\to\infty} x_k^{(n)}$, for all $k \in \mathbb{N}$. Is $x \in l_1$? Prove your answer, yes or no. Prove that $\left\|x^{(n)} - x\right\|_\infty \to 0$ as $n \to \infty$.

c) Is $x^{(n)}$ a Cauchy sequence in l_1? Prove your answer, yes or no.

d) Is $x^{(n)}$ a Cauchy sequence in the sup-norm? Prove your answers, yes or no.

e) Is l_1 complete in the sup-norm?

5.5 SERIES OF FUNCTIONS: THE WEIERSTRASS M-TEST

Definition 5.5.1 *Suppose for all $n \in \mathbb{N}$ that f_n is defined on the domain D. Let*

$$s_n(x) = \sum_{k=1}^{n} f_k(x).$$

If $s_n(x)$ converges to the number $s(x)$ for each $x \in D$ we say that $\sum_{k=1}^{\infty} f_k$ converges pointwise to s on D. However, if

$$\|s_n - s\|_{\text{sup}} \to 0$$

as $n \to \infty$ on D, then we say the series $\sum_{k=1}^{\infty} f_k$ converges uniformly to the sum s on D.

■ **EXAMPLE 5.7**

Let $f_k(x) = x^k$, for all $x \in D = [0, 1)$. Since $|x| < 1$,

$$\sum_{k=1}^{\infty} f_k(x) = \sum_{k=1}^{\infty} x^k = \frac{x}{1 - x}$$

pointwise convergent on D. However, we can see as follows that this series is *not* uniformly convergent on D.

Suppose it were uniformly convergent. Then the sequence s_n would be Cauchy in the sup-norm. Let $\epsilon = 1$, and there exists N such that $n > m \geq N$

implies $\|s_n - s_m\|_{\sup} < 1$. That is, $\sum_{k=m+1}^{n} x^k < 1$ for all $n > m \geq N$ and for all $x \in D$. Hence

$$\lim_{x \to 1-} \sum_{k=m+1}^{n} x^k \leq 1.$$

However, $\lim_{x \to 1-} \sum_{k=m+1}^{n} x^k = n - m - 1 > 1$ whenever $n > m + 2$. This is a contradiction. Hence the convergence is not uniform on $[0, 1)$.

Exercise 5.40 shows that the series in Example 5.7 does converge uniformly on a smaller domain. The next theorem shows that when we do have uniform convergence of an infinite series, this permits strong conclusions of interest in the calculus.

Theorem 5.5.1 *Let f_k be defined on a domain D for all $k \in \mathbb{N}$. Then we have the following conclusions.*

   i. *If $f_k \in \mathcal{C}(D)$ for all k and if $\sum_{k=1}^{\infty} f_k$ converges uniformly to f on D, then $f \in \mathcal{C}(D)$ as well.*

   ii. *If $f_k \in \mathcal{R}[a, b]$ for all k and if $\sum_{k=1}^{\infty} f_k$ converges uniformly to f on $[a, b]$, then $f \in \mathcal{R}[a, b]$ and*

$$\int_a^b f(x)\, dx = \sum_{k=1}^{\infty} \int_a^b f_k(x)\, dx.$$

   iii. *If $f_k \in \mathcal{C}^1[a, b]$ and if there exists $c \in [a, b]$ such that $\sum_{k=1}^{\infty} f_k(c)$ converges and if $\sum_{k=1}^{\infty} f_k'$ converges uniformly on $[a, b]$, then $\sum_{k=1}^{\infty} f_k$ converges uniformly to a differentiable function f on $[a, b]$ and*

$$f'(x) \equiv \sum_{k=1}^{\infty} f_k'(x).$$

Proof:

   i. For all $n \in \mathbb{N}$, $s_n = \sum_{k=1}^{n} f_k \in \mathcal{C}[a, b]$ and $\|s_n - f\|_{\sup} \to 0$ as $n \to \infty$. Thus $f \in \mathcal{C}[a, b]$.

   ii. This is Exercise 5.41.

   iii. By hypothesis, $s_n' = \sum_{k=1}^{n} f_k' \to g$ uniformly on $[a, b]$. And $s_n(c)$ converges for some $c \in [a, b]$. By the corresponding theorem on uniform limits and derivatives for *sequences*, we see that s_n converges uniformly to some differentiable function f and $f'(x) \equiv g(x) = \sum_{k=1}^{\infty} f_k'(x)$. ∎

Because uniform convergence is so useful, it is helpful to have a convenient test that can identify many (though *not* all) uniformly convergent series.

Theorem 5.5.2 (Weierstrass M-test) *Let f_k be defined on a domain D for all $k \in \mathbb{N}$. Let*

$$M_k = \|f_k\|_{\sup} < \infty$$

for all k. If $\sum_{k=1}^{\infty} M_k < \infty$ then $\sum_{k=1}^{\infty} f_k$ converges both absolutely and uniformly on D. Moreover,

$$\sum_{k=1}^{\infty} |f_k|$$

converges uniformly on D as well.

Proof: Let $\epsilon > 0$. By hypothesis, $\sigma_n = \sum_{k=1}^{n} M_k$ is a Cauchy sequence, so there exists $N \in \mathbb{N}$ such that $n > m \geq N$ implies $|\sigma_n - \sigma_m| < \epsilon$. Denoting

$$s_n = \sum_{k=1}^{n} f_k,$$

we see that

$$\|s_n - s_m\|_{\sup} = \left\| \sum_{k=m+1}^{n} f_k \right\|_{\sup}$$
$$\leq \sum_{k=m+1}^{n} \|f_k\|_{\sup}$$
$$= \sum_{k=m+1}^{n} M_k = |\sigma_n - \sigma_m| < \epsilon.$$

Thus s_n is Cauchy in the sup-norm, so $\sum_{k=1}^{\infty} f_k$ converges uniformly. The exact same argument as that just given still works if we replace f_k by $|f_k|$ in every step. This implies that $\sum_{k=1}^{\infty} |f_k|$ converges uniformly on D as well. ∎

■ **EXAMPLE 5.8**

Let

$$f(x) = \frac{1}{1-x}$$

for all $x \in (-1, 1)$. We know from the Geometric Series formula that

$$f(x) = \sum_{k=0}^{\infty} x^k,$$

provided that $|x| < 1$. We wish to express

$$f'(x) = \frac{1}{(1-x)^2}$$

as an infinite series in powers of x. Suppose we fix $x_0 \in (-1, 1)$ and select a real number r for which $|x_0| < r < 1$. We apply the Weierstrass M-Test by letting $M_k = kr^{k-1}$ to conclude that the series

$$\sum_{k=1}^{\infty} kx^{k-1}$$

of term-by-term derivatives converges uniformly on $[-r, r]$ by applying the Ratio Test (Theorem 5.2.3) to $\sum_{k=1}^{\infty} kr^{k-1}$. It follows then from Theorem 5.5.1 that $f'(x)$ exists and, moreover,

$$f'(x_0) = \sum_{k=1}^{\infty} kx^{k-1}.$$

Note that this formula has been established for all values of $x_0 \in (-1, 1)$.

EXERCISES

5.39   Let

$$f_k(x) = \frac{(-1)^{k+1}}{k} x^k.$$

  a) Prove: $\sum_{k=1}^{\infty} f_k(x)$ converges uniformly on $[0,1]$. (Hint: Use the alternating series test.)
  b) Find $\|f_k\|_{\sup}$ on $[0,1]$. Can the Weierstrass M-test be used to prove the uniform convergence of $\sum_{k=1}^{\infty} f_k(x)$ on $[0,1]$? Why or why not?

5.40   If $0 \le \alpha < 1$, prove that $\sum_{k=1}^{\infty} x^k$ converges uniformly on $[0, \alpha]$.

5.41   Prove part (ii) of Theorem 5.5.1.

5.42   If $\sum_{k=1}^{\infty} f_k$ converges uniformly on D, prove: $\|f_n\|_{\sup} \to 0$ as $n \to \infty$. Is the converse true? Prove or give a counterexample.

5.43   Determine whether or not each of the following series converges uniformly on the indicated domain.
  a) $\sum_{k=1}^{\infty} e^{-kx}$ on $[1, \infty)$.

  b) $\sum_{k=1}^{\infty} \frac{\sin kx}{k^2}$ on $\mathbb{R}$.   **c)** $\sum_{k=1}^{\infty} \sin^k x$ on $\left[0, \frac{\pi}{2}\right)$.
  d) $\sum_{k=1}^{\infty} \sin^k x$ on $[0, \delta]$, where   **e)** $\sum_{k=1}^{\infty} \tan^k x$ on $\left[0, \frac{\pi}{4}\right]$.
      $0 < \delta < \frac{\pi}{2}$.

5.44   Let

$$f(x) = \sum_{k=1}^{\infty} \frac{\sin kx}{k^3}$$

on $\mathbb{R}$. Prove that $f \in \mathcal{C}^1(\mathbb{R})$ and find an expression for $f'(x)$ in terms of an infinite series. Justify all your conclusions.

5.45   Suppose

$$\sum_{k=0}^{\infty} a_k \cos 2k\pi x$$

converges *uniformly* on $[0, 2\pi]$ to a function $f \in \mathcal{R}[0, 2\pi]$. Prove that

$$a_k = \frac{1}{\pi} \int_0^{2\pi} f(x) \cos 2k\pi x \, dx.$$

(Hint: First show how $\int_0^{2\pi} \cos 2j\pi x \cos 2k\pi x \, dx$ depends upon j and k, and apply a theorem about uniformly convergent series.)

5.6 POWER SERIES

Definition 5.6.1 *A power series in powers of* $(x - a)$ *is an infinite series*

$$\sum_{k=0}^{\infty} c_k (x - a)^k,$$

where c is a sequence of real coefficients and $a \in \mathbb{R}$ is the base point.

Given any power series, our first goal is to determine the set of values of the real variable x for which the power series converges. The next theorem shows that the set of points of convergence is always an interval (finite or infinite) centered about a. Note that the set of points of convergence of a power series is never empty, since $x = a$ is always a point of convergence.

Theorem 5.6.1 *Given any power series*

$$\sum_{k=0}^{\infty} c_k (x - a)^k,$$

there exists a radius of convergence $R \leq \infty$ *such that the series converges absolutely everywhere inside* $(a - R, a + R)$, *diverges everywhere outside* $[a - R, a + R]$, *and converges uniformly on every closed finite interval*

$$[\alpha, \beta] \subset (a - R, a + R).$$

Remark 5.6.1 If $R = \infty$, we interpret $(a - R, a + R)$ as $(-\infty, \infty)$. It is always a delicate question in any example to determine what happens at the endpoints $a \pm R$ of the interval of convergence, if $R < \infty$.

Proof: If we were to replace the variable x by $x' = x - a$, then we would be proving that the series $\sum_{k=0}^{\infty} c_k x'^k$ has the claimed convergence properties in $(-R, R)$. Then we would replace x' by $x - a$ and have the general result. So, without loss of generality, it is sufficient to prove the theorem for the case $a = 0$.

   i. We begin by proving the following claim: If the series $\sum_{k=0}^{\infty} c_k x^k$ converges for $x = x_1 \neq 0$ and if $[\alpha, \beta] \subset (-|x_1|, |x_1|)$, then $\sum_{k=0}^{\infty} c_k x^k$ converges both absolutely and uniformly on $[\alpha, \beta]$.

Pick $r > 0$ such that

$$-|x_1| < -r \le \alpha \le \beta \le r < |x_1|.$$

Thus for all $x \in [\alpha, \beta]$, we have $|x| \le r < |x_1|$, so that

$$\left| c_k x^k \right| = \left| c_k x_1^k \right| \cdot \left| \frac{x}{x_1} \right|^k \le \left| c_k x_1^k \right| \cdot \left| \frac{r}{x_1} \right|^k.$$

Since $\sum_{k=0}^{\infty} c_k x_1^k$ converges, $c_k x_1^k \to 0$, and so $\left| c_k x_1^k \right| \le B < \infty$ for all k and for some B. Since $0 \le r < |x_1|$, $\left| \frac{r}{x_1} \right| < 1$, and

$$\sum_{k=0}^{\infty} B \left| \frac{r}{x_1} \right|^k$$

is a convergent geometric series. But

$$\left\| c_k x^k \right\|_{\sup} \le B \left| \frac{r}{x_1} \right|^k$$

on $[\alpha, \beta]$, so $\sum_{k=0}^{\infty} \left\| c_k x^k \right\|_{\sup} < \infty$. Hence $\sum_{k=0}^{\infty} c_k x^k$ converges both absolutely and uniformly on $[\alpha, \beta]$ by the Weierstrass M-test. Observe that every $x \in (-|x_1|, |x_1|)$ lies in some such interval

$$[\alpha, \beta] \subset (-|x_1|, |x_1|).$$

Thus we have also proven that $\sum_{k=0}^{\infty} c_k x^k$ converges absolutely on the interval $(-|x_1|, |x_1|)$.

ii. Now let I be the set of all points at which $\sum_{k=0}^{\infty} c_k x^k$ converges, and let

$$R = \sup \left\{ |x_1| \,\middle|\, x_1 \in I \right\}.$$

If $[\alpha, \beta] \subset (-R, R)$, then there exists $x_1 \in I$ such that

$$[\alpha, \beta] \subset (-|x_1|, |x_1|).$$

Hence the series converges uniformly and absolutely on $[\alpha, \beta]$. And it follows that the series converges absolutely on (-R,R). On the other hand, if $x \notin [-R, R]$, then $|x|$ is too big to permit $x \in I$. Hence $x \notin [-R, R]$ implies $x \notin I$. That is, $(-R, R) \subseteq I \subseteq [-R, R]$.

∎

Remark 5.6.2 We observe that if $|x - a| < R$, then $\sum_{k=0}^{\infty} \left| c_k (x - a)^k \right|$ converges. But if $|x - a| > R$, then $\sum_{k=0}^{\infty} \left| c_k (x - a)^k \right|$ diverges. For this reason, we can determine the radius R of convergence of $\sum_{k=0}^{\infty} c_k (x - a)^k$ by testing the *positive-term* series $\sum_{k=0}^{\infty} \left| c_k (x - a)^k \right|$ for convergence. The ratio test is frequently useful for this purpose.

■ **EXAMPLE 5.9**

We will find the interval I of convergence of $\sum_{k=1}^{\infty} \frac{x^k}{k}$.

Here $a = 0$, and we begin by finding R. Applying the ratio test to the corresponding series of absolute values, we compute

$$\lim_{k \to \infty} \frac{\left| \frac{x^{k+1}}{k+1} \right|}{\left| \frac{x^k}{k} \right|} = \lim_{k \to \infty} \frac{k}{k+1} |x| = |x|.$$

Thus the series converges (absolutely) for $|x| < 1$ and diverges for $|x| > 1$. The endpoints must be tested separately. At $x = 1$ we get the harmonic series. But when $x = -1$, we get

$$\sum_{k=1}^{\infty} \frac{(-1)^k}{k}$$

which is -1 times the alternating harmonic series, which converges. Thus $I = [-1, 1)$.

■ **EXAMPLE 5.10**

Theorem 5.6.1 enables us to use the definite integral to obtain power series expansions for an assortment of interesting functions. Here we will obtain a power series expansion for $\tan^{-1} x$. We begin with the geometric series formula

$$\sum_{k=0}^{\infty} r^k = \frac{1}{1-r}$$

for all $r \in (-1, 1)$. Now set $r = -t^2$, and restrict t to the interval $(-1, 1)$ to insure $|r| < 1$ as required. Be sure to observe that Theorem 5.5.1 enables us to integrate both sides of the expansion

$$\frac{1}{1+t^2} = \sum_{k=0}^{\infty} \left(-t^2 \right)^k = \sum_{k=0}^{\infty} (-1)^k t^{2k},$$

but the integrals must by *definite integrals* (with upper and lower limits of integration). Thus, if $|x| < 1$, we have

$$\int_0^x \frac{1}{1+t^2} \, dt = \sum_{k=0}^{\infty} (-1)^k \int_0^x t^{2k} \, dt,$$

yielding

$$\tan^{-1} x = \sum_{k=0}^{\infty} \frac{(-1)^k}{2k+1} x^{2k+1}. \tag{5.3}$$

By our derivation, we know only that Equation (5.3) is valid for $x \in (-1, 1)$. Yet we know $\tan^{-1} x$ is a continuous function defined on [-1,1]. In Exercise

(5.46) the reader will show that Equation (5.3) remains valid at $x = \pm 1$, hence for all $x \in [-1, 1]$.

EXERCISES

5.46

    a) Show that the series on the right side of Equation (5.3) converges uniformly on $[-1, 1]$ so that if we let

$$g(x) = \sum_{k=0}^{\infty} \frac{(-1)^k}{2k + 1} x^{2k+1},$$

    then $g \in \mathcal{C}[-1, 1]$. (Hint: Use the alternating series test.)

    b) Use the preceding result to show that Equation (5.3) remains valid for all $x \in [-1, 1]$.

    c) Use the preceding result to find an infinite series the sum of which is π. (Hint: What is $\tan^{-1}(1)$?)

5.47

    a) Use the geometric series formula [Equation (5.1)] to derive a power series expansion of $\frac{1}{t}$ in powers of $(t-1)$, valid for $|t-1| < 1$. Does the resulting power series converge (to some real number) at $t = 2$? What about $t = 0$?

    b) Integrating from $t = 1$ to $t = x$, for all $x \in (0, 2)$, find a power series expansion in powers of $(x - 1)$ for $\log x$.

    c) Prove that the power series representation of $\log x$ found in (b) remains valid for $x = 2$. What about $x = 0$? Explain your conclusions.

5.48   Show that the series

$$\sum_{k=1}^{\infty} \frac{x^k}{k}$$

does *not* converge uniformly on $(-1, 1)$. (Hint: Show that the the sequence of partial sums is not Cauchy in the sup-norm.)

5.49   Find

$$\sum_{k=1}^{\infty} \frac{1}{k 2^k}.$$

(Hint: Begin with the series $\sum_{k=0}^{\infty} t^k$ and integrate between two appropriate limits.)

5.50   Find the interval of convergence for each of the following power series.

    a) $\sum_{k=0}^{\infty} \frac{x^k}{k!}$ **b)** $\sum_{k=0}^{\infty} \frac{(-1)^{k+1}}{k+1}(x+1)^k$

    c) $\sum_{k=1}^{\infty} \frac{(-1)^{k+1}(x-1)^{2k+1}}{2k+1}$ **d)** $\sum_{k=1}^{\infty} \frac{x^k}{k^k}$

    e) $\sum_{k=0}^{\infty} \frac{k!}{k^k} x^k$. Hint: For the endpoints, it may help to recall the power series expansion of e^x.

5.51   Use Exercise 5.17 to prove that the radius R of convergence of the power series $\sum_{k=0}^{\infty} a_k x^k$ is given by

$$R = \frac{1}{\limsup \sqrt[k]{|a_k|}}$$

provided that $\limsup \sqrt[k]{|a_k|}$ is a strictly positive real number. Give a suitable interpretation of this formula for R in the case that $\limsup \sqrt[k]{|a_k|} \in \{0, \infty\}$.

5.7   REAL ANALYTIC FUNCTIONS AND C^{∞} FUNCTIONS

If $f(x) = \sum_{k=0}^{\infty} c_k (x - a)^k$ for all $x \in (a - R, a + R)$, we would like to show that

$$f'(x) = \sum_{k=1}^{\infty} k c_k (x - a)^{k-1},$$

the sum of the term-by-term derivatives of the power series for f. In order to use Theorem 5.5.1 to prove this conclusion, we will need to know how the radius R' of convergence for

$$\sum_{k=1}^{\infty} k c_k (x - a)^{k-1}$$

compares with R. Because of the factor k multiplying c_k in the derived series, we would expect that $x - a$ should be smaller: that is, we expect that $R' \leq R$. Actually, we will prove the surprising fact that $R' = R$.

Theorem 5.7.1 *If R is the radius of convergence of*

$$\sum_{k=0}^{\infty} c_k (x - a)^k$$

and if R' is the radius of convergence of

$$\sum_{k=1}^{\infty} k c_k (x - a)^{k-1},$$

then $R = R'$.

Proof: As in the proof of Theorem 5.6.1, we could change variables to $x' = x - a$, and in the x'-variable the base point a would be replaced by 0. So without loss of generality we can give the proof under the assumption that $a = 0$.

   i. First we will prove that $R' \leq R$. For this it will suffice to show that if $|x| < R'$, then $|x| \leq R$ (See Exercise 5.52.) Since $|x| < R'$,

$$\sum_{k=1}^{\infty} k \left| c_k x^{k-1} \right| < \infty.$$

But then the series

$$\sum_{k=1}^{\infty} \left| c_k x^k \right| = |x| \sum_{k=1}^{\infty} \left| c_k x^{k-1} \right|$$

converges by the Comparison Test. Hence $\sum_{k=0}^{\infty} c_k x^k$ converges absolutely, so $R' \leq R$.

ii. Now we must prove the counterintuitive claim that $R \leq R'$. As in the first case, we will show that if $|x| < R$ then $|x| \leq R'$. Because of the delicate comparison needed for case (ii), we pick x_1 such that $|x| < |x_1| < R$. So we are seeking to prove that if

$$\sum_{k=0}^{\infty} \left| c_k x_1^k \right| < \infty, \quad \text{then} \quad \sum_{k=1}^{\infty} k \left| c_k x^{k-1} \right| < \infty.$$

But for all $k \geq 1$ such that $c_k \neq 0$, we have

$$\left| \frac{k c_k x^{k-1}}{c_k x_1^k} \right| = k \left| \frac{x}{x_1} \right|^{k-1} \frac{1}{|x_1|}$$

$$= k r^{k-1} \frac{1}{|x_1|} \overset{(1)}{\longrightarrow} 0$$

where $r = \left| \frac{x}{x_1} \right|$. Here we justify the limit (1) by means of the nth Term Test, since since $\sum_{k=1}^{\infty} k r^{k-1}$ converges by the ratio test (Exercise 5.13). Thus

$$\sum_{k=1}^{\infty} k \left| c_k x^{k-1} \right| < \infty$$

by the Limit Comparison Test (Exercise 5.8). Hence $|x| \leq R'$, as claimed. ∎

Definition 5.7.1 *If there exists $R > 0$ and if there exists a coefficient sequence c such that*

$$f(x) = \sum_{k=0}^{\infty} c_k (x - a)^k$$

for all $x \in (a - R, a + R)$, then we call f (real) analytic at $x = a$.

Theorem 5.7.2 *If f is (real) analytic at a and if R is as in Definition 5.7.1, then $f \in \mathcal{C}^\infty(a - R, a + R)$ and*

$$c_k = \frac{f^{(k)}(a)}{k!}$$

for all $k \geq 0$.

Remark 5.7.1 If $k = 0$, we interpret $f^{(0)}$ as being f itself, and recall the convention that $0! = 1$. This theorem states that no function can be equal to the sum of a power series on an open interval without being infinitely differentiable (i.e., without the derivative $f^{(k)}$ being continuous on the open interval for all k). Moreover, if f is analytic on $(a - R, a + R)$, then the coefficients of the power series must be the coefficients from Taylor's Theorem (Theorem 4.6.1). In other words, *if a function can be expressed as the sum of a series in powers of $(x - a)$, its coefficients are unique.*

Proof: Let $x \in (a - R, a + R)$, so there exist α and β such that

$$x \in (\alpha, \beta) \subset [\alpha, \beta] \subset (a - R, a + R).$$

Both the series $\sum_{k=0}^{\infty} c_k(x - a)^k$ and the series $\sum_{k=1}^{\infty} kc_k(x - a)^{k-1}$ converge uniformly and absolutely on $[\alpha, \beta]$. Thus $f'(x)$ exists and

$$f'(x) = \sum_{k=1}^{\infty} kc_k(x - a)^{k-1}.$$

Since the derived series has the same radius R of convergence as the original series, this argument can repeated as often as we like to prove that $f^{(p)}(x)$ exists and

$$f^{(p)}(x) = \sum_{k=p}^{\infty} k(k - 1) \cdots (k - [p - 1])c_k(x - a)^{k-p}$$

for each $p \in \mathbb{N}$. Since both $f^{(p)}$ and $f^{(p+1)}$ exist on $(a - R, a + R)$,

$$f^{(p)} \in \mathcal{C}(a - R, a + R),$$

so $f \in \mathcal{C}^{\infty}(a - R, a + R)$.

It remains to determine the coefficients c_k. We begin by substituting $x = a$ into

$$f(x) = \sum_{k=0}^{\infty} c_k(x - a)^k$$

yielding

$$c_0 = f(a) = \frac{f^{(0)}(a)}{0!}.$$

Next we substitute $x = a$ into

$$f'(x) = \sum_{k=1}^{\infty} kc_k(x - a)^{k-1},$$

obtaining

$$c_1 = \frac{f'(a)}{1!}.$$

In general, we substitute $x = a$ into

$$f^{(p)}(x) = \sum_{k=p}^{\infty} k(k - 1) \cdots (k - [p - 1])c_k(x - a)^{k-p},$$

showing that

$$c_p = \frac{f^{(p)}(a)}{p!}.$$

■

The reader may be surprised to learn that although every (real) analytic function at a must be in $\mathcal{C}^\infty(a - R, a + R)$ for some $R > 0$, there exist $f \in \mathcal{C}^\infty(\mathbb{R})$ such that f is *not* analytic at 0.

■ **EXAMPLE 5.11**

Let

$$f(x) = \begin{cases} e^{-\frac{1}{x^2}} & \text{if } x \neq 0 \\ 0 & \text{if } x = 0. \end{cases}$$

If $x \neq 0$, then $f'(x) = \frac{2}{x^3} f(x)$. By direct calculation we see that

$$f'(0) = \lim_{x \to 0} \frac{f(x)}{x} = 0.$$

(See Exercise 5.57.) If $x \neq 0$,

$$f''(x) = \frac{-6}{x^4} f(x) + \frac{4}{x^6} f(x).$$

In Exercise 5.57 the reader will show also that $f''(0) = 0$. The industrious reader may then devise a proof by the method of mathematical induction that $f^{(k)}(0)$ exists and $= 0$ for all $k \in \mathbb{N}$. If we construct the *Maclaurin* series

$$\sum_{k=0}^{\infty} \frac{f^{(k)}(0)}{k!} x^k,$$

this converges uniformly and absolutely to the sum 0 for all $x \in \mathbb{R}$. However, this sum does *not* equal $f(x)$ on any interval $(-R, R)$, with $R > 0$. Hence a $\mathcal{C}^\infty$ function need not be analytic.

Knowing that proving the convergence of a (Taylor) power series

$$\sum_{k=0}^{\infty} \frac{f^{(k)}(a)}{k!} (x - a)^k$$

is *insufficient* to establish that it converges to the given function $f(x)$, how can we establish (other than by the methods in Section 5.6) that a given function

$$f \in \mathcal{C}^\infty(a - R, a + R)$$

is analytic? In fact, the Nth partial sum of this series is just the nth Taylor Polynomial $P_n(x)$ of Theorem 4.6.1. To show that $P_n(x) \to f(x)$ as $n \to \infty$, we recall from the latter theorem that the Taylor Remainder

$$R_n(x) = \frac{f^{(n+1)}(\mu)}{(n + 1)!} (x - a)^{n+1}.$$

It is necessary and sufficient to show that $R_n(x) \to 0$, bearing in mind that we cannot control the location of μ between a and x.

■ EXAMPLE 5.12

Let $f(x) = e^x$ for all $x \in \mathbb{R}$. We claim that

$$f(x) = \sum_{k=0}^{\infty} \frac{f^{(k)}(0)}{k!} x^k = \sum_{k=0}^{\infty} \frac{x^k}{k!}$$

for all $x \in \mathbb{R}$. We need to show that

$$R_n(x) = \frac{f^{(n+1)}(\mu)}{(n+1)!} x^{n+1} \to 0$$

as $n \to \infty$ where μ is somewhere between 0 and x. But for our function f,

$$|R_n(x)| = \left| \frac{e^\mu}{(n+1)!} x^{n+1} \right| \le e^{|x|} \frac{|x|^{n+1}}{(n+1)!} \to 0$$

as $n \to \infty$ by virtue of the nth Term Test, since

$$\sum_{k=0}^{\infty} \frac{|x|^{n+1}}{(n+1)!}$$

converges by the ratio test. This proves the claim.

Theorem 5.7.3 *Suppose f and g are both real analytic at $x = a$. Then the product fg is also analytic at $x = a$.*

Proof: By hypothesis, there exist positive numbers R_1 and R_2 such that $|x| < R_1$ implies

$$f(x) = \sum_{0}^{\infty} a_k (x-a)^k$$

and $|x| < R_2$ implies

$$g(x) = \sum_{0}^{\infty} b_k (x-a)^k,$$

with both series being absolutely convergent. Let $R = \min\{R_1, R_2\}$. If

$$c_k (x-a)^k = \sum_{l=0}^{k} a_l (x-a)^l b_{k-l} (x-a)^{k-l}$$

is the kth term of the Cauchy product of the sequences $a_k(x-a)^k$ and $b_k(x-a)^k$, then $|x| < R$ implies

$$f(x)g(x) = \sum_{0}^{\infty} c_k (x-a)^k.$$

■

See Exercise 5.53.

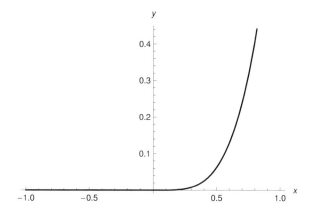

Figure 5.1 $f(x) = x^4$ if $0 < x < 1$.

EXERCISES

5.52   Suppose $|x| < R'$ implies $|x| \le R$. Prove that $R' \le R$.

5.53   Use Examples 5.10 and 5.12 with Theorem 5.7.3 to show that

$$h(x) = e^x \tan^{-1}(x)$$

is real analytic at 0. Find the coefficient of x^4 in the power series expansion of h.

5.54   Let $f(x) = \tan^{-1} x$. Find $f^{(100)}(0)$ and $f^{(101)}(0)$. [Hint: Use Theorem 5.7.2 and Equation (5.3).]

5.55   Prove that the following functions are *not* analytic at 0.
   a) $f(x) = |x|$
   b) For each $k \in \mathbb{N}$ let

$$f(x) = \begin{cases} x^k & \text{if } x > 0, \\ 0 & \text{if } x \le 0. \end{cases}$$

5.56   Let

$$f(x) = \begin{cases} x^4 & \text{if } 0 < x < 1, \\ 0 & \text{if } -1 < x \le 0. \end{cases}$$

Prove your answers to the following questions, and see Fig. 5.1.

   a) Is $f \in C^\infty(0,1)$?   **b)** Is $f \in C^\infty(-1,0)$?
   c) Is $f \in C^\infty(-1,1)$?

5.57   † Let

$$f(x) = \begin{cases} e^{-\frac{1}{x^2}} & \text{if } x \ne 0, \\ 0 & \text{if } x = 0. \end{cases}$$

Prove:

a) $f'(0)$ exists and equals 0. (Hint: Use L'Hôpital's Rule. See Exercises 4.8 and 4.46.)

b) $f''(0)$ exists and equals 0.

c) $f^{(k)}(0)$ exists and equals 0 for all $k \in \mathbb{N}$. (Hint: Use induction to prove that $f^n(x)$ has a useful form for each $n \in \mathbb{N}$.)

5.58   Prove that $\sin x$ is analytic at 0 by showing that its Taylor series about $a = 0$ converges *to* $\sin x$ for all $x \in \mathbb{R}$.

5.59   Let

$$f(x) = \begin{cases} \frac{\sin x}{x} & \text{if } x \neq 0, \\ c & \text{if } x = 0. \end{cases}$$

Prove that by selecting c suitably, f becomes a real analytic function at 0, with a power series expansion in powers of x that converges to $f(x)$ for all x.

5.60   Let the function $f(x)$ be as in Exercise 5.57. Define the function g by

$$g(x) = f(x)f(x-1).$$

Prove that $g \in \mathcal{C}^\infty(\mathbb{R})$, $g(x) \geq 0$ for all x, and $g^{(k)}(0) = 0 = g^{(k)}(1)$ for all $k = 0, 1, 2, \ldots$. (See Fig. 5.2.)

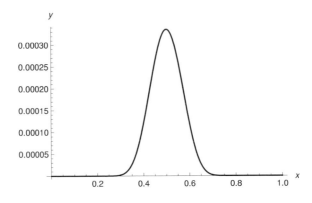

Figure 5.2 $g(x) = f(x)f(x-1)$.

5.61   ◇ Let

$$h(x) = \begin{cases} 0 & \text{if } x \leq 0, \\ \frac{\int_0^x g(t)\,dt}{\int_0^1 g(t)\,dt} & \text{if } 0 < x < 1, \\ 1 & \text{if } x \geq 1, \end{cases}$$

where g is as in Exercise 5.60 above. Prove that $h \in \mathcal{C}^\infty(\mathbb{R})$, $h(x) = 0$ for all $x \leq 0$, $h(x) = 1$ for all $x \geq 1$, $h^{(k)}(x) = 0$ at $x = 0$ and at $x = 1$ for all $k \in \mathbb{N}$, and $0 \leq h(x) \leq 1$ for all x.

5.62   † ◊[13] Let $[a, b]$ and $[c, d]$ denote any two closed finite intervals of the x- and y-axes, respectively. Construct a function $l \in C^\infty(\mathbb{R})$ such that $l(x) = c$ for all $x \leq a$, $l(x) = d$ for all $l \geq b$, and $c \leq l(x) \leq d$ for all $x \in [a, b]$. (Hint: Adjust the function h from Exercise 5.61 above. The reader can find extensions of these conclusions to $\mathbb{R}^n$ in [20].)

5.8 WEIERSTRASS APPROXIMATION THEOREM

We saw in Section 5.7 that if f is real analytic at a, then $f \in C^\infty(a - R, a + R)$ and

$$f(x) = \sum_{k=0}^\infty \frac{f^{(k)}(a)}{k!}(x - a)^k$$

for all $x \in (a - R, a + R)$, where $R > 0$ is the radius of convergence. Thus on every closed finite interval $[\alpha, \beta] \subset (a - R, a + R)$, the sequence of nth degree Taylor polynomials $P_n \to f$ uniformly. That is,

$$\|P_n - f\|_{\text{sup}} \to 0$$

as $n \to \infty$. On the other hand, we saw an example of a function $f \in C^\infty(\mathbb{R})$ for which the sequence P_n of Taylor polynomials converged everywhere to 0, *not* to f. The following theorem is not a theorem about power series, but it is located here to stand in contrast to the theorems of the preceding two sections. We will see that every continuous function on a closed, finite interval $[a, b]$ is the uniform limit of some sequence of polynomials. This includes the function of Example 5.11 on $[-1, 1]$ for example, though the polynomials in this case could not be the Taylor polynomials.

Theorem 5.8.1 (Weierstrass Approximation Theorem) *Let $a < b$, both a and b in $\mathbb{R}$. If $f \in C[a, b]$ then there exists a sequence of* polynomials p_n *such that*

$$\|f - p_n\|_{\text{sup}} \to 0$$

as $n \to \infty$.

Proof: There are *five* parts to the proof of this famous theorem. The first two parts are reductions to a somewhat simpler case.

   i. We claim it suffices to prove the theorem for $[0, 1]$.

     Suppose the theorem were true for $[0,1]$ and $f \in C[a, b]$. Let

$$u = \frac{x - a}{b - a}$$

---

[13]This exercise is used to prove Theorem 6.5.1, which establishes the convergence of Fourier series in the L^2-norm.

so that u goes from 0 to 1 as x goes from a to b. Define

$$g(u) = f(x) = f((b - a)u + a).$$

Thus $g \in \mathcal{C}[0, 1]$, and by hypothesis there exists a polynomial p such that $|g(u) - p(u)| < \epsilon$ for all $u \in [0, 1]$. Thus

$$\left| f(x) - p\left(\frac{x - a}{b - a}\right) \right| < \epsilon$$

for all $x \in [a, b]$. But

$$p\left(\frac{x - a}{b - a}\right)$$

is still a polynomial in x.

ii. We claim it suffices to prove the theorem only for $f \in \mathcal{C}[0, 1]$ such that $f(0) = 0 = f(1)$.

Suppose the theorem were known true when $f(0) = 0 = f(1)$. Let $f \in \mathcal{C}[0, 1]$ be arbitrary and let $L(x) = f(0) + x[f(1) - f(0)]$. Then $f - L$ is still continuous and vanishes at the endpoints. By hypothesis, there exists a polynomial p such that $\|f - L - p\|_{\sup} < \epsilon$. But $L + p$ is still a polynomial.

By virtue of parts (i) and (ii), we can assume henceforth without loss of generality that

$$f \in \mathcal{C}_0(\mathbb{R}) = \{f \in \mathcal{C}(\mathbb{R}) \mid x \notin (0, 1) \implies f(x) = 0\}.$$

Definition 5.8.1 *Suppose for all $n \in \mathbb{N}$ we have a sequence k_n in $\mathcal{R}[-1, 1]$ of nonnegative functions on $[-1, 1]$ such that*

(a) $\int_{-1}^{1} k_n(x)\, dx = 1$ for all n, and

(b) for all $0 < \delta < 1$ we have

$$\|k_n\|_{\sup} \to 0$$

on

$$\{x \mid \delta \leq |x| \leq 1\}.$$

Such a sequence of functions is called an approximate identity.

iii. We suppose for now that we have an approximate identity k_n and we let

$$f_n(x) = \int_{-1}^{1} f(x + t)k_n(t)\, dt$$

for all $f \in \mathcal{C}_0(\mathbb{R})$ and we *claim* that $\|f_n - f\|_{\sup} \to 0$ *on* $[0, 1]$ as $n \to \infty$. The function f_n is called the *convolution* of f with k_n. Each function k_n can be called also the *kernel* function of the integral operator that produces f_n given f and n.

Let $\epsilon > 0$. By uniform continuity of f (Exercise 5.66), there exists $\delta > 0$ such that $|t| \leq \delta$ implies $|f(x) - f(x + t)| \leq \epsilon/2$, for all $x \in [0, 1]$. Let $M = \|f\|_{\text{sup}}$ on $[0,1]$ and let $S_n = \|k_n\|_{\text{sup}}$ on $\{x \mid \delta \leq |x| \leq 1\}$. Thus $S_n \to 0$ as $n \to \infty$. Then, since $\int_{-1}^{1} k_n(x)\, dx = 1$,

$$|f(x) - f_n(x)| = \left| f(x) - \int_{-1}^{1} f(x+t)k_n(t)\, dt \right|$$

$$\leq \int_{-1}^{1} |f(x) - f(x+t)| k_n(t)\, dt$$

$$= \left(\int_{-1}^{-\delta} |f(x) - f(x+t)| k_n(t)\, dt \right.$$

$$+ \int_{-\delta}^{\delta} |f(x) - f(x+t)| k_n(t)\, dt$$

$$\left. + \int_{\delta}^{1} |f(x) - f(x+t)| k_n(t)\, dt \right)$$

$$< 2MS_n + \frac{\epsilon}{2} + 2MS_n$$

$$= \frac{\epsilon}{2} + 4MS_n < \frac{3\epsilon}{4}$$

if n is taken sufficiently big to make $S_n < \frac{\epsilon}{16M}$. For n this big,

$$\|f - f_n\|_{\text{sup}} \leq \frac{3\epsilon}{4} < \epsilon.$$

Thus $\|f - f_n\|_{\text{sup}} \to 0$ as $n \to \infty$.

iv. Suppose k_n and f_n are as above. Suppose also that each function k_n happens to be a polynomial. Then we claim that f_n is also a polynomial.

Letting $u = x + t$, and since $x \in [0, 1]$ and $f \equiv 0$ off $[0,1]$,

$$f_n(x) = \int_{x-1}^{x+1} f(u)k_n(u-x)\, du = \int_{0}^{1} f(u)k_n(u-x)\, du.$$

But $k_n(u - x)$ is a polynomial of some degree N in $(u - x)$ and can be written in the form $q_0(u) + q_1(u)x + \cdots + q_N(u)x^N$. Thus

$$f_n(x) = \int_{0}^{1} f(u)q_0(u)\, du + \cdots + \int_{0}^{1} f(u)q_N(u)\, du \cdot x^N,$$

which is a polynomial in x.

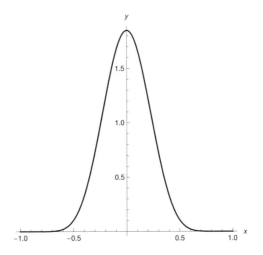

Figure 5.3 $Q_{10}(x)$.

v. Let

$$Q_n(x) = c_n \left(1 - x^2\right)^n,$$

a polynomial in x with c_n chosen so that

$$\int_{-1}^{1} Q_n(x)\, dx = 1.$$

We *claim* that Q_n is an approximate identity. Clearly, $Q_n(x) \geq 0$ on $[-1, 1]$ and $Q_n \in \mathcal{R}[0, 1]$. See Fig. 5.3.

(a) We claim that $\left(1 - x^2\right)^n \geq 1 - nx^2$, for all $x \in [0, 1]$. In fact, this was Exercise 4.15.

(b) We claim $c_n < \sqrt{n}$. To this end we calculate that

$$\int_{-1}^{1} \left(1 - x^2\right)^n\, dx = 2\int_{0}^{1} \left(1 - x^2\right)^n\, dx$$

$$\geq 2\int_{0}^{\frac{1}{\sqrt{n}}} \left(1 - x^2\right)^n\, dx$$

$$\overset{(I)}{\geq} 2\int_{0}^{1/\sqrt{n}} \left(1 - nx^2\right)\, dx$$

$$= \frac{4}{3\sqrt{n}} > \frac{1}{\sqrt{n}},$$

so that

$$c_n = \left(\int_{-1}^{1} \left(1 - x^2\right)^n\, dx\right)^{-1} < \sqrt{n}.$$

We have used the inequality from Exercise 4.15 to justify the step marked by (I).

(c) If $0 < \delta < 1$, then

$$Q_n(x) < \sqrt{n}\left(1 - \delta^2\right)^n$$

on $\delta \leq |x| \leq 1$. But $\sqrt{n}\left(1 - \delta^2\right)^n \to 0$ as $n \to \infty$ since

$$\sum \sqrt{n}\left(1 - \delta^2\right)^n$$

converges by the ratio test. Thus

$$\|Q_n\|_{\sup} \to 0$$

on $\{x \mid \delta \leq |x| \leq 1\}$ as $n \to \infty$. ∎

EXERCISES

5.63   Show that the conclusion of the Weierstrass Theorem would be false for $f(x) = \frac{1}{x}$ on $(0,1)$, thereby verifying the necessity of the use of a closed interval $[a, b]$.

5.64   Show that the Weierstrass Theorem would be false if we replaced [a,b] with the infinite interval $\mathbb{R}$. (Hint: Consider $f(x) = e^x$ on $\mathbb{R}$.)

5.65

a) Suppose $f \in \mathcal{C}[0, 1]$ has the property that

$$\int_0^1 f(x)x^k\,dx = 0$$

for all $k = 0, 1, 2, 3, \ldots$. Prove that $f(x) \equiv 0$ on $[0, 1]$. (Hint: Deduce a conclusion about the value of $\int_0^1 f(x)p(x)\,dx$ if p is any *polynomial*. Then apply the Weierstrass Approximation Theorem.)

b) Define $T_k(f) = \int_0^1 f(x)x^k\,dx$ for all $k = 0, 1, 2, 3, \ldots$. Prove: For each $k = 0, 1, 2, 3, \ldots$, T_k is a *bounded linear functional* on $\mathcal{C}[0, 1]$, equipped with the *sup-norm*.

c) Suppose $T_k(f) = T_k(g)$ for all $k = 0, 1, 2, 3, \ldots$, where f and g are in $\mathcal{C}[0, 1]$. Prove: $f(x) \equiv g(x)$ on $[0, 1]$.

5.66   † Prove that every $f \in \mathcal{C}_0(\mathbb{R})$ is uniformly continuous on $\mathbb{R}$.

5.67   Construct an approximate identity on $[-1, 1]$ consisting entirely of *step functions*, and justify your claims.

5.68   Let $f \in \mathcal{C}[0, 1]$ and let $\phi_n(x) = (n + 1)x^n$ on $[0, 1]$. Let

$$T_f(\psi) = \int_0^1 \psi(x)f(x)\,dx$$

for all $\psi \in \mathcal{C}[0, 1]$.

a) Prove: $\lim_{n \to \infty} T_f(\phi_n) = f(1)$ for all $f \in \mathcal{C}[0, 1]$. (Hint: Find

$$\lim_{n \to \infty} \int_0^1 nx^{n+k} \, dx$$

for each nonnegative integer k. What could you conclude if f were a polynomial?)

b) Prove that $\phi_n \to 0$ pointwise on $[0, 1] \setminus \{1\}$ but not uniformly.

c) Prove that T_f is a *bounded linear functional* on the Banach space $\mathcal{C}[0, 1]$, but $T_f(\phi_n)$ does not converge to $T_f(\lim_{n \to \infty} \phi_n)$.

5.69   Consider the vector space $\mathcal{P}$ from Exercise 3.44. Determine whether or not $\mathcal{P}$ is complete, with the norm from that exercise. Prove your conclusion.

5.9 TEST YOURSELF

EXERCISES

5.70   Determine *absolute convergence, conditional convergence, or divergence*:

a) $\sum_{k=1}^{\infty} \frac{(-1)^{k+1}}{k(k+1)}$

b) $\sum_{k=1}^{\infty} \frac{(-1)^{k+1}}{\sqrt{k}}$

5.71   Given an example of $x_k \to 0$ for which $\sum_{k=1}^{\infty} (-1)^{k+1} x_k$ *diverges*.

5.72   Let $f : \mathbb{R} \to \mathbb{R}$ be a differentiable function such that $\|f'\|_{\sup} = \frac{1}{2}$. True or False: There exists a real number $p \in \mathbb{R}$ such that $f(p) = p$.

5.73   True or False: The *conditionally summable* sequences of real numbers form a *vector space*.

5.74   Determine convergence or divergence:

a) $\sum_{k=2}^{\infty} \left(\frac{1}{k \log k} \right)$

b) $\sum_{k=1}^{\infty} kr^{k-1}$, where $|r| < 1$.

5.75   Find two sequences, x_k and y_k such that $\frac{x_{k+1}}{x_k} \to 1$ and $\frac{y_{k+1}}{y_k} \to 1$, but $\sum_{k=1}^{\infty} x_k$ converges, whereas $\sum_{k=1}^{\infty} y_k$ diverges.

5.76   Determine convergence or divergence:

a) $\sum_{k=0}^{\infty} \frac{k^k}{k!}$

b) $\sum_{k=2}^{\infty} \frac{1}{(\log k)^k}$

5.77   True or False: The series $\sum_{k=1}^{\infty} \frac{(-1)^{k+1}}{k^2}$ has a *rearrangement* that *diverges* to ∞.

5.78   Give an example of two *pairs* of sequences x_k and y_k such that

$$\sum_1^\infty x_k y_k \neq \left(\sum_1^\infty x_k\right)\left(\sum_1^\infty y_k\right)$$

and state the numerical values of *the sums* of each of the three series for each pair!

5.79   Let x_k and y_k be any two absolutely summable sequences, with the index $k \geq 1$. Write out the term c_5 of the *Cauchy product* of x and y.

5.80   For each $n \in \mathbb{N}$ define a sequence $x^{(n)} \in l_1$ by letting

$$x_k^{(n)} = \frac{n+1}{n3^k}$$

for all $k \in \mathbb{N}$. Calculate $\left\|x^{(n)}\right\|_1$ for each fixed $n \in \mathbb{N}$.

5.81   Find $M_k = \|f_k\|_{\text{sup}}$ and determine whether or not $\sum_1^\infty M_k$ converges, in each case.

    a) $\sum_{k=1}^\infty e^{-kx}$ on $[2, \infty)$.
    b) $\sum_{k=1}^\infty \sin^k x$ on $[0, \pi/2]$.

PART II

ADVANCED TOPICS IN ONE VARIABLE

CHAPTER 6

FOURIER SERIES [14]

Periodic (and nearly periodic) phenomena have played a large role in human activities since early recorded history. The cycles of day and night, the phases of the moon, the tides, the flooding of rivers, the seasons, and the migrations of birds and animals affect human beings' lives. Civilization as we know it is based upon agriculture, and this requires anticipation of the cycles in the world around us.

Almost three thousand years ago Babylonian astronomers successfully predicted the times of lunar and solar eclipses by expressing these complicated events as summations of numerous simpler periodic events. These remarkable predictions were accurate to the extent of predicting eclipses that would be visible at least from some part of the world.

Greek astronomers approximately two thousand years ago built a bronze device that predicted retrograde motions of the outer planets, which causes these planets to appear to reverse direction sometimes in their paths viewed against the background of distant stars. This device predicted the retrograde motions by compounding the effects of multiple periodic circular motions.

[14]This chapter is not required for any subsequent chapters.

In modern physics the color of light is determined by the frequency of oscillation of an electromagnetic wave, and the pitch of a musical note is determined by the frequency of oscillation of a vibrating string, membrane, or air column in a musical instrument. But the timbre of a musical instrument is determined by combining numerous oscillations of different frequencies, each with its own characteristic amplitude. Thus the recognizable differences of *concert tone A* on a flute, a piano, or a violin reflects in each case the summation of many different frequencies of relatively small amplitude added to the lowest frequency, which determines the perceived pitch.

In all the examples listed above, complex oscillations are analyzed as sums of many simple oscillations. This subject is known as *harmonic analysis*; it is known also as *Fourier analysis* after Joseph Fourier, who successively applied it to analyze the flow of heat through a metal rod in 1822. There are many distinguished books about Fourier series, though most presume knowledge of the Lebesgue integral. The author recommends especially the book by Dym and McKean [6]. It provides a useful summary of the needed properties of the Lebesgue integral in the beginning.

6.1   THE VIBRATING STRING AND TRIGONOMETRIC SERIES

We will begin with an important example from physics,[15] which explains why one might wish to express all functions from a broad class as the sums of infinite series of sine waves and cosine waves.

■ **EXAMPLE 6.1**

Suppose a piece of music wire (such as a violin, guitar, or piano string) is stretched taut and fastened down at the endpoints of the interval $[0, 1]$ on the x-axis. Now either a finger or an implement is used to stretch the string away from the axis, into some initial plane curve described by a displacement function $f(x)$. At time $t = 0$ the string is released, and, depending on the manner in which the plucking or hammering of the string was done, the string has an initial velocity at each point $x \in [0, 1]$ given by $g(x)$. This leads to the following partial differential equation governing the function $u(x, t)$, which is the displacement of the string at coordinate x and at time t.

$$\frac{\partial^2 u}{\partial t^2} = a^2 \frac{\partial^2 u}{\partial x^2}. \tag{6.1}$$

Here a is a constant reflecting the tension to which the string is tightened and the density of the steel in the string. Since our purpose here is to motivate a mathematical problem and not to keep track of the effect of physical constants on the solution, *we will take $a = 1$ henceforth.*

The boundary conditions on the string and the initial conditions on the displacement and velocity at time $t = 0$ are given as follows.

---

[15]Neither first-year physics nor differential equations (ordinary or partial) is required for this course. The Example is presented for physical motivation and historical perspective.

$$\begin{cases} u(0,t) &= 0, \\ u(1,t) &= 0, \\ u(x,0) &= f(x), \\ \frac{\partial u}{\partial t}(x,0) &= g(x). \end{cases} \tag{6.2}$$

What is needed is a technique that may be enable us to construct a function $u(x,t)$ which satisfies Equation (6.1) and conditions (6.2). A classic method of solution is to seek solutions having the special simplified form

$$u(x,t) = F(x)G(t),$$

where F and G must be determined. (This method is called *separation of variables*.) Substitution into Equation (6.1) yields

$$\frac{F''(x)}{F(x)} = \frac{G''(t)}{G(t)},$$

which proves that both sides must be equal to the same constant since the left side is independent of t and the right side is independent of x. If that constant were a *nonnegative* number λ^2, then we would have

$$F''(x) = \lambda^2 F(x), \tag{6.3}$$

which has the general solution

$$F(x) = c_1 e^{\lambda x} + c_2 e^{-\lambda x}.$$

But then there would be no solution for $F(x)$ other than the identically zero solution given the boundary conditions (6.2). (See Exercise 6.5.)

Thus we take the constant to be a *negative* number $-\lambda^2$, which yields the general solution

$$F_n(x) = c \sin 2n\pi x \tag{6.4}$$

for some integer $n \in \mathbb{Z}$ and an arbitrary constant c. Similar work for the function G yields a sequence of corresponding solutions

$$u_n(t) = (A_n \cos 2n\pi t + B_n \sin 2n\pi t) \sin 2\pi n x.$$

We observe that Equation (6.1) will be satisfied by any function of the form

$$u(x,t) = \sum_{n=0}^{\infty} (A_n \cos nt + B_n \sin nt) \sin nx, \tag{6.5}$$

provided that all questions of convergence of the series itself and its twice derived series can be managed.

The goal of this example is to satisfy both the *boundary* conditions and the *initial* conditions in (6.2). In order to achieve this, we will need to be able to express both f and g as sums of infinite *trigonometric series*:

$$f(x) = \sum_{n=1}^{\infty} A_n \sin 2n\pi x$$

and

$$g(x) = \sum_{n=1}^{\infty} 2n\pi B_n \cos 2n\pi x.$$

Trigonometric Series

Thus the problem of describing the motion of a vibrating string leads us to the importance of determining which functions f and g can be expressed as *convergent* infinite series of sine and cosine functions of various periods. And we will need a method of computation of the necessary coefficients of such expansions.

As a problem of pure mathematics, we ask for which functions f defined on $[0, 1]$ it is possible to expand

$$f(x) = \sum_{n=0}^{\infty} a_n \cos 2n\pi x + b_n \sin 2n\pi x \tag{6.6}$$

in a suitably convergent manner. (We have included the index $n = 0$ in the summation in order to allow for nonzero constant functions.) Also, we need to ask whether given such a function f we can find the coefficients a_n and b_n.

Theorem 6.1.1 If *the series in Equation (6.6) converges* uniformly *on* $[0, 1]$ *to some function* f, *then* $f \in \mathcal{C}[0, 1]$,

$$\begin{cases} a_n = & 2 \int_0^1 f(x) \cos 2n\pi x \, dx, \\ b_n = & 2 \int_0^1 f(x) \sin 2n\pi x \, dx \end{cases} \tag{6.7}$$

for each $n > 0$, *and* $a_0 = \int_0^1 f(x) \, dx$. *Moreover, if the series*

$$\sum_{n=0}^{\infty} (-2n\pi a_n \sin 2n\pi x + 2n\pi b_n \cos 2n\pi x)$$

converges uniformly *on* $[0, 1]$, *then* f *is differentiable and*

$$f'(x) = \sum_{n=0}^{\infty} -2n\pi a_n \sin 2n\pi x + 2n\pi b_n \cos 2n\pi x. \tag{6.8}$$

Proof: Since each summand on the right side of Equation (6.6) is a continuous function on $[0, 1]$, the uniform convergence of this series guarantees the continuity of the sum, f, by Theorem 5.5.1. We can prove Equation (6.7) by using Theorem 5.5.1 to compute the integrals on the right sides as sums of infinite series of integrals, and showing that all but one integral vanishes. (See Exercise 6.7.) There will be no loss of generality if we take $b_0 = 0$ always. Equation (6.8) can be proven similarly, and this is left to Exercise 6.8. ∎

Definition 6.1.1 *A function* $f : \mathbb{R} \to \mathbb{R}$ *is called* periodic with period T, *provided that* $f(x + T) = f(x)$ *for all* $x \in \mathbb{R}$.

If the trigonometric series in Equation (6.6) converges at least pointwise to a function f on $[0, 1]$, then the same series converges for all $x \in \mathbb{R}$ to a function of period 1 since each summand of the series has period 1. Thus any function f given as in Equation (6.6) can be viewed as being a periodic function of period 1 on the entire real line. We observe that if $\phi : [0, 1) \to \mathbb{R}$ is any function defined on the left-closed and right-open unit interval, then ϕ can be extended to a function ϕ_e of period 1 on the real line by letting $\phi_e(x) = \phi(x - \lfloor x \rfloor)$, where the *floor function* $\lfloor x \rfloor$ denotes the greatest integer that does not exceed x.

A periodic function on the real line that is not constantly zero cannot be Riemann integrable, since its support is unbounded. However, we have the following concept.

Definition 6.1.2 *If* $f : \mathbb{R} \to \mathbb{R}$ *is periodic with period* T, *we call* f Riemann integrable on the circle *if its restriction to any closed interval of length* T *is Riemann integrable.*

The reader should note that if f has period T, then $f(a) = f(a + T)$ for each real number a. Exercise 6.4 establishes that the integral of a periodic integrable function on the circle is independent of the choice of the interval $[a, a + T]$ over which the integration takes place.

An interesting way to picture the domain of a periodic function geometrically (or topologically) is as follows. Imagine a closed interval $[a, a + T]$ of length T. Picture the interval bent into a circle with the endpoints joined together to form a single point. This is a circle of circumference T. One can also picture this circle algebraically as the quotient group of the *additive group* of real numbers modulo the *additive* subgroup of *integer multiples* of T. That is, the circle of circumference T can be pictured as $\mathbb{R}/T\mathbb{Z}$.

EXERCISES

6.1 Let $a < b$ and $\phi : [a, b) \to \mathbb{R}$.

    a) Prove that ϕ can be extended to a periodic function ϕ_e defined on the entire real line, with period $T = b - a$.

    b) Let $f(x) = \phi_e(a + (b - a)x)$ for each $x \in \mathbb{R}$. Prove that f has period 1.

    c) *If* f can be expanded into a series according to Equation (6.6), derive a similar series expansion for ϕ_e.

6.2 Prove that the series

$$\sum_{n=0}^{\infty} \frac{\cos nx}{n^3}$$

converges uniformly to a periodic function f of period 2π in $C^1(\mathbb{R})$, the continuously differentiable functions on the real line.

6.3 Prove that the series

$$\sum_{n=0}^{\infty} \frac{\sin 2n\pi x}{2^n}$$

converges uniformly to a periodic function f of period 1 in $C^{\infty}(\mathbb{R})$, the infinitely differentiable functions on the real line.

6.4 † Suppose $f : \mathbb{R} \to \mathbb{R}$ has period T and suppose that the restriction of f to the interval $[0, T]$ is Riemann integrable. Prove that f is Riemann integrable on every closed interval of length T and that

$$\int_a^{a+T} f(x)\,dx = \int_0^T f(x)\,dx$$

for all $a \in \mathbb{R}$.

6.5 In Equation (6.3) prove that there is no nonconstant solution for F if we take λ negative. (Hint: Use the boundary conditions in Equation (6.2).)

6.6 Explain why *negative* integers n are not needed in Equation (6.5).

6.7 Prove Equation (6.7).

6.8 Prove Equation (6.8).

6.2 EULER'S FORMULA AND THE FOURIER TRANSFORM

We will see that it is helpful both computationally and conceptually to recast the concept of trigonometric series so that it applies to complex-valued functions as well as to real-valued ones. First we remind the reader of some elementary properties of the set $\mathbb{C}$ of complex numbers. The properties are listed in Table 6.1.

The first six axioms listed are identical to the field axioms for the real numbers. Axiom 7 applies to $\mathbb{C}$ but not to $\mathbb{R}$. Because of the fact that squares of complex numbers can be negative, there is no order relation for $\mathbb{C}$ and of course it follows that there is no Archimedean property for $\mathbb{C}$ either. Similarly, there is no concept of positivity or negativity for nonreal complex numbers, though these features are retained by the subfield of real numbers. We regard $\mathbb{R}$ as being a subset of $\mathbb{C}$ consisting of all those complex numbers having the form

$$x + i0 = x + 0 = x.$$

Definition 6.2.1 *If $z = x + iy \in \mathbb{C}$, then we define the* conjugate $\bar{z}$ *of z by*

$$\bar{z} = x - iy$$

and the modulus *of z to be the* nonnegative real *number $|z|$ given by*

$$|z|^2 = z\bar{z} = x^2 + y^2.$$

Table 6.1   Field of Complex Numbers

The set
$$\mathbb{C} = \{x + iy \mid x \in \mathbb{R},\ y \in \mathbb{R}\}$$
of all complex numbers is a field with two operations, called addition and multiplication. These satisfy the following properties:

i. *Closure:* If w and z are elements of $\mathbb{C}$, then $w + z \in \mathbb{C}$ and $wz \in \mathbb{C}$.

ii. *Commutativity:* If w and z are elements of $\mathbb{C}$, $w + z = z + w$ and $wz = zw$.

iii. *Associativity:* If v, w, and z are elements of $\mathbb{C}$, $v + (w + z) = (v + w) + z$ and $v(wz) = (vw)z$.

iv. *Distributivity:* If v, w, and z are elements of $\mathbb{C}$, $v(w + z) = vw + vz$.

v. *Identity:* There exist elements 0 and 1 in $\mathbb{C}$ such $0 + z = z$ and $1z = z$, for all $z \in \mathbb{C}$. Moreover, $0 \neq 1$.

vi. *Inverses:* If $z \in \mathbb{C}$, there exists $-z \in \mathbb{C}$ such that $-z + z = 0$. Also, for all $z \neq 0$, there exists $z^{-1} = \frac{1}{z} \in \mathbb{C}$ such that $z\frac{1}{z} = 1$.

vii. The number $i^2 = -1$, the additive inverse of the number 1.

We call x and y, respectively, the real *and the* imaginary *parts of $z = x + iy$ and these are denoted by*
$$x = \Re(z),\ y = \Im(z).$$

We remark that $z\bar{z}$ is a nonnegative real number for all $z \in \mathbb{C}$. The complex numbers can be modeled conveniently in a geometrical manner by identifying $z = x + iy$ with the point (x, y) in the Cartesian plane. We think of the x-axis as the *real axis* and the y-axis as the *imaginary axis* in this picture.

The plane can be equipped with polar coordinates (r, θ) as well as with rectangular coordinates (x, y). The relationship between these two systems of coordinates is that $x = r\cos\theta$ and $y = r\sin\theta$. Thus each complex number z can be written in the *polar form*
$$z = r(\cos\theta + i\sin\theta).$$

Definition 6.2.2 *A sequence of complex numbers z_n is said to converge to $z \in \mathbb{C}$ if and only if $|z_n - z| \to 0$ as $n \to \infty$.*

It is easily verified that $z_n \to z$ if and only *both* $\Re z_n \to \Re z$ and $\Im z_n \to \Im z$. (See Exercise 6.10.)

Definition 6.2.3 *An infinite series $\sum_{n=0}^{\infty} z_n$ of complex numbers is said to converge to a complex number S if and only if the sequence*

$$S_N = \sum_{n=0}^{N} z_n$$

converges to S.

We leave it to Exercise 6.11 for the reader to prove that $\sum_{n=0}^{\infty} z_n$ converges to a complex number S if and only *both* $\sum_{n=0}^{\infty} \Re z_n$ converges to $\Re S$ and $\sum_{n=0}^{\infty} \Im z_n$ converges to $\Im S$.

Definition 6.2.4 *For each complex number $z \in \mathbb{C}$ we define*

$$e^z = \sum_{n=0}^{\infty} \frac{z^n}{n!}, \tag{6.9}$$

which is shown to converge in Theorem 6.2.1.

Theorem 6.2.1 *For each complex number z, the series in Equation (6.9) converges. Moreover, for each real number x, we have* Euler's formula*:*

$$e^{ix} = \sum_{n=0}^{\infty} \frac{(ix)^n}{n!} = \cos x + i \sin x. \tag{6.10}$$

Proof: We note first that $\sum_{n=0}^{\infty} \frac{|z^n|}{n!}$ converges since the real exponential series converges absolutely for every real number. But it is immediate that $|\Re(z^n)| \leq |z^n|$ and also that $|\Im(z^n)| \leq |z^n|$. Thus the sum of the real parts in Equation (6.9) converges absolutely, as does the sum of the imaginary parts. Now we apply Exercise 6.11 to the series appearing in Equation (6.9) in order to conclude that the series expansion that defines e^z converges for every $z \in C$.

For Euler's formula, we observe that

$$(ix)^n = \begin{cases} (-1)^k x^{2k} & \text{if } n = 2k \\ (-1)^k i x^{2k+1} & \text{if } n = 2k+1. \end{cases}$$

It follows that

$$e^{ix} = \sum_{k=0}^{\infty} (-1)^k \frac{x^{2k}}{(2k)!} + i \sum_{k=0}^{\infty} (-1)^k \frac{x^{2k+1}}{(2k+1)!}$$

$$= \cos x + i \sin x.$$

∎

Corollary 6.2.1 *For each real number x, we have*

$$\cos x = \frac{e^{ix} + e^{-ix}}{2}, \tag{6.11}$$

$$\sin x = \frac{e^{ix} - e^{-ix}}{2i}. \tag{6.12}$$

Proof: Apply Euler's formula. ∎

We observe next that the Nth partial sum of the trigonometric series in Equation (6.6) can be rewritten using Corollary 6.2.1 in the form

$$S_N(f) = \sum_{n=0}^{N} (a_n \cos 2\pi nx + b_n \sin 2\pi nx) \tag{6.13}$$

$$= \sum_{n=-N}^{N} c_n e^{2\pi i n x},$$

where $c_0 = a_0$ and

$$c_n = \frac{a_{|n|} - i \operatorname{sgn}(n) b_{|n|}}{2}$$

if $n \neq 0$. (Here sgn denotes the signum function.) We would like to convert the formulae in Equation (6.7) for a_n and b_n into direct formulae for the calculation of c_n. In order to do this, we will need a suitable definition for the integral of a complex-valued function of a real variable.

Definition 6.2.5 *Suppose that $f : [a, b] \to \mathbb{C}$, so that*

$$f(x) = \Re f(x)) + i \Im f(x)$$

for each $x \in [a, b]$. If the two real-valued functions $\Re f(x)$ and $\Im f(x)$ are both Riemann integrable on $[a, b]$, then we say that $f \in \mathcal{R}([a, b], \mathbb{C})$ and

$$\int_a^b f(x)\, dx = \int_a^b \Re f(x)\, dx + i \int_a^b \Im f(x)\, dx.$$

Moreover, for each $k \in \mathbb{Z}$ we define the kth character by

$$\chi_k(x) = e^{2\pi i k x}. \tag{6.14}$$

Remark 6.2.1 Each function $\chi_k : \mathbb{R} \to \mathbb{R}$ in such a way that $\chi_k : \mathbb{Z} \to \{1\}$. In the language of abstract algebra, this means that each function χ_k is well-defined on the additive *quotient group* $\mathbb{R}/\mathbb{Z}$, which is interpreted geometrically as a circle of circumference 1. Note that $\chi_{-k}(x) = \bar{\chi}_k(x)$, the *complex conjugate* of $\chi_k(x)$.

Corollary 6.2.2 *If f is a Riemann integrable function on $[0,1]$ with uniformly convergent trigonometric series, then the Nth partial sum of that series can be expressed in the form*

$$S_N = \sum_{-N}^{N} c_n \chi_n(x),$$

where

$$c_n = \int_0^1 f(x)\overline{\chi}_n(x)\, dx$$

for each integer n, positive, negative, or zero.

We leave the proof of this corollary to the reader in Exercise 6.13.

The reader should note that as $N \to \infty$ the summation that defines S_N leads to what is naturally expressed as a sum *from* $-\infty$ *to* ∞. This leads us to the formulation of the general concept of the *Fourier series* of any Riemann integrable function f on $[0,1]$, without making any claims initially regarding convergence of that series to such a general function.

Definition 6.2.6 *Let $f \in \mathcal{R}([0,1],\mathbb{C})$. We define the* Fourier series *of f to be the (doubly) infinite series*

$$S(f) = \sum_{n=-\infty}^{\infty} \widehat{f}(n)\chi_n(x), \tag{6.15}$$

where we define the numbers $\widehat{f}(n)$, called the Fourier coefficients *of f, by*

$$\widehat{f}(n) = \int_0^1 f(x)\overline{\chi}_n(x)\, dx \tag{6.16}$$

for each $n \in \mathbb{Z}$. The function $\widehat{f} : \mathbb{Z} \to \mathbb{C}$ is called the Fourier transform *of f. Even if the doubly infinite summation in Equation (6.15) were not convergent in any sense, we would call*

$$S_N(f) = \sum_{-N}^{N} \widehat{f}(n)\chi_n(x)$$

the Nth partial sum of the Fourier series of the Riemann integrable function f.

We note in advance that in Theorem 6.5.1 we will prove that the Fourier series of every Riemann integrable function does in fact converge to f in the sense of the L^2-norm, which will be defined later. For the case of pointwise convergence, we observe that since the summation in Equation (6.15) is over a countable set of indices, it will be independent of order, provided that it is absolutely [16] convergent. Thus

---

[16] An infinite series of *complex* numbers c_n is called *absolutely convergent*, provided that the sum of $|c_n|$ is finite, where the symbol that looks like an absolute value actually means *modulus* in this context. The theorems concerning independence of order of summation apply to absolutely convergent series of complex numbers as well as to real series. The proofs are identical, even down to the use of the same absolute value symbol for the modulus.

there will be no loss of generality if we understand that the double summation in Equation (6.15) means the limit of S_N as $N \to \infty$, whether the convergence be pointwise, uniform, or in the sense of L^2-norm, which remains to be defined. In the next section we will begin to investigate the convergence of S_N. First we will need to know that the standard computational properties of the exponential function carry over to the case of complex exponents.

Theorem 6.2.2 *If w and z are complex numbers, then $e^{w+z} = e^w e^z$.*

Remark 6.2.2 In Exercise 5.27 the reader showed that the multiplicative property of the exponential function

$$e^x e^y = e^{x+y}$$

for all real numbers x and y follows from the binomial theorem. We will *not* require that exercise for what follows. The proof below of the multiplicative property of e^z for all complex numbers z is based on the multiplicative property of e^x for all real numbers x, which we assume to be well known to the reader.

Proof: Because

$$e^x e^y = e^{(x+y)}$$

the Cauchy product of the series expansions on the left-hand side yields

$$\sum_{n=0}^{\infty} \sum_{k=0}^{n} \frac{x^k y^{(n-k)}}{k!(n-k)!} = \sum_{n=0}^{\infty} \frac{(x+y)^n}{n!}.$$

Since real convergent power series are unique, it follows that

$$\sum_{k=0}^{n} \frac{x^k y^{(n-k)}}{k!(n-k)!} = \frac{(x+y)^n}{n!},$$

which can be observed to follow alternatively from the binomial theorem.

The complex series

$$e^z = \sum_{n=0}^{\infty} \frac{z^n}{n!}$$

is said to *converge absolutely*, though for complex series this means that the sum of the *moduli* converges. It follows that the Cauchy product can be applied to the expansions of e^w and e^z, and this results in the conclusion of the theorem. ∎

Definition 6.2.7 *If $f(x) = u(x) + iv(x)$ maps an interval I to $\mathbb{C}$ where u and v are real-valued, we say that f is differentiable on I if u and v are differentiable and then we define $f'(x) = u'(x) + iv'(x)$.*

In Exercise 6.12 the reader will show that

$$\frac{de^{(a+ib)x}}{dx} = (a+ib)e^{(a+ib)x}, \tag{6.17}$$

$$\int_c^x e^{(a+ib)t}\, dt = \frac{e^{(a+ib)t}}{(a+ib)}\Big|_{t=c}^x \tag{6.18}$$

for any c and x in I, except that the Equation (6.18) requires $a + ib \neq 0$.

EXERCISES

6.9   Let $z = x + iy$ and $z' = x' + iy'$ be complex numbers. Show that

    a) $\overline{zz'} = \bar{z}\bar{z}'$.

    b) $|zz'| = |z||z'|$.

    c) $|z + z'| \le |z| + |z'|$

    d) $\Re(zz') = xx' - yy'$.

    e) $\Im(zz) = xy' + x'y$.

    f) $\chi_{-n}(x) = \overline{\chi}_n(x)$

6.10   † Use Definition 6.2.2 to prove that a complex sequence $z_n \to z$ if and only both $\Re z_n \to \Re z$ and $\Im z_n \to \Im z$.

6.11   † Use Definition 6.2.3 to prove that a complex series $\sum_{n=0}^{\infty} z_n$ converges to a complex number S if and only *both*

$$\sum_{n=0}^{N} \Re(z_n) \to \Re S \ \text{ and } \ \sum_{n=0}^{N} \Im(z_n) \to \Im S$$

as $N \to \infty$.

6.12   † Prove Equations (6.17) and (6.18) for any real numbers a and b and for any c and x in I. For the second formula assume $a + ib \ne 0$. (Hint: Use Equation (6.9), or else use Euler's formula (Theorem 6.2.1) together with the multiplicative property of the exponential function.)

6.13   Prove Corollary 6.2.2.

6.14   Use Euler's formula together with Theorem 6.2.2 to prove that

$$\cos(x + y) = \cos x \cos y - \sin x \sin y$$

and

$$\sin(x + y) = \sin x \cos y + \cos x \sin y.$$

6.15   Give an example of $f \in R([a, b], \mathbb{C})$ for which $\int_a^b f(x)^2 \, dx < 0$.

6.16   † Show that $\left| \int_a^b f(x) \, dx \right| \le \int_a^b |f(x)| \, dx$ even for $f \in R([a, b], \mathbb{C})$. (See Definition 6.2.5.)

6.17   † Use the result of Exercise 6.12 to prove that for all integers n and m we have

$$\int_0^1 \chi_n(x)\overline{\chi}_m(x) \, dx = \begin{cases} 0 & \text{if } n \ne m, \\ 1 & \text{if } n = m. \end{cases}$$

This equation establishes that the functions $\chi_n(x)$ are what are called *mutually orthogonal characters*. The set $\{\chi_k \mid k \in \mathbb{Z}\}$ is called an *orthonormal set*.

6.18   † Suppose $f \in R[0, 1]$ has period 1, and let $a \in \mathbb{R}$. Define the *a-translate* f_a of f by $f_a(x) = f(x + a)$. Let $k \in \mathbb{Z}$ and define the *k-rotation* $f\chi_k$ of f by

$(f\chi_k)(x) = e^{2\pi ikx}f(x)$. Prove that *the Fourier transform converts translations into rotations, and rotations into translations* in the following sense.

a) $\widehat{f_a}(n) = \chi_n(a)\widehat{f}(n)$.

b) $\widehat{f\chi_k}(n) = \widehat{f}(n-k)$.

6.19   Calculate the Fourier transform $\widehat{f}(n)$ for each $n \in \mathbb{Z}$ if f is a periodic function of period 1 defined *on an interval* of length 1 as shown below. Write also the formal Fourier series $S(f)$ for each given function. (Integration by parts is sometimes helpful for these problems.)

a) Let

$$f(x) = 1_{[a,b]}(x) = \begin{cases} 1 & \text{if } x \in [a,b], \\ 0 & \text{if } x \in [0,1) \setminus [a,b], \end{cases}$$

where $0 \le a < b < 1$.

b) $f(x) = 1 + \cos 2\pi nx + \sin 4\pi nx$ on $[0,1)$.

c) $f(x) = x$ on $[0,1)$.

d) $f(x) = x$ on $\left[-\frac{1}{2}, \frac{1}{2}\right]$.

e) $f(x) = x^2$ on $\left[-\frac{1}{2}, \frac{1}{2}\right)$.

f) $f(x) = e^x$ on $[0,1)$.

6.20   Suppose that f is an even function on $\left[-\frac{1}{2}, \frac{1}{2}\right]$ and g is an odd function on the same domain. Prove that

a) $\widehat{f}$ is even.

b) $\widehat{g}$ is odd.

6.21   For the periodic function of period 1 given by $f(x) = x$ for $0 \le x < 1$, find

$$\lim_{N\to\infty} S_N(x) = \sum_{n=-\infty}^{\infty} \widehat{f}(n)\chi_n(x)$$

at the point $x = 0$. Is the value of this sum the same as $f(0)$? (You can use the data from Exercise 6.19.)

6.22   Suppose that the function $f \in \mathcal{R}[0,1]$ is a periodic function with period 1, and suppose that f is real-valued. Prove that

a) $\widehat{f}(-n) = \overline{\widehat{f}(n)}$.

b) $S_N(f)$ is real-valued for each $N \in \mathbb{N}$.

6.23   The functions $\chi_n(x)$ defined in Equation (6.14) are called *characters*, for which reason the Greek letter χ is used to represent them.[17] Show that $\chi_n(x)$ is a homomorphism of the additive group $\mathbb{R}/\mathbb{Z}$ onto the multiplicative group of complex numbers of modulus 1. That is, show that

$$\chi_n(x+y) = \chi_n(x)\chi_n(y).$$

---

[17]This exercise is only for students who have studied some abstract algebra. It is not necessary for any subsequent topics.

6.3   BESSEL'S INEQUALITY AND l_2

Bessel's inequality is very important in the study of Fourier series, especially with regard to convergence. In order to introduce this inequality, we need the structure of an inner product space for the space of *complex-valued* Riemann integrable functions. In Definition 6.2.5 we defined the integral of any complex-valued function f for which $\Re f$ and $\Im f$ are both Riemann integrable. We say that such functions f are in the space $\mathcal{R}([a, b], \mathbb{C})$. And in Definition 3.4.1 we defined the *inner product* (or *scalar product*) of any two *real-valued* Riemann integrable functions as

$$\langle f, g \rangle = \int_a^b f(x)g(x)\, dx.$$

This definition will not suffice for the purpose of making the Riemann integrable complex-valued functions into an inner product space. The difficulty is that

$$\int_a^b f(x)^2\, dx$$

can be negative and is thus not the square of a norm. We need to show how one can define a special kind of inner product (called a *Hermitian* inner product) on the vector space $\mathcal{R}([a, b], \mathbb{C})$ over the complex numbers. (See Exercise 6.24.)

Definition 6.3.1 *In any complex vector space V (Table 2.1 on page 59, using complex scalars), we call a function*

$$\langle \cdot, \cdot \rangle : V \times V \to \mathbb{C}$$

a Hermitian scalar product[18] *if and only if it has the following three properties.*

  i. $\langle a\mathbf{x} + \mathbf{y}, \mathbf{z} \rangle = a\langle \mathbf{x}, \mathbf{z} \rangle + \langle \mathbf{y}, \mathbf{z} \rangle$ *for all* $a \in \mathbb{C}$ *and for all* $\mathbf{x}$ *and* $\mathbf{y}$ *in V.* (Linearity in the First Variable)

  ii. $\langle \mathbf{x}, \mathbf{y} \rangle = \overline{\langle \mathbf{y}, \mathbf{x} \rangle}$ *for all* $\mathbf{x}$ *and* $\mathbf{y}$ *in V.* (Conjugate Symmetry)

  iii. $\langle \mathbf{x}, \mathbf{x} \rangle \geq 0$ *for all* $\mathbf{x} \in V$ *and* $\langle \mathbf{x}, \mathbf{x} \rangle = 0 \Leftrightarrow \mathbf{x} = \mathbf{0} \in V.$ (Positive Definiteness)

  We show next how to introduce the structure of a Hermitian scalar product in $\mathcal{R}([a, b], \mathbb{C})$.

Definition 6.3.2 *In the vector space $\mathcal{R}([a, b], \mathbb{C})$, we define*

$$\langle f, g \rangle = \int_a^b f(x)\overline{g(x)}\, dx \tag{6.19}$$

---

[18]This may be called also a Hermitian inner product.

and we define the L^2-norm [19] *of $f \in \mathcal{R}([a,b], \mathbb{C})$ by*

$$\|f\|_2 = \sqrt{\langle f, f \rangle} = \left(\int_a^b |f(x)|^2 \, dx \right)^{\frac{1}{2}}. \tag{6.20}$$

We will see shortly that this product is a Hermitian scalar product, with one subtle proviso to be described momentarily. We see that

$$\langle f, f \rangle = \int_a^b |f(x)|^2 \, dx \geq 0$$

since $f(x)\overline{f(x)} = |f(x)|^2$, where the vertical bars in $|f(x)|$ connote the modulus of the complex value of $f(x)$. Also, we see that in the special case in which f and g happen to be real-valued, then the Hermitian inner product $\langle f, g \rangle$ is the same as the inner product defined for real-valued integrable functions in Section 3.4. Properties (i) and (ii) of the definition of a Hermitian inner product are left to the reader to verify in Exercise 6.25. The one property that is not satisfied by $\mathcal{R}([a,b], \mathbb{C})$ is that we could have $\langle f, f \rangle = 0$ without having $f = 0$. We can remedy this deficiency if we agree to understand that two Riemann integrable functions are to be considered *equivalent* provided that the integral of the absolute value of their difference is zero. Thus it is really the *equivalence classes* that are the vectors in the Hermitian inner product space $\mathcal{R}([a,b], \mathbb{C})$.

Definition 6.3.3 *Let f and g be in the space $\mathcal{R}([a,b], \mathbb{C})$. We say that f is equivalent to g, written as $f \sim g$, provided that $\int_a^b |f(x) - g(x)| \, dx = 0$.*

See Exercise 6.28.

Theorem 6.3.1 *In a complex vector space V equipped with a Hermitian scalar product as defined in Definition 6.3.1, we define*

$$\|\mathbf{x}\| = \sqrt{\langle \mathbf{x}, \mathbf{x} \rangle} \tag{6.21}$$

for all vectors $\mathbf{x} \in V$. The function $\| \cdot \|$ as defined in Equation (6.21) is a norm, as in Definition 2.4.4, where scalars c are taken as complex and $|c|$ is interpreted as the modulus of c. Moreover the Cauchy–Schwarz Inequality *is satisfied:*

$$|\langle \mathbf{x}, \mathbf{y} \rangle| \leq \|\mathbf{x}\| \|\mathbf{y}\|.$$

Proof: To prove the Cauchy–Schwarz inequality, we fix $\mathbf{x}$ and $\mathbf{y}$, and we proceed as follows. If $\mathbf{x} = \mathbf{0}$ the Cauchy–Schwarz inequality is trivial. So suppose $\mathbf{x} \neq \mathbf{0}$.

---

[19] This norm is named for Henri Lebesgue, inventor of the Lebesgue integral. Here we use the norm only in the context of the Riemann integral, however.

For all $c \in \mathbb{C}$, observe that $\langle c\mathbf{x} + \mathbf{y}, c\mathbf{x} + \mathbf{y} \rangle \geq 0$ for all c. By linearity of the scalar product in the first variable and conjugate-linearity in the second variable we see that

$$\|\mathbf{x}\|^2 |c|^2 + 2\Re(c\langle \mathbf{x}, \mathbf{y} \rangle) + \|\mathbf{y}\|^2 \geq 0$$

for all $c \in \mathbb{C}$. Let

$$c = -\frac{\overline{\langle \mathbf{x}, \mathbf{y} \rangle}}{\|\mathbf{x}\|^2}.$$

An easy calculation shows that $|\langle \mathbf{x}, \mathbf{y} \rangle|^2 \leq \|\mathbf{x}\|^2 \|\mathbf{y}\|^2$.

The first two conditions of Definition 2.4.4 are easily verified for $\| \cdot \|$. The third condition, the triangle inequality, is left for Exercise 6.27.   ∎

We are ready to introduce Bessel's inequality. The reader should recall from Exercise 6.17 that the characters χ_k with $k \in \mathbb{Z}$ comprise an *orthonormal* set with respect to the Hermitian inner product. Moreover, the Fourier coefficients of a Riemann integrable function f are defined by Equation (6.16) in such a way that

$$\widehat{f}(k) = \langle f, \chi_k \rangle$$

using the Hermitian inner product.

Theorem 6.3.2 (Bessel's inequality) *Let $f \in \mathcal{R}([0, 1], \mathbb{C})$ be a periodic function of period 1. Then we have* Bessel's inequality

$$\sum_{k=-\infty}^{\infty} \left| \widehat{f}(k) \right|^2 \leq \|f\|_2^2, \tag{6.22}$$

with the right side being necessarily finite. Moreover, we have

$$\left\| f - \sum_{k=-n}^{n} \widehat{f}(k)\chi_k \right\|_2^2 = \|f\|_2^2 - \sum_{k=-n}^{n} \left| \widehat{f}(k) \right|^2,$$

so that equality holds in Bessel's inequality if and only if

$$\left\| f - \sum_{k=-n}^{n} \widehat{f}(k)\chi_k \right\|_2 \to 0 \tag{6.23}$$

as $n \to \infty$.

Remark 6.3.1 We will show in Theorem 6.5.1 that the limit shown in (6.23) exists and is zero. Then we will have established *equality* in Bessel's inequality. The resulting equality is called the *Plancherel identity*.

Proof: We calculate *carefully* the following nonnegative Hermitian scalar product:

$$\left\langle f - \sum_{-n}^{n} \widehat{f}(k)\chi_k, f - \sum_{-n}^{n} \widehat{f}(k)\chi_k \right\rangle = \|f\|_2^2 - \sum_{-n}^{n} \overline{\widehat{f}(k)}\widehat{f}(k)$$

$$= \|f\|_2^2 - \sum_{-n}^{n} \left| \widehat{f}(k) \right|^2 \geq 0.$$

Thus the partial sums of the terms $\left|\widehat{f}(k)\right|^2$ are all bounded above by $\|f\|_2^2$, and this yields Bessel's inequality. ∎

Definition 6.3.4 *We define the space l_2 of square-summable complex double sequences as follows:*

$$l_2 = \left\{ c : \mathbb{Z} \to \mathbb{C} \,\middle|\, \sum_{k=-\infty}^{\infty} |c_k|^2 < \infty \right\}.$$

Remark 6.3.2 We should observe that the Fourier transform, denoted by $\mathcal{F} : f \to \widehat{f}$, is a mapping of $\mathcal{R}([0, 1], \mathbb{C})$ *into* l_2. One might hope that the Fourier transform maps $\mathcal{R}([0, 1], \mathbb{C})$ *onto* l_2, but this is not true. It is commonly proven in a graduate course in harmonic analysis that the Fourier transform can be extended to the much larger space of all *Lebesgue measurable square-integrable functions*, which is denoted L^2, and that the Fourier transform is an invertible map from L^2 *onto* l_2. However, the development of the Lebesgue integral requires approximately one semester at the graduate level.

The space l_2 of square-summable double sequences has many interesting properties.

Theorem 6.3.3 *The space l_2 is a vector space over $\mathbb{C}$ and for each pair of elements c and d in l_2, we can define a Hermitian inner product by the formula*

$$\langle c, d \rangle = \sum_{k=-\infty}^{\infty} c_k \bar{d}_k \tag{6.24}$$

and a norm by the formula

$$\|c\|_2 = \sqrt{\langle c, c \rangle} = \left(\sum_{-\infty}^{\infty} |c_k|^2 \right)^{\frac{1}{2}}$$

that satisfies the Cauchy–Schwarz inequality

$$|\langle c, d \rangle| \leq \|c\|_2 \|d\|_2.$$

Proof: The first task is to show that the sum in Equation (6.24) is convergent. In fact, we will show that

$$\left| \sum_{-\infty}^{\infty} c_k \bar{d}_k \right| \leq \sum_{-\infty}^{\infty} |c_k \bar{d}_k| \leq \|c\|_2 \|d\|_2 \tag{6.25}$$

and this will establish absolute convergence. For each $c \in l_2$, define the *truncation*

$$_n c = (c_{-n}, \ldots, c_0, \ldots, c_n) \in \mathbb{C}^{2n+1}.$$

For these truncated sequences there is no convergence problem and it is easy to see that we define a Hermitian scalar product by the formula

$$\langle {}_nc, \, {}_nd \rangle = \sum_{-n}^{n} c_k \bar{d}_k.$$

Theorem 6.3.1 can be applied to the truncated sequences to prove that

$$|\langle {}_nc, \, {}_nd \rangle| \leq \sum_{-n}^{n} |c_k||d_k|$$

$$\leq \sqrt{\sum_{-n}^{n} |c_k|^2} \sqrt{\sum_{-n}^{n} |d_k|^2} \leq \|c\|_2 \|d\|_2$$

for each value of $n \in \mathbb{N}$. This establishes Equation (6.24) and it establishes the Cauchy–Schwarz inequality

$$|\langle c, d \rangle| \leq \|c\|_2 \|d\|_2$$

for l_2 at the same time. It remains only to prove that l_2 is a vector space, which we leave to the reader in Exercise 6.30.   ∎

EXERCISES

6.24   Prove that $\mathcal{R}([a, b], \mathbb{C})$ satisfies all the axioms to be a vector space over the field of complex numbers.

6.25   Prove that $\mathcal{R}([a, b], \mathbb{C})$ satisfies properties (i) and (ii) of Definition 6.3.1.

6.26   † Prove that in a Hermitian inner product space we have
a) $\langle f, cg \rangle = \bar{c}\langle f, g \rangle$, where $c \in \mathbb{C}$.
b) $\langle f, g + h \rangle = \langle f, g \rangle + \langle f, h \rangle$.

6.27   † If $\|x\|$ is defined by means of a Hermitian inner product, prove the Triangle Inequality:

$$\|x + y\| \leq \|x\| + \|y\|.$$

(Hint: Use the Cauchy–Schwarz inequality.)

6.28   Suppose that f, f_1, g, and g_1 are all Riemann integrable complex-valued functions on $[a, b]$ such that $f \sim f_1$ and $g \sim g_1$ as in Definition 6.3.3. Prove that

$$\langle f, g \rangle = \langle f_1, g_1 \rangle.$$

(Hint: Use the Cauchy–Schwarz inequality.)

6.29 † $\diamondsuit^{20}$ Suppose that $f \in \mathcal{R}([0,1], \mathbb{C})$ has period 1 and that $c_1, \ldots, c_n$ are any n complex constants. Then

$$\left\| f - \sum_{k=-n}^{n} \widehat{f}(k)\chi_k \right\|_2 \leq \left\| f - \sum_{k=-n}^{n} c_k \chi_k \right\|_2$$

with equality holding if and only if $c_k = \widehat{f}(k)$ for each k. Hint: Write the difference inside the right-side norm as the sum of two sums of differences, one of these involving $\left(\widehat{f}(k) - c_k\right)\chi_k$.

6.30 † Complete the proof of Theorem 6.3.3 by proving that if c and d are in l_2 and if $\alpha \in \mathbb{C}$ then $\|\alpha c + d\|_2^2 < \infty$. (Hint: Write the square of the norm as a Hermitian scalar product and apply the Cauchy–Schwarz inequality.)

6.31 † $\diamondsuit^{21}$ Prove that l_2 is complete, in the sense that each Cauchy sequence in the l_2-norm converges to a square-summable complex sequence. (Hint: Prove that if $c^{(n)} \in l_2$ is a Cauchy sequence, then $c_k^{(n)}$ converges to some $c_k \in \mathbb{C}$ for each $k \in \mathbb{Z}$. Then prove that $\sum_{-\infty}^{\infty} |c_k|^2 < \infty$ and that $\|c^{(n)} - c\|_2 \to 0$ as $n \to \infty$.)

6.4 UNIFORM CONVERGENCE & RIEMANN LOCALIZATION

In this section we will prove that if a periodic function $f \in C^p(\mathbb{R})$ has period 1, then the Fourier series $S(f)$ *converges uniformly* to f, provided that $1 \leq p \leq \infty$. In other words a periodic function on the circle will have a uniformly convergent Fourier series provided that f has at least a first-order continuous derivative.

Before we state formally and prove the main result, which is Theorem 6.4.2, it is important to remark that if f is not differentiable this theorem can fail. It is known that if f is only continuous but not differentiable, then the Fourier series of f can *diverge* for uncountably many values of x. Divergence is even more erratic behavior than convergence to a value different from $f(x)$, a phenomenon the reader has seen in Exercise 6.21.

The main theorem will require a preliminary result that introduces the *Dirichlet kernel*.

Theorem 6.4.1 *If $f \in \mathcal{R}[0,1]$ is a periodic function of period 1, the nth partial sum S_n of its Fourier series is given by*

$$S_n(x) = \int_{-\frac{1}{2}}^{\frac{1}{2}} f(t)D_n(x-t)\,dt, \tag{6.26}$$

---

[20]This exercise is used in the proof of Theorem 6.5.1, establishing the convergence of Fourier series in the L^2-norm.
[21]This exercise is cited in Remark 6.5.2.

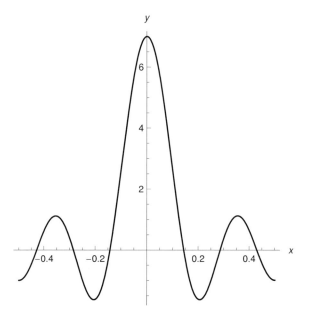

Figure 6.1   Dirichlet kernel D_n for $n = 3$.

where the Dirichlet kernel D_n *is defined by*

$$D_n(x) = \begin{cases} \frac{\sin(2n+1)\pi x}{\sin \pi x} & \text{if } x \notin \mathbb{Z}, \\ 2n + 1 & \text{if } x \in \mathbb{Z} \end{cases} \tag{6.27}$$

and is shown in Fig. 6.1. Also, $\int_{-\frac{1}{2}}^{\frac{1}{2}} D_n(x)\,dx = 1$ for each $n \in \mathbb{N}$.

Remark 6.4.1 The Dirichlet kernel does not converge to zero for x bounded away from the origin. It depends for its work on rapid oscillations in sign to produce cancelations, together with most of its integral being nearly 1 over a small interval around the origin.

Proof: We observe that

$$S_n(x) = \sum_{k=-n}^{n} \left(\int_0^1 f(t) \bar{\chi}_k(t)\,dt \right) \chi_k(x) = \sum_{k=-n}^{n} \int_0^1 f(t) \chi_k(x - t)\,dt$$

$$= \int_0^1 f(t) \sum_{k=-n}^{n} \chi_k(x - t)\,dt = \int_{-\frac{1}{2}}^{\frac{1}{2}} f(t) \left(\sum_{k=-n}^{n} e^{2\pi i k(x - t)}\,dt \right)$$

since the integrand has period 1 and can be integrated with the same result on any interval of length 1 (Exercise 6.4). It will suffice to prove that the sum inside the integrand is the Dirichlet kernel evaluated at $x - t$. We reason as follows using Euler's formula and the sum of a geometric series.

$$\sum_{k=-n}^{n} e^{2\pi i k x} = e^{-2\pi i n x} \sum_{k=0}^{2n} e^{2\pi i k x} = e^{-2\pi i n x} \frac{1 - e^{2\pi i (2n+1) x}}{1 - e^{2\pi i x}}$$

$$= \frac{e^{-2\pi i n x} - e^{2\pi i (n+1) x}}{1 - e^{2\pi i x}} = \frac{e^{-i(2n+1)\pi x} - e^{i(2n+1)\pi x}}{e^{-i\pi x} - e^{i\pi x}}$$

$$= \frac{\sin(2n+1)\pi x}{\sin \pi x} = D_n(x),$$

provided that the denominator in the geometric series formula is not zero, which is equivalent to $x \notin \mathbb{Z}$. If $x \in \mathbb{Z}$, then the sum is clearly $2n + 1$.

Finally, since we have shown above that $D_n(x) = \sum_{k=-n}^{n} e^{2\pi i k x}$, it follows readily that $\int_{-\frac{1}{2}}^{\frac{1}{2}} D_n(x) \, dx = 1$ for each $n \in \mathbb{N}$. ∎

The following famous lemma is very useful.

Lemma 6.4.1 (Riemann–Lebesgue Lemma) *If $f \in \mathcal{R}([0, 1], \mathbb{C})$, then*

$$\widehat{f}(n) \to 0$$

as $|n| \to \infty$.

The proof of this lemma is left to Exercise 6.32.

Definition 6.4.1 *If $f \in \mathcal{R}([a, b], \mathbb{C})$, we define the L^1-norm of f by the formula*

$$\|f\|_1 = \int_a^b |f(x)| \, dx,$$

where the vertical bars indicate the modulus of the complex-valued function f.

If we identify functions f and g as equivalent if $\int_a^b |f(x) - g(x)| \, dx = 0$, then $\| \cdot \|_1$ is a legitimate norm[22] on $\mathcal{R}([a, b], \mathbb{C})$. (See Exercise 6.33.)

Lemma 6.4.2 *Let $f \in C^p(\mathbb{R})$ have period 1, and suppose $1 \leq p < \infty$. Then*

$$\left| \widehat{f}(n) \right| \leq \frac{\left\| f^{(p)} \right\|_1}{(2\pi |n|)^p}. \tag{6.28}$$

---

[22]However, it can be shown that the normed vector space $\mathcal{R}([a, b], \mathbb{C})$ is not complete in the L^1-norm. (See Exercise 6.50.b.) The completion of this space with respect to the L^1-norm requires the Lebesgue integral.

Proof: We begin by applying integration by parts to

$$\widehat{f'}(n) = \int_0^1 f'(x)\overline{\chi_n(x)}\,dx$$

$$= f(x)\overline{\chi_n(x)}\Big|_0^1 - \int_0^1 f(x)\overline{\chi_n}'(x)\,dx$$

$$= 2\pi i n \widehat{f}(n).$$

We iterate this argument a total of p times obtaining

$$\widehat{f}(n) = \frac{\widehat{f^{(p)}}(n)}{(2\pi i n)^p}. \tag{6.29}$$

Finally, we observe that for each function $g \in \mathcal{R}([0,1], \mathbb{C})$ we have

$$|\widehat{g}(n)| = \left|\int_0^1 g(x)\overline{\chi_n}(x)\,dx\right| \le \int_0^1 |g(x)\overline{\chi_n}(x)|\,dx = \|g\|_1.$$

■

Theorem 6.4.2 *Let $f \in C^p(\mathbb{R})$ have period 1, and suppose $1 \le p < \infty$. Then the Nth partial sum*

$$S_N = \sum_{n=-N}^{N} \widehat{f}(n)\chi_n(x)$$

converges uniformly to f on the real line. Moreover,

$$\|S_N - f\|_{\sup} \le KN^{\frac{1}{2}-p}$$

for some constant K that is independent of N but dependent upon f and p.

Proof: It would suffice for the first part of the theorem to give a proof for $p = 1$, but the the first part follows from the inequality that is the second part, and that is what we will prove. Our first step will be to prove that the sequence S_n is Cauchy in the sup-norm. For each $n \in \mathbb{N}$ and for each $m \ge n$ we have

$$|S_m(x) - S_n(x)| \le \sum_{|k| \ge n} \left|\widehat{f}(k)\right|$$

$$\overset{(1)}{\le} \left(\sum_{|k|>n} \left|\widehat{f^{(p)}}(k)\right|^2\right)^{\frac{1}{2}} \left(\sum_{|k|>n} \frac{1}{(2\pi k)^{2p}}\right)^{\frac{1}{2}}$$

$$\overset{(2)}{\le} \left\|f^{(p)}\right\|_2 \left(2\int_n^\infty (2\pi x)^{-2p}\,dx\right)^{\frac{1}{2}}$$

$$= \left\|f^{(p)}\right\|_2 \frac{(2\pi)^{\frac{1}{2}-p}}{\sqrt{\pi(2p-1)}} n^{\frac{1-2p}{2}} \to 0$$

as $n \to \infty$. For inequality (1) we have used Equation (6.29) and the Cauchy–Schwarz inequality for l_2. For inequality (2) we have used both Bessel's inequality and the integral test for infinite series of positive terms. This proves that

$$\|S_m - S_n\|_{\sup} \to 0$$

and S_m is uniformly convergent to some continuous function ϕ. Letting $m \to \infty$, we see that

$$\|\phi - S_n\|_{\sup} \to 0$$

as $n \to \infty$ and that the convergence takes place at the rate claimed.

It remains still to prove that $S_n \to f$ or, in other words, that $\phi = f$. Since uniform convergence is established already, we need prove only pointwise convergence to f. We fix x arbitrarily and observe that

$$S_n(x) - f(x) = \int_{-\frac{1}{2}}^{\frac{1}{2}} f(x+y)D_n(y)\,dy - f(x)\int_{-\frac{1}{2}}^{\frac{1}{2}} D_n(y)\,dy$$

$$= \int_{-\frac{1}{2}}^{\frac{1}{2}} \frac{f(x+y) - f(x)}{\sin \pi y} \sin \pi(2n+1)y\,dy$$

$$= \int_{-\frac{1}{2}}^{\frac{1}{2}} Q(y)\frac{e^{i\pi(2n+1)y} - e^{-i\pi(2n+1)y}}{2i}\,dy,$$

where we have used Euler's formula, and we define

$$Q(y) = \begin{cases} \frac{f(x+y)-f(x)}{\sin \pi y} & \text{if } y \neq 0, \\ \frac{f'(x)}{\pi} & \text{if } y = 0. \end{cases}$$

Next we define

$$Q_+(y) = Q(y)e^{i\pi y} \text{ and } Q_-(y) = Q(y)e^{-i\pi y} \tag{6.30}$$

and the reader will show in Exercise 6.37 that each of these functions is continuous and thus Riemann integrable. Finally, we see that

$$S_n(x) - f(x) = \frac{-i}{2}\left(\widehat{Q_+}(-n) - \widehat{Q_-}(n)\right) \to 0$$

as $n \to \infty$ by the Riemann–Lebesgue lemma. ∎

The following theorem provides very important information about the behavior of the pointwise convergence of Fourier series.

Theorem 6.4.3 (Riemann Localization) *Suppose that f and g are Riemann integrable functions of period 1, and suppose there exists a subinterval $(a, b) \subset [0, 1]$ such that $f(x) = g(x)$ for all $x \in (a, b)$. Then $S_n(f) - S_n(g) \to 0$ uniformly on each interval $[c, d] \subset (a, b)$.*

Remark 6.4.2 The Fourier coefficients $\widehat{f}(n)$ of a Riemann integrable function of period 1 are called *global* objects because they result from an integral of f against a character over the whole domain $[0, 1]$, which we can think of as being the entire circle $\mathbb{R}/\mathbb{Z}$. The word *global* is used to connote the fact that the behavior of f over its entire domain is reflected in the calculation of $\widehat{f}(n)$ for each value of n. On the other hand, a convergence property of the Fourier series $S(f)$ in a very small interval that reflects the behavior of f only in that small interval is called *local* in nature. The *Riemann Localization* Theorem is a striking example of *local-global duality*. The reader should note that the integrable functions f and g could differ as much as we like outside the possibly minute interval (a, b), and also that *neither $S(f)$ nor $S(g)$ need be convergent even pointwise on (a, b)*, and yet $S_n(f) - S_n(g)$ must converge uniformly to zero on any closed proper subinterval of (a, b).

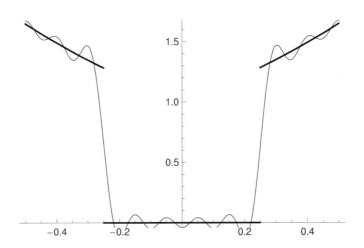

Figure 6.2 S_{10} for $F(x) = e^{|x|}[1 - 1_{(-.25,.25)}]$ on $[-.5, .5)$ with period 1.

Proof: We begin by making a few simplifications that incur no loss of generality. Observe first that $S_n(f) - S_n(g) = S_n(h)$, where $h = f - g$ and $h(x) \equiv 0$ on (a, b). Let $x_0 = \frac{a+b}{2}$ and $r = \frac{b-a}{2}$, so that $(a, b) = (x_0 - r, x_0 + r)$. Our goal is to prove that $S_n(h) \to 0$ uniformly on each $[c, d] \subset (x_0 - r, x_0 + r)$. As a final simplification, we would like to represent the circle $\mathbb{R}/\mathbb{Z}$ as $[-\frac{1}{2}, \frac{1}{2})$ and it would be convenient to have $x_0 = 0$. To this end, define $F(x) = h(x + x_0) = h_{x_0}(x)$. Thus $F \equiv 0$ on $(-r, r)$. If $[c, d] \subset (a, b)$ then there exists $\delta \in (0, r)$ such that $[x_0 - \delta, x_0 + \delta] \supseteq [c, d]$. If we can prove that $S_n(F) \to 0$ uniformly on $[-\delta, \delta]$ then $S_n(h) \to 0$ uniformly on $[c, d]$ since $S_n(h)$ is simply an x_0-translation of $S_n(F)$. (An example of a function such as F in this proof is shown in Fig. 6.2 together with the tenth partial sum of its Fourier series.)

We proceed with the proof of the Riemann Localization theorem. Recall that

$$S_n(x) = \int_{-\frac{1}{2}}^{\frac{1}{2}} F(x+y)D_n(y)\,dy$$

$$= \int_{-\frac{1}{2}}^{\frac{1}{2}} \frac{F(x+y)}{\sin \pi y} \sin(2n+1)\pi y\,dy.$$

We define *two* functions F^+ and F^- by

$$F^{\pm}(y) = \frac{F(x+y)e^{\pm i\pi y}}{2i \sin \pi y}$$

for all $x \in \left[-\frac{1}{2}, \frac{1}{2}\right)$. If $|x| \leq \delta$, then since $F \equiv 0$ on $(-r, r)$, we have $F^{\pm}(y) \equiv 0$ if $|y| < r - \delta = \delta'$. It follows that $F^{\pm} \in \mathcal{R}\left(\left[-\frac{1}{2}, \frac{1}{2}\right], \mathbb{C}\right)$. We calculate from Euler's formula that

$$S_n(x) = \widehat{F^+}(-n) - \widehat{F^-}(n) \to 0$$

as $n \to \infty$ by the Riemann–Lebesgue lemma.

It remains to prove that the convergence of S_n to 0 is uniform on $[-\delta, \delta]$. If x_1 and x_2 are in $[-\delta, \delta]$, then

$$|S_n(x_2) - S_n(x_1)| \leq \int_{\frac{\delta'}{2} \leq |y| \leq \frac{1}{2}} |F(x_2+y) - F(x_1+y)||D_n(y)|\,dy$$

$$\leq \frac{1}{\sin \frac{\delta'\pi}{2}} \|F_{x_2} - F_{x_1}\|_1$$

independent of n. Moreover, the right-hand side is uniformly continuous as a function of x_1 and/or x_2 in the circle, $\mathbb{R}/\mathbb{Z}$, which is compact. (Here we use Exercise 6.39.) Thus for each $\epsilon > 0$ there exists $\eta > 0$ such that $|x_2 - x_1| < \eta$ implies

$$\|F_{x_2} - F_{x_1}\|_1 < \frac{\epsilon}{2} \sin \frac{\delta'\pi}{2}$$

for all x_1 and x_2 in the circle $\mathbb{R}/\mathbb{Z}$. Now we lay out an η-*net* of points

$$-\delta = x_1 < x_2 < \cdots < x_n = \delta$$

such that $x_k - x_{k-1} < \eta$ for each k. We pick a value of $n \in \mathbb{N}$ large enough so that $n \geq N$ implies $S_n(x_k) < \frac{\epsilon}{2}$ for each $k = 1, \ldots, n$. Then $n \geq N$ implies that $|S_n(x)| < \epsilon$ for all x such that $|x| < \delta$. ∎

Remark 6.4.3 In Exercise 6.40 the reader will use the Riemann localization theorem to show that if an integrable function of period 1 is smooth on an interval but not globally smooth, there will still be uniform convergence on a closed subinterval of the integral of smoothness, provided one keeps at least some positive distance $\delta > 0$ away from the endpoints. See Fig. 6.3. Notice that just before and just after each

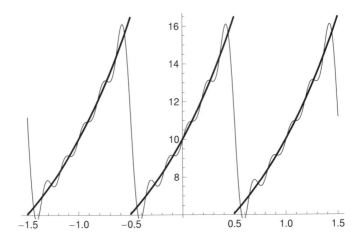

Figure 6.3 S_{10} for $f(x) = 10e^x$, $-0.5 \le x < 0.5$, with period 1.

long vertical jump, the Fourier approximation S_5 seems to overshoot the function by a larger amount than is the case well within the interval of smoothness. That overshoot near the jump is known as *Gibbs' phenomenon*. By using a Fourier approximation S_n with larger values of n, one can narrow the support of that large overshoot, confining it closer in the x-coordinate to the jump point, but one cannot eliminate the greater overshoot near that jump point. Gibbs' phenomenon is visible also in Fig. 6.2.

EXERCISES

6.32   † Prove Lemma 6.4.1. (Hint: The result follows quickly from the fact that $\widehat{f} \in l_2$, because of Bessel's inequality.)

6.33   Prove that $\| \cdot \|_1$ in Definition 6.4.1 has all the properties required to be a norm on the vector space of equivalence classes of Riemann integrable functions $\mathcal{R}([a, b], \mathbb{C})$.

6.34   If f and g are functions of period 1 in $C^1([0, 1], \mathbb{C})$ and if $\widehat{f}(k) = \widehat{g}(k)$ for all $k \in \mathbb{Z}$, prove that $f(x) = g(x)$ for all x.

6.35   Find the Fourier series that converges uniformly to the periodic function $f(x) = \cos 2\pi x + \sin 4\pi x \sin 6\pi x$ of period 1.

6.36   Prove that an absolutely convergent Fourier series must be uniformly convergent as well.

6.37   † Show that the functions Q_+ and Q_- in Equation (6.30) are continuous.

6.38   If f is a Riemann integrable function of period 1, prove that $f \in C^\infty([0, 1], \mathbb{C})$ if and only if $k^n \widehat{f}(k) \to 0$ as $k \to \infty$ for all $n \in \mathbb{N}$.

6.39 † If $f \in \mathcal{R}([0, 1), \mathbb{C})$ is a function of period 1, and if we denote the t-translate of f by $f_t(x) = f(x+t)$ then the function $\phi(t) = \|f_t - f\|_1$ is a continuous function of $t \in \mathbb{R}/\mathbb{Z}$, the circle. (Hint: Let $\epsilon > 0$. Show that there is a step function σ such that $\|f - \sigma\|_1 < \frac{\epsilon}{3}$. Next use a double application of the triangle inequality for the L^1-norm.)

6.40 ◇ Suppose $f \in C^1((a, b), \mathbb{C})$ for some $(a, b) \subset [0, 1]$ and that f is a Riemann integrable function of period 1. Prove that the Fourier series $S(f)$ converges uniformly to f on each proper closed subinterval $[c, d] \subset (a, b)$. See Fig. 6.3. Note that in the figure $a = -0.5$ and $b = 0.5$ and $[a, b)$ is also a valid domain for the study of functions of period 1. (Hint: Use Exercise 5.62 to create a smooth periodic function that agrees with f on (a, b). Apply Theorem 6.4.3.)

6.5 L^2-CONVERGENCE & THE DUAL OF l^2

We have shown in Equation (6.22) that if $f \in \mathcal{R}([0, 1], \mathbb{C})$ is a function of period 1, then the double sequence $\widehat{f}(n)$ is in l^2, and

$$\sum_{-\infty}^{\infty} \left|\widehat{f}(n)\right|^2 \leq \|f\|_2^2$$

with equality holding if and only if $\|S_n(f) - f\|_2 \to 0$ as $n \to \infty$. The case of equality in Bessel's inequality is called the *Plancherel identity*, which we will establish with Theorem 6.5.1. The Plancherel identity can be interpreted as an infinite-dimensional version of the Pythagorean Theorem for $\mathcal{R}[0, 1]$, which we interpret as an infinite-dimensional Hermitian inner product space, utilizing the equivalence relation $f \sim g$ in $\mathcal{R}[0, 1]$ if and only if $\int_0^1 |f(x) - g(x)|^2 \, dx = 0$. Some applications of this identity are given in Exercise 6.44.

Theorem 6.5.1 *If $f \in \mathcal{R}([0, 1], \mathbb{C})$ is a function of period 1, then*

$$\|S_n(f) - f\|_2 \to 0$$

as $n \to \infty$. Consequently we have the Plancherel identity:

$$\sum_{-\infty}^{\infty} \left|\widehat{f}(n)\right|^2 = \|f\|_2^2. \tag{6.31}$$

Remark 6.5.1 A celebrated theorem of Lennart Carleson [5] established that the Fourier series of any square-integrable Lebesgue measurable function must converge to $f(x)$ pointwise except on a set of Lebesgue measure zero,[23] and that theorem does apply to all the functions covered by our theorem above. However, a set of points

---

[23]The reader will not need to know about Lebesgue measure here. The definition of a set of Lebesgue measure zero—known also as a *Lebesgue null set*—is, however, provided in this book as Definition 11.2.1.

can have Lebesgue measure zero and still be an uncountably infinite set. There are examples known of continuous functions f for which $S_n(f)$ is actually *divergent* for infinitely many values of x. And there is an example of a Lebesgue integrable function f for which the Fourier series diverges at each point x! The extraordinary pathologies of Fourier series in regard to pointwise convergence, even for continuous functions, make theorems like the one we are about to prove very interesting and useful.

Proof: Suppose first that f is real-valued. By Exercise 3.29, there exist step functions $\sigma(x) \le f(x) \le \sigma'(x)$ for all $x \in [a, b]$, so that

$$\int_a^b \sigma(x)\,dx \le \int_a^b f(x)\,dx \le \int_a^b \sigma'(x)\,dx$$

such that $|\int_a^b \sigma'(x)\,dx - \int_a^b \sigma(x)\,dx| < \epsilon$, implying also that

$$\int_a^b |f(x) - \sigma(x)|\,dx < \epsilon \quad \text{and} \quad \int_a^b |f(x) - \sigma'(x)|\,dx < \epsilon.$$

This follows immediately from the use of the upper sums and the lower sums from the Darboux integrability criterion. These sums are integrals of step functions with heights corresponding to the infimum and the supremum of f on the intervals of a partition.

Consider next a step function having only two values, each on a subinterval of strictly positive length. Let p be the point at which the single jump discontinuity occurs. By Exercise 5.62 we know that for each $\delta > 0$ there exists a function $\phi \in C^\infty[0, 1]$ such that $\phi(x) = \sigma(x)$ except on an interval $(p - \delta, p + \delta)$. In effect, we are connecting the two steps with a smooth (C^∞) curve which departs from the lower step very near the point of jump discontinuity and joins the upper step smoothly only slightly to the opposite side of the jump discontinuity. This permits us to make $\|\phi - \sigma\|_1$ as small as we like.

For general step functions we can iterate the process just described at each of the finitely many jump discontinuities of σ. Moreover, we can do this keeping $\|\phi\|_{\sup} = \|\sigma\|_{\sup} \le \|f\|_{\sup}$. (If f is complex-valued, the same approximations can be produced by working separately with the real and imaginary parts, $\Re(f)$ and $\Im(f)$.) In this manner we establish that there exists a sequence of functions $\phi_n \in C^\infty[0, 1]$ with period readily adjusted to be 1, and having the properties

$$\|f - \phi_n\|_1 \to 0, \text{ and } \|\phi_n\|_{\sup} \le \|f\|_{\sup} = M < \infty$$

for all n. It follows that

$$\|f - \phi_n\|_2^2 = \int_0^1 |f - \phi_n|^2\,dx \le 2M\|f(x) - \phi_n(x)\|_1 \to 0$$

as $n \to \infty$. By Theorem 6.4.2 we know that $S_k(\phi_n) \to \phi_n$ uniformly on $[0, 1]$ as $k \to \infty$. By Exercise 6.41 it follows that $S_k(\phi_n) \to \phi_n$ as $k \to \infty$ in the

L^2-norm. Thus for each $\epsilon > 0$ there exists a number K such that $k \geq K$ implies that $\|f - S_k(\phi)\|_2 < \epsilon$. By Exercise 6.29, the Fourier coefficients of f provide optimal L^2 approximation to f. Thus

$$\|f - S_k(f)\|_2 \leq \|f - S_k(\phi)\|_2 < \epsilon$$

for all $k \geq K$. This implies the theorem. ∎

Remark 6.5.2 Thanks to Theorem 6.5.1, we know that if f and g are in $\mathcal{R}([0, 1], \mathbb{C})$ and if $\hat{f} = \hat{g}$, then $\|f - g\|_2 = 0$. (See Exercise 6.5.2.) And Exercise 6.43 tells us that the mapping $\mathcal{F} : f \to \hat{f}$ is a linear map of L^2-equivalence classes that preserves the Hermitian scalar product. It can be shown, however, that $\mathcal{R}([0, 1], \mathbb{C})$ is *not* complete in the L^2-norm. (See Exercises 6.48, 6.49, and 6.50.) On the other hand, the reader has shown in Exercise 6.31 that l^2 is complete in its l^2-norm. In a graduate course in Lebesgue measure and integration, it is customary to prove that the space of square-integrable Lebesgue measurable functions is a complete Hermitian inner product space, called $L^2([0, 1], \mathbb{C})$. The latter space is very much larger than the space $\mathcal{R}([0, 1], \mathbb{C})$, although the Lebesgue integral agrees with the Riemann integral for every Riemann integrable function. What is different is that the Lebesgue integral is capable of integrating many functions that are not Riemann integrable.[24]

We will prove in Theorem 6.5.2 that l^2 is a *self-dual* complete Hermitian inner product space. The space $L^2([0, 1], \mathbb{C})$ can be proven to be a self-dual complete Hermitian inner product space as well, and $\mathcal{F}$ is an isomorphism of $L^2([0, 1], \mathbb{C})$ onto l^2. Complete Hermitian inner product spaces are called *Hilbert spaces* and they are always self-dual. Hilbert spaces play an especially important role in Fourier analysis. In Exercise 8.15 the reader will show that finite-dimensional Euclidean spaces are always self-dual. This means that l^2 (and L^2) are natural infinite-dimensional analogues of Euclidean space.

Theorem 6.5.2 *A function $T : l^2 \to \mathbb{C}$ is a bounded linear functional if and only if there is an element $d \in l^2$ such that $Tc = \langle c, d \rangle$ for each $c \in l^2$. Moreover, $\|T\| = \|d\|_2$.*

Proof: In Theorem 6.3.3 we have proven that *if $Tc = \langle c, d \rangle$ for each $c \in l^2$* then $|Tc| \leq \|d\|_2 \|c\|_2$. This shows that T is a bounded linear functional, with $\|T\| \leq \|d\|_2$. Now we will prove the converse, and the equality of norms.

For each $j \in \mathbb{Z}$, define the vector $e^{(j)} \in l^2$ by letting its kth coordinate be

$$e_k^{(j)} = \delta_{jk} = \begin{cases} 1 & \text{if } j = k, \\ 0 & \text{if } j \neq k, \end{cases}$$

---

[24]Theorem 6.5.1 can be considered a version of the Riesz-Fischer Theorem, restricted according to the limitations of the Riemann integral. As stated originally by Riesz, this theorem said that an l_2-sequence of coefficients corresponds to an L^2-convergent series of functions in an expansion using an orthonormal basis, such as the *trigonometric* basis $\{e^{2\pi i n x}\}$ used for Fourier series.

where δ_{jk} is called the Kronecker delta. If $c \in l^2$, then $c = \sum_{k=-\infty}^{\infty} c_k e^{(k)}$ because

$$\left\| c - \sum_{|k| \leq n} c_k e^{(k)} \right\|_2 = \left\| \sum_{|k| > n} c_k e^{(k)} \right\|_2 \rightarrow 0$$

as $n \rightarrow \infty$ for any square summable sequence c.

Now let T be any bounded linear functional on l^2, and define $d_n = \overline{T\left(e^{(n)}\right)}$ for each n. It follows from the linearity and the continuity of T that

$$T\left(\sum_{|k| \leq n} c_k e^{(k)} \right) = \sum_{|k| \leq n} c_k \bar{d}_k \rightarrow \sum_{k=-\infty}^{\infty} c_k \bar{d}_k = T(c)$$

as $n \rightarrow \infty$. Hence

$$T(c) = \langle c, d \rangle \tag{6.32}$$

for all $c \in l^2$. We need to prove that $d \in l^2$. The nth *truncation* of d is given by

$$_n d = \sum_{|k| \leq n} d_k e^{(k)} \in l^2$$

for each n. Hence $|T(_n d)| \leq \|T\| \|_n d\|_2$, which implies by Equation (6.32) that

$$\|_n d\|_2^2 \leq \|T\| \|_n d\|_2 \leq \|T\| \|d\|_2$$

for each n. Thus

$$\|d\|_2^2 \leq \|T\| \|d\|_2,$$

which implies that $\|d\|_2 \leq \|T\| < \infty$. Hence $d \in l^2$. Moreover,

$$|Tc| = |\langle c, d \rangle| \leq \|c\|_2 \|d\|_2,$$

by the Cauchy–Schwarz inequality, which implies that $\|T\| \leq \|d\|_2$, which implies that

$$\|T\| = \|d\|_2.$$

$\blacksquare$

EXERCISES

6.41   Let $S_k(\phi)$ denote the kth partial sum of the Fourier series for a Riemann integrable function ϕ. Prove that if $\|S_k(\phi) - \phi\|_{\sup} \rightarrow 0$ as $k \rightarrow \infty$, then

$$\|S_k(\phi) - \phi\|_2 \rightarrow 0$$

as $k \rightarrow \infty$, thereby completing the proof of Theorem 6.5.1.

6.42   If f and g in $\mathcal{R}([0, 1], \mathbb{C})$ and if $\hat{f} = \hat{g}$, then $\|f - g\|_2 = 0$.

6.43   Prove the following generalization of the Plancherel identity, called *Parseval's identity*: for each f and g in $\mathcal{R}([0,1], \mathbb{C})$ we have

$$\langle f, g \rangle = \langle \widehat{f}, \widehat{g} \rangle.$$

Parseval's identity shows that the Fourier transform preserves the Hermitian scalar product. (Hint: Apply the Plancherel identity to $\langle f + g, f + g \rangle$ to prove that

$$\Re\langle f, g \rangle = \Re\langle \widehat{f}, \widehat{g} \rangle.$$

Then do something similar with if to obtain the desired conclusion.)

6.44   Use Exercise 6.19 together with the Plancherel identity [Equation (6.31)] to find the sums of the following infinite series.

a) $\sum_{n \in \mathbb{N}} \frac{1}{n^2}$

b) $\sum_{n \in \mathbb{N}} \frac{1}{(2n)^2}$

c) $\sum_{n \in \mathbb{N}} \frac{1}{(2n-1)^2}$

6.45   Suppose that $f \in \mathcal{R}[0,1]$ is a function of period 1. If $\langle f, \chi_n \rangle = 0$ for each $n \in \mathbb{Z}$, prove that $\| f \|_2 = 0$ so that f is equivalent to zero in the normed vector space of equivalence classes in $\mathcal{R}[0,1]$ with the L^2-norm.

6.46   Let c be a nonzero element of l^2. Denote

$$c^{\perp} = \left\{ d \in l^2 \mid \langle d, c \rangle = 0 \right\},$$

the orthogonal complement of the one-dimensional subspace $\mathbb{C}c$ of l^2.

a) Prove that $c^{\perp}$ is a vector subspace of $\mathcal{R}[0,1]$.

b) Prove that each $x \in l^2$ has a *unique* decomposition $x = zc + d$, where $z \in \mathbb{C}$ and $d \in c^{\perp}$.

c) Prove that there exists a bounded linear functional T on l^2 such that $T(c) = 1$ and $T : c^{\perp} \to \{0\}$.

d) Suppose now that c and d are any two nonzero elements of l^2, and suppose further that neither element is a scalar multiple of the other. (That is, the two are linearly independent.) Prove that there exists a bounded linear functional $T : l^2 \to \mathbb{C}$ such that $T(c) \neq T(d)$.

6.47   ◇ It is known that the Fourier series of a continuous, periodic function need not converge uniformly: in fact it can diverge for an uncountably infinite set of values of x, though it is not easy to give an example and we do not do so here. On the other hand, every uniformly convergent Fourier series must converge to a continuous function. The following steps will establish Fejer's Theorem: Every continuous function of period 1 is the uniform limit of the Cesaro means (Exercise 1.56) of the partial sums of its Fourier series.

a) Define the nth Fejer kernel F_n to be the average of the first n Dirichlet kernels. (See Fig. 6.4.) Fill in the missing steps in the following calculation:

$$F_n(x) = \frac{1}{n} \sum_{k=0}^{n-1} D_k(x) = \frac{1}{n} \sum_{0}^{n-1} \frac{\sin(2k+1)\pi x}{\sin \pi x}$$

$$= \frac{1}{n \sin \pi x} \Im \left(\sum_{0}^{n-1} \left(e^{i\pi x} \right)^{2k+1} \right)$$

$$= \frac{1}{n} \left[\frac{\sin n\pi x}{\sin \pi x} \right]^2.$$

b) Prove that

$$\int_{-\frac{1}{2}}^{\frac{1}{2}} F_n(x) \, dx = 1.$$

c) Prove that if $0 < \delta < \frac{1}{2}$, then $F_n \to 0$ uniformly on the domain

$$\delta \leq |x| \leq \frac{1}{2}.$$

d) Define

$$\sigma_n(x) = \frac{1}{n} \sum_{k=0}^{n-1} S_k(x)$$

and prove that if f is continuous of period 1, then

$$\sigma_n(x) - f(x) = \int_{-\frac{1}{2}}^{\frac{1}{2}} [f(x+y) - f(x)]F_n(y) \, dy \to 0$$

uniformly in x as $n \to \infty$.

e) Use Fejer's theorem to give an alternative proof for the Weierstrass Polynomial Approximation Theorem (Theorem 5.8.1).

6.48   Let

$$f_n(x) = 1_{\left[\frac{1}{n},1\right]}(x)\frac{1}{\sqrt[4]{x}}$$

for each n so that $f_n \in \mathcal{R}[0,1]$. Prove that the sequence f_n is Cauchy in the L^2-norm. Prove that if $f \in \mathcal{R}[0,1]$, then $\|f - f_n\|_2$ fails to converge to zero, so that $\mathcal{R}[0,1]$ is not complete in the L^2-norm.

6.49   ◇ Let x_n be a sequence of all the rational numbers in $(0,1)$. Thus the sequence x_n is dense in $[0,1]$. Let

$$f_n(x) = \begin{cases} \dfrac{1}{\sqrt[4]{|x-x_n|}}, & x \neq x_n, \\ 0, & x = x_n \end{cases}$$

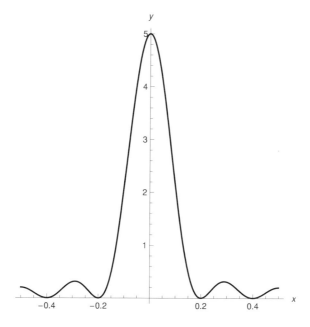

Figure 6.4   Fejer kernel $F_5(x)$.

and define

$$F_N(x) = \min \left(N, \sum_{n=1}^{N} \frac{1}{2^n} f_n(x) \right)$$

for each $N \in \mathbb{N}$. Prove that F_N is Riemann integrable on $[0, 1]$ for each N, and $\{F_N\}$ is a Cauchy sequence in the L^2-norm, but that there is no $f \in \mathcal{R}[0, 1]$ such that $\|F_N - f\|_2 \to 0$. Hence $\mathcal{R}[0, 1]$ is not complete in the L^2-norm.

6.50   ◇ Following Example 1.15, construct an open *dense* set

$$O = \bigcup_{n \in \mathbb{N}} O_n, \text{ with } O_n = \bigcup_{k=1}^{n} (a_k, b_k) \subset [0, 1]$$

such that

$$\sum_{k \in \mathbb{N}} |b_k - a_k| < \frac{1}{2}.$$

Let $f_n = 1_{O_n}$.

    a) Prove that f_n is a sequence of Riemann integrable functions that is also a Cauchy sequence in the L^2-norm, but that there is no $f \in \mathcal{R}[0, 1]$ such that $\|f_n - f\|_2 \to 0$ as $n \to \infty$. Hence $\mathcal{R}[0, 1]$ is not complete in the L^2-norm.

b) Show also that $\mathcal{R}[0, 1]$ is not complete in the L^1-norm.

6.6   TEST YOURSELF

EXERCISES

6.51   Suppose f is an *even* Riemann integrable function on the real line with period equal to 1. If

$$\int_{-\frac{1}{2}}^{1} f(x)\,dx = 3,$$

then find

a) $\int_{\frac{1}{2}}^{1} f(x)\,dx$

b) $\int_{0}^{1} f(x)\,dx$ **c)** $\int_{0}^{1} f(x)\sin 2\pi x\,dx$

6.52   If

$$f(x) = \sum_{n\in\mathbb{N}} \frac{\sin 2\pi n x}{n^2},$$

then find

a) $\int_{0}^{1} f(x)\cos 6\pi x\,dx$

b) $\int_{0}^{1} f(x)\sin 6\pi x\,dx$

6.53

a) Find the numerical value of

$$\frac{e^{\frac{i\pi}{3}} + e^{-\frac{i\pi}{3}}}{2}.$$

b) Find the numerical value of

$$e^{\frac{i\pi}{6}} - e^{-\frac{i\pi}{6}}.$$

c) Find the numerical value of

$$\Im\left(e^{\frac{i\pi}{3}} - e^{\frac{-i\pi}{6}}\right).$$

6.54   Find the *Hermitian* inner product (Definition 6.19) $\langle f, g\rangle$ if $f(x) = x$ and $g(x) = ix$ for all $x \in [0, 1)$, both functions extended to $\mathbb{R}$ so as to be periodic with period 1.

6.55   True or False: If $\langle \cdot, \cdot \rangle$ is a Hermitian inner product on a complex vector space V, then

$$\langle \mathbf{x}, c\mathbf{y} + \mathbf{z}\rangle = c\langle \mathbf{x}, \mathbf{y}\rangle + \langle \mathbf{x}, \mathbf{z}\rangle$$

for all $c \in \mathbb{C}$.

6.56 Suppose that $f \in \mathcal{R}\left[-\frac{1}{2}, \frac{1}{2}\right]$. Decide which of the terms *real-valued, pure imaginary-valued,*[25] *even, and odd* apply correctly to $\widehat{f}(n)$ if

    a) f is both real-valued and even.

    b) f is both real-valued and odd.

6.57 Let $f(x) = \cos^3 2\pi x$.

    a) Find the trigonometric Fourier series expansion (in terms of $\cos 2\pi n x$ and $\sin 2\pi n x$) for the function $f(x)$.

    b) Find $\int_0^1 f(x)\cos 2\pi x\, dx$.

6.58 Suppose the sequence of partial sums

$$S_n = \sum_{k=-n}^{n} c_n e^{2\pi i n x}$$

converges uniformly. True or False: The sum $\sum_{n\in\mathbb{Z}} |c_n|^2 < \infty$.

6.59 Use Exercise 6.19 together with the Plancherel identity [Equation (6.31)] to find the numerical value of the sum of each of the following infinite series:

    a)

$$\sum_{1}^{\infty} \frac{1}{n^4}.$$

    b)

$$\sum_{n\in\mathbb{N}\backslash 3\mathbb{N}} \frac{1}{n^4},$$

    where $3\mathbb{N} = \{3n \mid n \in \mathbb{N}\}$.

---

[25]Being pure imaginary-valued means being a real number times i. This is a special subset of the complex numbers, often denoted $i\mathbb{R}$.

CHAPTER 7

THE RIEMANN–STIELTJES INTEGRAL[26]

The Riemann integral of a function f provides a continuous analog of the process of summation of numerical values $f(\bar{x}_i)$, with each such value weighted by the *width* Δx_i of the interval $[x_{i-1}, x_i]$ from which $\bar{x}_i$ is selected. There are many reasons for generalizing this concept to allow for the weighting of the numerical values $f(\bar{x}_i)$ by numbers different from Δx_i. The Riemann–Stieltjes integral allows for the replacement of Δx_i by $\Delta g_i = g(x_i) - g(x_{i-1})$, where g is a function of *bounded variation*. The concept of bounded variation is explained in the first section.

A good example to have in mind would be the probabilistic expectation of a game of chance in which there is a winning given by $f(x)$ if a random number turns out to be x. However, x may be more likely to be in some intervals than in others, and this difference is measured by the function g.

The Riemann integral is an example of a bounded linear functional on the normed vector space $\mathcal{C}[a, b]$. The Riesz Representation Theorem will establish that every such bounded linear functional comes from a Riemann–Stieltjes integral with respect to a suitable function g of bounded variation.

---

[26]This chapter is not required for any subsequent chapters.

7.1   FUNCTIONS OF BOUNDED VARIATION

The idea behind the concept of the *total variation* of a function f defined on $[a, b]$ is to measure the extent to which the values of f oscillate up and down. One can think of it as being an *odometer* measurement of the amount of vertical travel, with all distances counted as positive.

Definition 7.1.1 *Let* $f : [a, b] \to \mathbb{R}$.

   i. If $P = \{a = x_0 < x_1 < \cdots < x_n = b\}$ *is any partition of* $[a, b]$, *we define*

$$P(f) = \sum_{k=1}^{n} |f(x_k) - f(x_{k-1})|.$$

   ii. We define the total variation *of* f *on* $[a, b]$ *by*

$$V_a^b(f) = \sup_{P} \{P(f)\}$$

   where the supremum is taken over all partitions *of* $[a, b]$.

   iii. Since f *is also defined on the interval* $[a, x]$ *for all* $x \in [a, b]$, *we define* $V_a^x(f)$, *the* total variation function, *to be the total variation of* f *on the interval* $[a, x]$. *This is understood to be a function of* x.

   iv. We say f *has* bounded variation *on* $[a, b]$, *denoted by* $f \in \mathcal{BV}[a, b]$, *if and only if* $V_a^b(f) < \infty$.

■ **EXAMPLE 7.1**

Suppose f is *monotone* on [a,b]. Then we claim $V_a^b(f) = |f(b) - f(a)|$, so that $f \in \mathcal{BV}[a, b]$.

We prove this first for f increasing on $[a, b]$. Then for all P we have

$$P(f) = \sum_{k=1}^{n} |f(x_k) - f(x_{k-1})| = \sum_{k=1}^{n} [f(x_k) - f(x_{k-1})]$$
$$= f(b) - f(a) = |f(b) - f(a)|.$$

Theorem 7.1.1 *Suppose* $f \in \mathcal{C}[a, b]$ *and* f' *is* bounded *at least on* (a, b). *Then* $f \in \mathcal{BV}[a, b]$ *and* $V_a^b(f) \le \|f'\|_{\sup}(b - a)$.

Proof: We apply the Mean Value Theorem for derivatives as follows. For all P, we have

$$|P(f)| = \sum_{k=1}^{n} |f(x_k) - f(x_{k-1})| = \sum_{k=1}^{n} |f'(\bar{x}_k)| \Delta x_k$$
$$\le \|f'\|_{\sup} \sum_{k=1}^{n} \Delta x_k = \|f'\|_{\sup}(b - a).$$

Here the Mean Value Theorem guarantees the *existence* of suitable points $\bar{x}_k$ in $[x_{k-1}, x_k]$. ∎

■ EXAMPLE 7.2

Let

$$f(x) = \begin{cases} x^2 \sin\left(\frac{\pi}{x}\right) & \text{if } x \neq 0, \\ 0 & \text{if } x = 0. \end{cases}$$

We claim $f \subset \mathcal{BV}[0, 1]$. See Fig. 7.1.

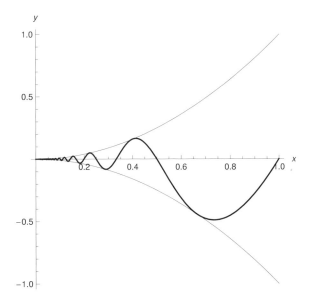

Figure 7.1 $f(x) = x^2 \sin\left(\frac{\pi}{x}\right)$, with envelope $u(x) = x^2$, $l(x) = -x^2$.

In fact, we showed in Exercise 4.7 that $f'(0) = 0$ and that for all $x \neq 0$ we have $f'(x) = 2x \sin\left(\frac{\pi}{x}\right) - \pi \cos\left(\frac{\pi}{x}\right)$. Thus on $[0, 1]$ we have

$$\|f'\|_{\sup} \leq 2 + \pi < \infty$$

so $f \in \mathcal{BV}[0, 1]$, as claimed.

Theorem 7.1.2 *Let $f \in \mathcal{BV}[a, b]$, and let $a \leq x \leq y \leq b$. Then*

$$V_a^y(f) = V_a^x(f) + V_x^y(f).$$

Proof:

i. First we will prove that

$$V_a^y(f) \leq V_a^x(f) + V_x^y(f).$$

So let P be any partition of $[a, y]$. It is possible that $x \notin P$, so let $P^* = P \cup \{x\}$. Then $P^* = P' \cup P''$, where P' is a partition of $[a, x]$ and P'' is a partition of $[x, y]$. By Exercise 7.4,

$$P(f) \le P^*(f) = P'(f) + P''(f)$$
$$\le V_a^x(f) + V_x^y(f).$$

Now we take the supremum over all P to obtain

$$V_a^y(f) \le V_a^x(f) + V_x^y(f).$$

ii. We will complete the proof by showing that

$$V_a^y(f) \ge V_a^x(f) + V_x^y(f).$$

For this we let P_1 be any partition of $[a, x]$ and P_2 any partition of $[x, y]$, and let $P = P_1 \cup P_2$, a particular partition of $[a, y]$. Then

$$P_1(f) + P_2(f) = P(f) \le V_a^y(f)$$

for all P_1 and P_2. We take the supremum *first* over all P_1, showing that

$$V_a^x(f) \le V_a^y(f) - P_2(f).$$

Then we take the supremum over all P_2, and we find that

$$V_a^x(f) + V_x^y(f) \le V_a^y(f).$$

∎

■ **EXAMPLE 7.3**

Let

$$f(x) = \begin{cases} x \sin\left(\frac{\pi}{x}\right) & \text{if } x \ne 0 \\ 0 & \text{if } x = 0. \end{cases}$$

We claim that $V_0^1(f) = \infty$, so that $f \notin \mathcal{BV}[0, 1]$.

It will help the reader to sketch the graph of this function, indicating its infinitely many oscillations as $x \to 0+$. It is helpful to sketch the lines $y = x$ and $y = -x$ as helper lines. See Fig. 4.1. We observe that the graph of f touches the two helper lines wherever

$$\frac{\pi}{x_k} = \frac{(2k+1)\pi}{2}$$

an *odd* multiple of $\frac{\pi}{2}$. Although these are not extreme points of f, because $f'(x_k) \ne 0$, we can still see that

$$V_{x_{k+1}}^{x_k}(f) \ge |f(x_{k+1}) - f(x_k)|.$$

Moreover,

$$|f(x_k)| = x_k = \frac{2}{2k+1},$$

with the *sign* of $f(x_k)$ *alternating.* Thus

$$V^{x_k}_{x_{k+1}}(f) \geq \frac{2}{2k+1} + \frac{2}{2k+3} = \frac{8k+8}{4k^2+8k+3}.$$

By repeated application of Theorem 7.1.1, we have for all $N \in \mathbb{N}$,

$$V^1_0(f) \geq V^{x_{N-1}}_{x_N}(f) + V^{x_{N-2}}_{x_{N-1}}(f) + \cdots + V^{x_1}_{x_2}(f)$$

$$\geq \sum_{k=2}^{N} \frac{8k+8}{4k^2+8k+3} \to \infty$$

as $N \to \infty$. Hence $V^1_0(f) = \infty$, so $f \notin \mathcal{BV}[0,1]$.

Theorem 7.1.3 *The function $f \in \mathcal{BV}[a,b]$ if and only if there exist two monotone functions, g and h are increasing, such that $f = g - h$ on $[a,b]$.*

Proof:

i. We prove the *if* implication (from right to left) first. In this case, since g and h are monotone, $g, h \in \mathcal{BV}[a,b]$, so $f = g - h \in \mathcal{BV}[a,b]$ as well, by Exercise 7.6.

ii. Now we prove the *only if* part (from left to right). Now suppose $f \in \mathcal{BV}[a,b]$. If we let $g(x) = V^x_a(f)$, Theorem 7.1.1 implies that g is increasing on $[a,b]$. Let $h(x) = g(x) - f(x)$, so $f = g - h$. It suffices to prove that h is increasing on $[a,b]$ as well. So let $x < y$ in $[a,b]$, and we need to show that $h(x) \leq h(y)$. That is, we need to show

$$V^x_a(f) - f(x) \leq V^y_a(f) - f(y),$$

which is equivalent to showing that

$$f(y) - f(x) \leq V^y_a(f) - V^x_a(f) = V^y_x(f)$$

by Theorem 7.1.1 again. However, $P = \{x, y\}$ is a partition of $[x, y]$, so

$$f(y) - f(x) \leq |f(y) - f(x)| = P(f) \leq V^y_x(f).$$

∎

Remark 7.1.1 Theorem 7.1.3 can be called a *representation theorem* for $\mathcal{BV}[a,b]$. What this means is that the theorem shows that a function f lies in the set $\mathcal{BV}[a,b]$ if and only if it is the difference between two monotone increasing functions. In some sense it is easier to understand the concept of a function being monotone increasing than it is to grasp the concept of a function having bounded variation. Thus this representation theorem for $BV[a, b]$ expresses every such function as the difference between two simpler and seemingly more familiar objects. It asserts also that every such difference has bounded variation. However, one may assume too easily that monotone increasing functions are easy to understand! Consider Exercise 7.10, in which the reader will construct a bounded monotone increasing function that has a jump discontinuity at each rational number on the x-axis. It is very difficult to picture the graph. See Exercise 2.12 for the definition of a jump discontinuity.

EXERCISES

7.1   † Suppose f is decreasing on $[a, b]$. Prove $V_a^b(f) = |f(b) - f(a)|$, so that $f \in BV[a, b]$.

7.2   Prove: If $f \in BV[a, b]$, then f is *bounded* on $[a, b]$.

7.3   Let $f(x) = \sin(x^{100})$. Prove that $f \in BV[0, 10]$.

7.4   † Let P be any partition of $[a, b]$, $x' \in [a, b]$, and $f : [a, b] \to \mathbb{R}$. Let $P^* = P \cup \{x'\}$. Prove: $P(f) \le P^*(f) \le V_a^b(f)$.

7.5   Let P be any partition of $[a, b]$, $a \le b < x'$, and let $P' = P \cup \{x'\}$, a partition of $[a, x']$. Prove: $P(f) \le P'(f) \le V_a^{x'}(f)$.

7.6   Prove that $V_a^b(cf + g) \le |c|V_a^b(f) + V_a^b(g)$, so that $BV[a, b]$ is a *vector space*. (Hint: Consider $P(cf + g)$ and apply the triangle inequality.)

7.7   Let

$$f(x) = \begin{cases} x^2 \sin\left(\frac{\pi}{x^2}\right) & \text{if } x \ne 0, \\ 0 & \text{if } x = 0. \end{cases}$$

See Fig. 7.2.
    a) Prove that $f \notin BV[0, 1]$.
    b) Prove that f' exists on $[0, 1]$. Is $\|f'\|_{\sup} < \infty$? Explain.

7.8   Prove that $f \in BV[a, b]$ if and only if f is the difference of two *monotone* functions.

7.9   If $f, g \in BV[a, b]$, prove that $fg \in BV[a, b]$. (Caution: the product of two monotone functions need not be monotone.)

7.10   Let the rational numbers in $(0, 1)$ be listed in a sequence x_k, $k \in \mathbb{N}$. Let

$$f_k(x) = \begin{cases} 0 & \text{if } 0 \le x < x_k, \\ \frac{1}{2^k} & \text{if } x_k \le x \le 1. \end{cases}$$

Let

$$f(x) = \sum_{k=1}^{\infty} f_k(x).$$

Prove:

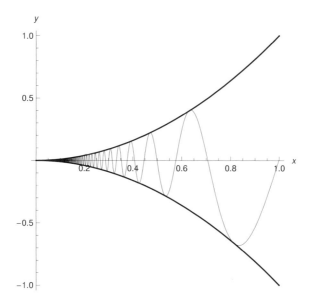

Figure 7.2 $f(x) = x^2 \sin\left(\frac{\pi}{x^2}\right)$, with envelope $u(x) = x^2$, $l(x) = -x^2$.

a) The series $\sum_{k=1}^{\infty} f_k$ converges *uniformly* on $[0, 1]$.
b) f is increasing on $[0, 1]$.
c) $f \in \mathcal{R}[0, 1]$.
d) f has a *jump discontinuity* at every rational point x_k in $(0, 1)$. (This means

$$\lim_{x \to x_k-} f(x) \neq \lim_{x \to x_k+} f(x)$$

although both one-sided limits exist.)
e) f is continuous at each irrational value of x in $[0, 1]$.

7.11 Prove or give a counterexample:
a) If $f \in \mathcal{BV}[a, b]$, then $f \in \mathcal{R}[a, b]$.
b) If $f \in \mathcal{R}[a, b]$, then $f \in \mathcal{BV}[a, b]$.

7.12 † Let $f \in \mathcal{BV}[a, b]$ and suppose f is *continuous* at $x_0 \in [a, b]$. Prove: $V_a^x(f)$ is also continuous at $x = x_0$. (Hint: Use Theorem 7.1.1, Example 7.1, and Exercise 7.6 above. You will need to use Exercises 2.8–2.12 to show that we can represent $f = \phi_1 - \phi_2$ with ϕ_1 and ϕ_2 increasing *and* with ϕ_1 and ϕ_2 *continuous* at x_0.)

7.13 ◇ Follow the steps below to construct a function $f \in \mathcal{BV}[0, 1]$ that is monotone increasing on $[0, 1]$ and differentiable with $f'(x) \geq 0$ for all $x \in [0, 1]$, yet f' is *unbounded* on $[0, 1]$, which implies also that $f' \notin \mathcal{C}[0, 1]$.

a) Let $x_n = \frac{1}{2^n}$ for all $n \in \mathbb{N}$. Thus $x_n \searrow 0$ as $n \to \infty$. Now use the function l from Exercise 5.62, to define $f \in C^\infty[x_n, x_{n-1}]$ so that

$$f(x_n) = \frac{1}{4^{n+1}} = (x_{n+1})^2.$$

We require that $f'(x) \geq 0$ for all x and that f' vanish at x_n, at x_{n-1}, and at all x in the interval

$$\left[x_n, x_{n-1} - \frac{3}{n4^{n+1}} \right].$$

b) Now link together the segments of the graph of f smoothly for all the intervals indexed by n let $f(0) = 0$. Prove that f' is unbounded although $f'(x)$ exists for all $x \in [0, 1]$, including $x = 0$. (Hint: Use the Mean Value Theorem.)

7.14   Suppose $f'(x)$ exists for all $x \in [a, b]$, and suppose $f' \in \mathcal{R}[a, b]$. Use the Fundamental Theorem of Calculus to prove that $f \in \mathcal{BV}[a, b]$ and

$$V_a^b(f) \leq \int_a^b |f'(x)|\, dx.$$

7.15   Let

$$g(x) = \begin{cases} \sin\left(\frac{\pi}{x}\right) & \text{if } 0 < x \leq 1, \\ 0 & \text{if } x = 0. \end{cases}$$

Let $f(x) = \int_0^x g(t)\, dt$. Use Theorem 7.1 to prove $f \in \mathcal{BV}[0, 1]$ (Hint: Use Exercise 3.27.)

7.16   ◇ Let $\sigma[a, b]$ denote the family of step functions defined on $[a, b]$ (Exercise 3.8), and let $\bar{\sigma}[a, b]$ denote the set of all uniform limits of step functions.

a) Prove: $\bar{\sigma}[a, b]$ is a complete normed vector space equipped with the sup-norm.

b) Prove: $\mathcal{BV}[a, b]$ is not a complete normed vector space in the sup-norm.

c) Prove: $\bar{\sigma}[a, b] \supsetneq C[a, b]$.

d) Prove: $\bar{\sigma}[a, b] \supsetneq \mathcal{BV}[a, b]$.

e) Prove: $\bar{\sigma}[a, b] \subsetneq \mathcal{R}[a, b]$.

f) Let V be any sup-norm complete vector space of functions that contains $\mathcal{BV}[a, b]$. Prove that $V \supseteq \bar{\sigma}[a, b]$. (Because of this result, we call $\bar{\sigma}[a, b]$ the *sup-norm completion* of $\mathcal{BV}[a, b]$. (In general, the *completion of a normed vector space* V can be defined as the intersection of all those complete normed vector spaces that contain V as a normed vector subspace.)

g) Prove: $f \in \bar{\sigma}[a, b]$ if and only if $f(x+)$ and $f(x-)$ both exist for each $x \in (a, b)$ and $f(a+)$ and $f(b-)$ both exist. (See Exercises 2.8 and 2.9.) (Hint: Use the Heine–Borel Theorem for the implication from right to left.)

7.2 RIEMANN–STIELTJES SUMS AND INTEGRALS

The Riemann integral

$$\int_a^b f(x)\, dx = \lim_{\|P\|\to 0} P(f, \mu)$$

$$= \lim_{\|P\|\to 0} \sum_{k=1}^n f(\mu_k)\Delta x_k$$

where $\mu = \{\mu_k \in [x_{k-1}, x_k] \mid k = 1, 2, \ldots, n\}$ is a set of arbitrary *evaluation points* for f and Δx_k measures the length of the kth subinterval determined by the partition P of $[a, b]$. The Riemann–Stieltjes integral differs in one important way from the latter concept. Instead of weighting each subinterval by its length Δx_k as in the Riemann sums, we will use the changes of a second function g on that interval serve as the weight of the interval. There are many reasons for making such an extension of the concept of the integral.

For example, the interval $[a, b]$ might be the space of possible outcomes of a probabilistic experiment. Then $\Delta g_k = g(x_k) - g(x_{k-1})$ could represent the probability of the outcome landing in the interval $[x_{k-1}, x_k]$ of possibilities, and the function f could be the *value* in some sense of such an outcome. In this illustration, $\int_a^b f\, dg$ would be a probabilistically expected value to result from running the experiment.

Another reason for extending the concept of integration in this way is that it will enable us to give a complete description of the *dual space* $\mathcal{C}'[a, b]$ of the Banach space $\mathcal{C}[a, b]$ of continuous functions on $[a, b]$.

Definition 7.2.1 *Suppose* $f, g : [a, b] \to \mathbb{R}$ *and* $P = \{x_0, x_1, \ldots, x_n\}$ *is any partition of* $[a, b]$. *Let*

$$\mu = \{\mu_k \in [x_{k-1}, x_k] \mid k = 1, 2, \ldots, n\}$$

and let $\Delta g_k = g(x_k) - g(x_{k-1})$, *for all* $k = 1, 2, \ldots, n$. *Define the Riemann–Stieltjes sum*

$$P(f, g, \mu) = \sum_{k=1}^n f(\mu_k)\Delta g_k.$$

We say that f *is Riemann–Stieltjes integrable with respect to* g *on* $[a, b]$ *if and only if there exists* $L \in \mathbb{R}$ *such that for all* $\epsilon > 0$ *there exists* $\delta > 0$ *such that*

$$\|P\| < \delta \implies |P(f, g, \mu) - L| < \epsilon,$$

independent of the choice of P *and of* μ *subject to the stipulations above. If this condition holds, we write*

$$\int_a^b f\, dg = \lim_{\|P\|\to 0} P(f, g, \mu) = L$$

and $f \in \mathcal{RS}([a, b], g)$, *the class of* Riemann–Stieltjes integrable *functions on* $[a, b]$ *with respect to* g. f *is called the* integrand *and* g *is called the* integrator.

■ **EXAMPLE 7.4**

Let $f \in C[a, b]$ and let

$$g(x) = \begin{cases} 0 & \text{if } a \leq x < t, \\ p & \text{if } t \leq x \leq b, \end{cases}$$

where t is some fixed real number. Let P be any partition of $[a, b]$. In order that $\Delta g_k \neq 0$, it is necessary and sufficient that $x_{k-1} < t \leq x_k$. Thus

$$P(f, g, \mu) = f(\mu_k)\Delta g_k = f(\mu_k)p \rightarrow f(t)p$$

since $\mu_k \rightarrow t$ as $\|P\| \rightarrow 0$. Similar reasoning could be applied to any step function.

Theorem 7.2.1 *Let $f \in \mathcal{R}[a, b]$ and suppose $g \in C^1[a, b]$, so that g' is continuous. Then $\int_a^b f \, dg$ exists and*

$$\int_a^b f \, dg = \int_a^b f(x)g'(x) \, dx,$$

which is a Riemann *integral.*

Remark 7.2.1 Notice that in the special case in which $g(x) \equiv x$, $g' \equiv 1$ and $\int_a^b f \, dg = \int_a^b f(x) \, dx$, the ordinary Riemann integral of f.

Proof: We apply the Mean Value Theorem for derivatives to write

$$\Delta g_k = g(x_k) - g(x_{k-1}) = g'(\bar{x}_k)\Delta x_k$$

for some $\bar{x}_k \in [x_{k-1}, x_k]$. Thus

$$P(f, g, \mu) = \sum_{k=1}^n f(\mu_k)g'(\bar{x}_k)\Delta x_k \tag{7.1}$$

$$= \sum_{k=1}^n f(\mu_k)g'(\mu_k)\Delta x_k + \sum_{k=1}^n f(\mu_k)[g'(\bar{x}_k) - g'(\mu_k)]\Delta x_k. \tag{7.2}$$

Since f and g' are in $\mathcal{R}[a, b]$, so is their product, and the first sum in Equation (7.2) converges to $\int_a^b f(x)g'(x) \, dx$, as $\|P\| \rightarrow 0$. Thus it suffices to prove that the second sum in Equation (7.2) converges to 0 as $\|P\| \rightarrow 0$. We prove this as follows. Let $\epsilon > 0$. Since g' is *uniformly continuous* on $[a, b]$, and since f is *bounded*, there exists $\delta > 0$ such that $\|P\| < \delta$ implies $|\bar{x}_k - \mu_k| < \delta$ which implies

$$|g'(\bar{x}_k) - g(\mu_k)| < \frac{\epsilon}{(b - a)\|f\|_{\sup}},$$

making the second sum in Equation (7.2) less than ϵ. (Note that if $b - a = 0$ or if $\|f\|_{\sup} = 0$, the claim about the second sum is trivial.)   ∎

Theorem 7.2.2 *Suppose $f_i \in \mathcal{RS}([a, b], g_j)$ for i and j in $\{1, 2\}$. Let $c \in \mathbb{R}$. Then*

i. $\int_a^b (cf_1 + f_2)\, dg_1$ *exists and*

$$\int_a^b (cf_1 + f_2)\, dg_1 = c \int_a^b f_1\, dg_1 + \int_a^b f_2\, dg_1.$$

ii. $\int_a^b f_1 d(cg_1 + g_2)$ *exists and*

$$\int_a^b f_1\, d(cg_1 + g_2) = c \int_a^b f_1\, dg_1 + \int_a^b f_1\, dg_2.$$

Remark 7.2.2 This theorem says that $\int_a^b f\, dg$ is *separately linear* in each of its two variables f and g. It also establishes that $\mathcal{RS}([a, b], g)$ is always a *vector space*.

Proof: We will prove the first part here. (See Exercise 7.18 for the second part.)

$$\left| P(cf_1 + f_2, g_1, \mu) - \left[c \int_a^b f_1\, dg_1 + \int_a^b f_2\, dg_1 \right] \right|$$

$$\leq |c| \left| P(f_1, g_1, \mu) - \int_a^b f_1\, dg_1 \right| + \left| P(f_2, g_1, \mu) - \int_a^b f_2\, dg_1 \right|$$

$$\rightarrow |c|0 + 0 = 0$$

as $\|P\| \rightarrow 0$.   ∎

In some ways the Riemann–Stieltjes integral has surprisingly different properties from those of the Riemann integral. Consider the following example.

■ **EXAMPLE 7.5**

Let

$$f(x) = \begin{cases} 0 & \text{if } x \in [0, 1], \\ 1 & \text{if } x \in (1, 2] \end{cases}$$

and let

$$g(x) = \begin{cases} 0 & \text{if } x \in [0, 1), \\ 1 & \text{if } x \in [1, 2]. \end{cases}$$

Then we make the following observations.

i. $\int_0^1 f\, dg = f(1) \cdot 1 = 0$, since $f \in \mathcal{C}[0, 1]$ and g is a step function on $[0, 1]$.

ii. $\int_1^2 f\, dg = 0$, since on $[1, 2]$ we have $g'(x) \equiv 0$, so

$$\int_1^2 f\, dg = \int_1^2 f(x)g'(x)\, dx = 0.$$

iii. $\int_0^2 f\, dg$ *does not exist* (that is, $f \notin \mathcal{RS}([0, 2], g)$), in stark contrast to the properties of the Riemann integral. Let us prove this claim as follows. No matter how small we make $\|P\|$, we can still have $x_{k-1} < 1 < x_k$, so that $\Delta g_k = 1$, and $f(\mu_k)$ can be either 0 or 1 depending upon how we choose μ_k. Thus $P(f, g, \mu)$ can be either 0 or 1, and cannot be forced to converge to a limit merely by requiring $\|P\| \to 0$. Note that f being Riemann–Stieltjes integrable on *both* $[0, 1]$ and $[1, 2]$ with respect to g fails to force f to be in $\mathcal{RS}([0, 2], g)$.

Theorem 7.2.3 *Let* $a < b < c$. *If* $\int_a^b f\, dg$, $\int_b^c f\, dg$, *and* $\int_a^c f\, dg$ *all exist, then*

$$\int_a^c f\, dg = \int_a^b f\, dg + \int_b^c f\, dg.$$

Proof: Let $\epsilon > 0$. There exists $\delta_1 > 0$ such that if P_1 is a partition of $[a, b]$ with $\|P_1\| < \delta_1$, then

$$\left| P_1(f, g, \mu) - \int_a^b f\, dg \right| < \frac{\epsilon}{3}.$$

And there exists $\delta_2 > 0$ such that if P_2 is a partition of $[b, c]$ with $\|P_2\| < \delta_2$, then

$$\left| P_2(f, g, \mu) - \int_b^c f\, dg \right| < \frac{\epsilon}{3}.$$

And there exists $\delta_3 > 0$ such that if P_3 is a partition of $[a, c]$ with $\|P_3\| < \delta_3$, then $|P_3(f, g, \mu) - \int_a^c f\, dg| < \frac{\epsilon}{3}$. So let $\delta = \min\{\delta_1, \delta_2, \delta_3\} > 0$, and let P_1 and P_2 be any partitions of $[a, b]$ and of $[b, c]$, respectively, with $\|P_i\| < \delta$, $i = 1, 2$. And let $P = P_1 \cup P_2$, a partition of $[a, c]$ with $\|P\| < \delta$ also. Then we have

$$\left| \int_a^c f\, dg - \left(\int_a^b f\, dg + \int_b^c f\, dg \right) \right| = \left| \left(\int_a^c f\, dg - P(f, g, \mu) \right) \right.$$

$$\left. + \left(P(f, g, \mu) - \left[\int_a^b f\, dg + \int_b^c f\, dg \right] \right) \right|$$

$$\leq \left| \int_a^c f\, dg - P(f, g, \mu) \right| + \left| P_1(f, g, \mu) - \int_a^b f\, dg \right| + \left| P_2(f, g, \mu) - \int_b^c f\, dg \right|$$

$$< \frac{\epsilon}{3} + \frac{\epsilon}{3} + \frac{\epsilon}{3} = \epsilon.$$

Example 7.5 is a special case of the following theorem that the reader should bear in mind when dealing with the Riemann–Stieltjes integral.

Theorem 7.2.4 *If f and g are both discontinuous at the same point $c \in [a, b]$, then $f \notin \mathcal{RS}([a, b], g)$.*

Remark 7.2.3 Note that in Example 7.5, when we restrict f and g to $[0, 1]$, only one of these two functions has a discontinuity. Similarly, if we restrict f and g to $[1, 2]$, again only one of the two functions has a discontinuity. But when viewed on $[0, 2]$, *both* functions have a discontinuity at 1!

Proof: Since f is discontinuous at c, either f is *right-discontinuous* (meaning $\lim_{x \to c+} f(x)$ fails either to exist or to be $f(c)$), or f is *left-discontinuous*. The same applies to g.

We will treat the case in which f and g are both discontinuous from the right. Then there exists $\epsilon_f > 0$ and there exists $\epsilon_g > 0$ such that for all $\delta > 0$ there exists μ', x' with

$$c < \mu' < x' < c + \delta, \ |g(x') - g(c)| \geq \epsilon_g$$

and

$$|f(\mu') - f(c)| \geq \epsilon_f.$$

No matter how small we make $\|P\|$, the value of $P(f, g, \mu)$ can fluctuate by at least the fixed positive amount $\epsilon_f \epsilon_g$ by choosing P with x' in it and then choosing μ' as the evaluation point for f in the interval between c and x'. Thus $\int_a^b f \, dg$ fails to exist. The other cases are very similar.   ∎

EXERCISES

7.17   Suppose that $f \in \mathcal{C}[a, b]$, $t \in (a, b)$, and

$$g(x) = \begin{cases} c_1 & \text{if } a \leq x < t, \\ c & \text{if } x = t, \\ c_2 & \text{if } t < x \leq b. \end{cases}$$

Prove that $f \in \mathcal{RS}([a, b], g)$ and $\int_a^b f \, dg = f(t)(c_2 - c_1)$.

7.18   Prove the second part of Theorem 7.2.2.

7.19   Find $\int_1^2 x \, d(\log x)$.

7.20   Find $\int_1^2 (x + x^3) \, d(\tan^{-1} x)$

7.21   Find $\int_0^3 x \, d\lfloor x \rfloor$ (Note: The *floor function* $\lfloor x \rfloor$ denotes the greatest integer that does not exceed x.)

7.22   Let

$$f(x) = \begin{cases} 1 & \text{if } x \in \mathbb{Q} \cap [a, b], \\ 0 & \text{if } x \in [a, b] \setminus \mathbb{Q}. \end{cases}$$

Prove that $\int_a^b f\,dg$ exists if and only if g is a constant function.

7.23 Let the *continuous* function $f \in \mathcal{RS}([a,b],g)$, $p \in (a,b)$, and $h(x) = g(x)$ for all $x \in [a,b] \setminus \{p\}$. Prove: $f \in \mathcal{RS}([a,b],h)$ and $\int_a^b f\,dh = \int_a^b f\,dg$.

7.3 RIEMANN–STIELTJES INTEGRABILITY THEOREMS

The following theorem shows a remarkable symmetry between the roles of the integrand and the integrator functions in the concept of the Riemann–Stieltjes integral.

Theorem 7.3.1 (Integration by Parts) *If $f \in \mathcal{RS}([a,b],g)$ then $g \in \mathcal{RS}([a,b],f)$ and*

$$\int_a^b f\,dg + \int_a^b g\,df = f(b)g(b) - f(a)g(a) = (fg)\Big|_a^b.$$

Remark 7.3.1 This theorem appears for ordinary Riemann integration in the more familiar formula for integration by parts, written as follows:

$$\int_a^b f\,dg = (fg)\Big|_a^b - \int_a^b g\,df.$$

Proof: It will suffice to show $\lim_{\|P\|\to 0} P(g,f,\mu)$ exists and

$$\lim_{\|P\|\to 0} P(g,f,\mu) = (fg)|_a^b - \int_a^b f\,dg.$$

Equivalently, it will suffice to show

$$(fg)|_a^b - P(g,f,\mu) \to \int_a^b f\,dg$$

as $\|P\| \to 0$. In fact,

$$(fg)\Big|_a^b - P(g,f,\mu) = f(b)g(b) - f(a)g(a)$$
$$- \Big\{ g(\mu_1)[f(x_1) - f(x_0)] + g(\mu_2)[f(x_2) - f(x_1)]$$
$$+ \cdots + g(\mu_n)[f(x_n) - f(x_{n-1})] \Big\}$$
$$= f(a)[g(\mu_1) - g(a)] + f(x_1)[g(\mu_2) - g(\mu_1)]$$
$$+ \cdots + f(b)[g(b) - g(\mu_n)]$$
$$\to \int_a^b f\,dg$$

as $\|P\| \to 0$ since $\{a, \mu_1, \mu_2, \ldots, \mu_n, b\}$ is a partition P' of $[a,b]$ and

$$\|P'\| \le 2\|P\| \to 0.$$

■

■ **EXAMPLE 7.6**

We evaluate $\int_{-1}^{2} x\, d|x| = 2|2| - (-1)| - 1| - \int_{-1}^{2} |x|\, dx = 4 + 1 - \frac{5}{2}$, where the right-hand integral can be read directly from a graph.

Theorem 7.3.2 *If $f \in C[a, b]$ and $g \in BV[a, b]$, then $f \in RS([a, b], g)$.*

Remark 7.3.2 By Theorem 7.3.1, this theorem implies also that

$$g \in RS([a, b], f).$$

Proof: Since $g \in BV[a, b]$, $g = g_1 - g_2$, where g_1 is increasing and g_2 is increasing on $[a, b]$. Thus it will suffice to prove the claim in the theorem for the case in which g is increasing on $[a, b]$. The proof will be very similar in concept to the proof of Riemann integrability of each $f \in C[a, b]$. Let P be any partition of $[a, b]$, let $M_k = \max_{x \in [x_{k-1}, x_k]} f(x)$ and $m_k = \min_{x \in [x_{k-1}, x_k]} f(x)$. Then

$$U(f, g, P) = \sum_{k=1}^{n} M_k \Delta g_k \text{ and } L(f, g, P) = \sum_{k=1}^{n} m_k \Delta g_k.$$

Clearly,

$$L(f, g, P) \leq P(f, g, \mu) \leq U(f, g, P)$$

for all P and μ. It is easy to show, just as we did for Riemann sums in Chapter 3, that

$$P' \supseteq P \implies L(f, g, P) \leq L(f, g, P') \leq U(f, g, P') \leq U(f, g, P).$$

Thus for all P and P' we have $L(f, g, P) \leq U(f, g, P')$. We define the upper integral $\overline{\int_a^b} f\, dg$ to be the infimum of all the upper sums, and the lower integral $\underline{\int_a^b} f\, dg$ to be the supremum of all the lower sums, again just as for Riemann integration. Since every lower sum is less than or equal to every upper sum, we have

$$\underline{\int_a^b} f\, dg \leq \overline{\int_a^b} f\, dg$$

and we claim that these two are actually equal. Let $\epsilon > 0$. It would suffice to show that

$$\overline{\int_a^b} f\, dg - \underline{\int_a^b} f\, dg < \epsilon.$$

For this it would be sufficient to prove there exists $\delta > 0$ such that $\|P\| < \delta$ implies $U(f, g, P) - L(f, g, P) < \epsilon$. By uniform continuity of f, there exists $\delta > 0$ such that $|x - x'| < \delta$ implies

$$|f(x) - f(x')| < \frac{\epsilon}{g(b) - g(a)}.$$

Thus if $\|P\| < \delta$, then $M_k - m_k < \frac{\epsilon}{g(b)-g(a)}$ for all k, and this implies

$$U(f,g,P) - L(f,g,P) < \epsilon.$$

Thus

$$\underline{\int_a^b} f\,dg = L = \overline{\int_a^b} f\,dg$$

which defines the number L. Hence $\|P\| < \delta$ implies both L and $P(f,g,\mu)$ must lie between $\underline{\int_a^b} f\,dg$ and $\overline{\int_a^b} f\,dg$ and hence within ϵ of each other. That is, $\|P\| < \delta$ implies $|P(f,g,\mu) - L| < \epsilon$, and the theorem is proven. ∎

Theorem 7.3.3 *For each $g \in \mathcal{BV}[a,b]$, define $T_g : \mathcal{C}[a,b] \to \mathbb{R}$ by*

$$T_g(f) = \int_a^b f\,dg.$$

Then T_g is a continuous linear functional on $\mathcal{C}[a,b]$ and

$$\|T_g\| \le V_a^b(g).$$

Remark 7.3.3 The symbol $\|T_g\|$ is called the *norm* of T_g. Norms of bounded linear functionals were defined in Remark 5.4.2.

Proof: By Theorem 7.3.2 T_g is a linear functional. It suffices to prove T_g is *bounded*. However, for all partitions P of $[a,b]$, we have

$$|P(f,g,\mu)| = \left| \sum_{k=1}^n f(\mu_k)\Delta g_k \right| \le \sum_{k=1}^n |f(\mu_k)||\Delta g_k|$$

$$\le \|f\|_{\sup} \sum_{k=1}^n |\Delta g_k| = \|f\|_{\sup} P(g)$$

$$\le \|f\|_{\sup} V_a^b(g).$$

Thus

$$|T_g(f)| = \left| \lim_{\|P\|\to 0} P(f,g,\mu) \right| \le \|f\|_{\sup} V_a^b(g)$$

for all $f \in \mathcal{C}[a,b]$. It follows that $\|T_g\| \le V_a^b(g)$. ∎

EXERCISES

7.24   Evaluate $\int_{-1}^1 x\,d(|x| + [x])$.

7.25   Evaluate $\int_0^{\pi/2} x\,d(\cos x)$.

7.26   Let $p \in [a,b]$ and define $T : \mathcal{C}[a,b] \to \mathbb{R}$ by $T(f) = f(p)$, a so-called *point evaluation*. Prove that T is a bounded linear functional on $\mathcal{C}[a,b]$ with $\|T\| = 1$.

7.27   Let T be as in Exercise 7.26. Find a function $g \in \mathcal{BV}[a, b]$ such that $T_g(f) \equiv T(f)$, where T_g is defined in Theorem 7.3.3. Can you find g in this exercise in such a way that $V_a^b(g) = 1 = \|T\|$? Explain.

7.28   Let $g \in \mathcal{BV}[a, b]$ and suppose $h(x) = g(x)$ except at one point $x = p \in (a, b)$. Show that $h \in \mathcal{BV}[a, b]$ and $T_h \equiv T_g$. Must $V_a^b(h) = V_a^b(g)$? If yes, prove it. If no, give a counterexample.

7.29   Let $g \in \mathcal{BV}[0, 2]$ such that

$$g(x) = \begin{cases} c_1 & \text{if } 0 \le x < 1, \\ c_2 & \text{if } x = 1, \\ c_3 & \text{if } 1 < x \le 2. \end{cases}$$

Prove that for all $f \in \mathcal{C}[0, 2]$, $T_g(f)$ depends *only* upon the difference between c_1 and c_3, and is *independent* of c_2.

7.30   Give an example of $g \in \mathcal{BV}[0, 2]$ for which $\|T_g\| < V_0^2(g)$.

7.31   Let $f(x)$ be defined as in Example 7.3. Let $1_{\left[\frac{1}{n}, 1\right]}$ be the indicator function of the interval $\left[\frac{1}{n}, 1\right]$, and let $f_n(x) = f(x) 1_{\left[\frac{1}{n}, 1\right]}(x)$ for all $x \in [0, 1]$. Prove:

    a) $f_n \in \mathcal{BV}[0, 1]$ for all $n \in \mathbb{N}$.

    b) $f_n \to f$ uniformly on $[0, 1]$.

    c) Prove or give a counterexample: the uniform limit of a sequence of functions of bounded variation must be of bounded variation.

7.32   Let

$$g_n(x) = \begin{cases} x^2 & \text{if } \frac{1}{n} \le x \le 1, \\ 0 & \text{if } 0 \le x < \frac{1}{n} \end{cases}$$

and let

$$f(x) = \begin{cases} \frac{1}{x} & \text{if } 0 < x \le 1, \\ 0 & \text{if } x = 0. \end{cases}$$

Prove:

    a) The sequence g_n converges uniformly on $[0, 1]$ to $g(x) = x^2$ as $n \to \infty$.

    b) $f \in \mathcal{RS}([0, 1], g_n)$ for all $n \in \mathbb{N}$.

    c) $f \notin \mathcal{RS}([0, 1], g)$.

7.4 THE RIESZ REPRESENTATION THEOREM[27]

---

[27] The Riesz Representation Theorem is not required for any other part of this book. It is included here because it is very important, depends only upon advanced calculus topics presented earlier in this book, and provides an excellent introduction to advanced, graduate level analysis.

We have seen in Theorem 7.3.3 that to each $\alpha \in \mathcal{BV}[a, b]$, there corresponds a bounded linear functional $T_\alpha : \mathcal{C}[a, b] \to \mathbb{R}$ by

$$T_\alpha(f) = \int_a^b f \, d\alpha$$

and that $\|T_\alpha\| \leq V_a^b(\alpha)$ for all α. Thus $T_\alpha \in \mathcal{C}'[a, b]$, the *dual space* of the Banach space $\mathcal{C}[a, b]$. Our next theorem establishes that *every* $T \in \mathcal{C}'[a, b]$ can be represented as being T_α for some suitable $\alpha \in \mathcal{BV}[a, b]$. We adapt here the constructive method of proof presented in the classic book *Functional Analysis* by Frigyes Riesz and Béla Sz.-Nagy [18] in 1955. The proof is a considerably larger undertaking than those appearing earlier in the present advanced calculus text. It requires however only theorems with which the reader is familiar already.[28]

Theorem 7.4.1 (Riesz Representation Theorem) *Let* $T \in \mathcal{C}'[a, b]$. *Then there exists* $\alpha \in \mathcal{BV}[a, b]$ *such that* $T = T_\alpha$. *That is,*

$$T(f) = T_\alpha(f) = \int_a^b f \, d\alpha$$

for all $f \in \mathcal{C}[a, b]$. *Moreover,* α *can be selected so that* $V_a^b(\alpha) = \|T\|$ *and so that* $\alpha(a) = 0$.

Proof: Since the proof is substantial, we present the intuitive idea that motivates it first. We are given some $T \in \mathcal{C}'[a, b]$, so for all $f \in \mathcal{C}[a, b]$, $T(f) \in \mathbb{R}$, and T is both bounded and linear. Let
$$M = \|T\|,$$
for convenience. Note that the only kind of function we have a right to apply T to is a continuous function f. For example, we would need to justify application of T to such a function as

$$1_{[a,t)}(x) = \begin{cases} 1 & \text{if } a \leq x < t, \\ 0 & \text{if } t \leq x \leq b \end{cases}$$

since the latter function is not continuous at t. Observe that $1_{[a,a)}$ is the indicator function of the empty set, which is therefore identically zero. We are going to prove that it is possible to *extend* the domain of definition of T to a larger space that includes all functions such as $1_{[a,t)}$ while keeping $\|T\| = M$ even on this larger space.[29] Then

---

[28]Note that there is more than one theorem in the subject of functional analysis with the name Riesz Representation Theorem. Another famous one, known also as the Riesz-Fischer Theorem, describes the relationship between l_2 and L^2, using orthonormal bases. It can be interpreted also as describing the convergence of Fourier series in the square-norm, which the reader has met in Remark 6.5.2. Both theorems will likely be encountered again in graduate courses.

[29]The construction of the necessary extension of T occupies most of the lengthy proof of the Riesz Representation Theorem. A much shorter proof [8] exists, in which the extension is available automatically from the Hahn-Banach Theorem, which is beyond the scope of this book. Although the shorter, more advanced proof exists, there is intellectual merit in the constructive proof because the Hahn-Banach Theorem depends upon the Axiom of Choice, whereas the constructive proof is independent of that axiom.

we will define

$$\alpha(t) = \begin{cases} T(1_{[a,t)}) \text{ if } t \in [a,b), \\ T(1_{[a,b]}) \text{ if } t = b. \end{cases}$$

Observe that $\alpha(a) = T(0) = 0$. We will prove that $\alpha \in \mathcal{BV}[a,b]$ and that $T \equiv T_\alpha$.

The intuitive motivation comes from the following formal observation, in which we pretend to be able to compute $\int_a^b 1_{[a,t)} d\alpha$ for some not yet known $\alpha \in \mathcal{BV}[a,b]$. (This is a pretense since this Riemann–Stieltjes integral cannot even exist unless α happens to be continuous at t.) We take a partition P and calculate

$$P(1_{[a,t)}, \alpha, \mu) = \sum_{k=1}^{n} 1_{[a,t)}(\mu_k)[\alpha(x_k) - \alpha(x_{k-1})]$$

$$= \sum_{k=1}^{l} [\alpha(x_k) - \alpha(x_{k-1})]$$

$$= \alpha(x_l) - \alpha(a) = \alpha(x_l),$$

where x_l is the last partition point for which μ_l lies inside [0,t]. So we can hope that $P(1_{[a,t)}, \alpha, \mu) \to \alpha(t)$ as $\|P\| \to 0$. That is, $\int_a^b 1_{[a,t)} d\alpha = T(1_{[a,t)})$. Now we proceed to the rigorous arguments.

We have $T \in \mathcal{C}'[a,b]$ and $\|T\| = M < \infty$. We are going to show how to extend T to every *bounded* function f having the property that there exists a sequence $f_k \in \mathcal{C}[a,b]$ such that for all $x \in [a,b]$ we have $f_k(x) \nearrow f(x)$. Notice that f need not be continuous, so Dini's theorem does *not* apply here, and that f could be a function such as $1_{[a,t)}$, for all $t \in [a,b]$ (Exercise 7.33).

The proof will proceed in five parts.

i. Let $B^+[a,b]$ denote the set of all bounded functions f for which there exists $\{f_k\} \subset \mathcal{C}[a,b]$ with $f_k(x) \nearrow f(x)$ for all $x \in [a,b]$. We claim that if $f \in B^+[a,b]$ and f_k is as just described then the sequence $T(f_k)$ is convergent. Moreover, if $g_k \in \mathcal{C}[a,b]$ such that for all $x \in [a,b]$ we have $g_k(x) \nearrow f(x)$, then

$$\lim_{k \to \infty} T(g_k) = \lim_{k \to \infty} T(f_k).$$

This will enable us to *define* $T(f) = \lim_{k \to \infty} T(f_k)$. So let $f \in B^+[a,b]$ and let $B = \|f\|_{\sup} < \infty$. (We remark that of course $\mathcal{C}[a,b] \subset B^+[a,b]$.)

In order to show that the sequence $T(f_n)$ converges, it will suffice to show that

$$T(f_n) = T(f_1) + \sum_{k=2}^{n} [T(f_k) - T(f_{k-1})]$$

converges. We will show that

$$T(f_1) + \sum_{k=2}^{\infty} [T(f_k) - T(f_{k-1})]$$

converges absolutely. Let

$$\sigma_k = \operatorname{sgn}[T(f_k) - T(f_{k-1})],$$

where sgn denotes the signum function. We have

$$\sum_{k=2}^{n} |T(f_k) - T(f_{k-1})| = T\left(\sum_{k=2}^{n} \sigma_k(f_k - f_{k-1})\right)$$
$$\leq M(B + \|f_1\|_{\sup})$$

for all n since

$$\left|\sum_{k=2}^{n} \sigma_k(f_k - f_{k-1})(x)\right| \leq \sum_{k=2}^{n} (f_k - f_{k-1})(x)$$
$$= (f_n - f_1)(x) \leq B + \|f_1\|_{\sup} < \infty,$$

which means that

$$\left\|\sum_{k=2}^{n} \sigma_k(f_k - f_{k-1})\right\|_{\sup} \leq B + \|f_1\|_{\sup}.$$

Thus $\sum_{k=2}^{\infty}[T(f_k) - T(f_{k-1})]$ is absolutely convergent, and $T(f_n)$ converges. In order to complete the first part, we need to show that if in addition to f_k we have also a sequence $g_k \in C[a,b]$ such that $g_k(x) \nearrow f(x)$ for all $x \in [a,b]$, then

$$\lim_{k \to \infty} T(g_k) = \lim_{k \to \infty} T(f_k).$$

We know that $f_k(x) \nearrow f(x)$ and $g_k(x) \nearrow f(x)$ for all x as well. It follows that $f_k - \frac{1}{k} \uparrow f$ and $g_k - \frac{1}{k} \uparrow f$ pointwise as well, these two sequences being *strictly* increasing at each x. And

$$T\left(f_k - \frac{1}{k}\right) = T(f_k) - T\left(\frac{1}{k}\right) \to \lim_{k \to \infty} T(f_k)$$

since $\left\|\frac{1}{k}\right\|_{\sup} \to 0$ implies $T\left(\frac{1}{k}\right) \to 0$. So it would suffice to show that $f_k \uparrow f$ and $g_k \uparrow f$ pointwise on $[a,b]$ implies

$$\lim_{k \to \infty} T(f_k) = \lim_{k \to \infty} T(g_k).$$

What we know from the first part of this proof is that $T(f_k)$ and $T(g_k)$ both converge. We claim that for each k there exists j such that $f_k < g_j$ for all x. If that were false for some k, then consider x_j such that $f_k(x_j) \geq g_j(x_j)$, for all j. Since $x_j \in [a,b]$, the Bolzano–Weierstrass theorem implies there exists a subsequence $x_{j_i} \to p \in [a,b]$. By continuity of f_k, $f_k(p) \geq g_j(p)$ for all j. But $g_j(p) \uparrow f(p)$, so $f(p) \leq f_k(p)$. Yet $f_k(p) \uparrow f(p)$ is a *strictly increasing* sequence. This is a contradiction. Hence we see that f_k and g_k have subsequences such that

$$f_{k_1} < g_{j_1} < \cdots < f_{k_l} < g_{j_l} < \cdots,$$

and this sequence also increases strictly at each point to f. But that means T of this sequence converges, and so each subsequence of this sequence converges to the *same limit*. That is,

$$\lim_{l \to \infty} T(f_{k_l}) = \lim_{l \to \infty} T(g_{j_l}),$$

which implies in turn that $\lim_{k \to \infty} T(f_k) = \lim_{k \to \infty} T(g_k)$.

ii. Now we are able to define

$$T(f) = \lim_{k \to \infty} T(f_k)$$

for all $f \in B^+[a, b]$. If $c > 0$, clearly $cf_k \nearrow cf$ at each x too, so the set $B^+[a, b]$ is closed under multiplication by positive scalars. Similarly, f_1 and $f_2 \in B^+[a, b]$ implies $f_1 + f_2 \in B^+[a, b]$ (Exercise 7.34). Now let

$$B[a, b] = \{f = f_1 - f_2 \mid f_1, f_2 \in B^+[a, b]\}.$$

Then $B[a, b]$ is a *vector space* of functions (Exercise 7.6). Moreover, we claim that we can define

$$T(f) = T(f_1) - T(f_2)$$

for all $f \in B[a, b]$. For this extension of T to be well-defined, we need to know that if

$$f = f_1 - f_2 = g_1 - g_2$$

are two representations of f, where

$$f_1, f_2, g_1, g_2 \in B^+[a, b],$$

then

$$T(f_1) - T(f_2) = T(g_1) - T(g_2)$$

and T so-defined is linear on $B[a, b]$ (Exercise 7.36).

iii. Next, we wish to show that T, as extended above, remains bounded on the vector space $B[a, b]$. We will show that

$$|T(f)| \le M\|f\|_{\text{sup}},$$

where M is still the norm of T as given initially on $\mathcal{C}[a.b]$. So let $f, g \in B^+[a, b]$ and suppose $f_k, g_k \in \mathcal{C}[a, b]$ such that for all $x \in [a, b]$ we have $f_k(x) \nearrow f(x)$ and $g_k(x) \nearrow g(x)$. Although $f_k - g_k \to f(x) - g(x)$, it is *not* necessarily true that

$$[f_k(x) - g_k(x)] \nearrow [f(x) - g(x)].$$

What is more serious for our purposes is that we may have

$$\|f_k - g_k\|_{\sup} > \|f - g\|_{\sup} = K.$$

But we can fix this last problem by a method known as *truncation* as follows. We define a new sequence of functions $\phi_n \in C[a, b]$ (see Exercise 7.37) by

$$\phi_n(x) = \begin{cases} f_n(x) & \text{if } |f_n(x) - g_n(x)| \le K, \\ g_n(x) + K & \text{if } f_n(x) - g_n(x) > K, \\ g_n(x) - K & \text{if } f_n(x) - g_n(x) < -K. \end{cases} \tag{7.3}$$

Also,

$$\|\phi_n - g_n\|_{\sup} \le K = \|f - g\|_{\sup}.$$

We need to know that $\phi_n(x) \nearrow f(x)$ for all $x \in [a, b]$. This can be seen from the geometrical meaning of the truncation defined above in Equation (7.3) as follows. Consider the band in the plane trapped between the graphs of $g_n(x) + K$ and $g_n(x) - K$, over the interval $[a, b]$. The whole band moves upwards as n increases strictly since $g_n(x)$ increases. If the graph of f_n slips either over the top or under the bottom of the band, then we truncate the graph of f_n with the upper or lower boundary curve, respectively. This produces ϕ_n. It is clear that

$$\phi_n(x) \le \max\{f_n(x), g_n(x)\} \le f(x)$$

for all n and for all x. To see that $\phi_n(x) \nearrow$, consider the fact that for each x and for each n, $\phi_n(x)$ must be either the middle, the upper, or the lower value permitted by Equation (7.3). The only way it is conceivable that $\phi_n(x) \ge \phi_{n+1}(x)$ is if $\phi_{n+1}(x)$ is a lower value among the three possibilities than is $\phi_n(x)$. For example, if $\phi_n(x) = g_n(x) + K$ and $\phi_{n+1}(x) = g_{n+1}(x) - K$, then

$$\begin{aligned} \phi_{n+1}(x) = g_{n+1}(x) - K &> f_{n+1}(x) \\ &\ge f_n(x) > g_n(x) + K \\ &= \phi_n(x). \end{aligned}$$

On the other hand, if $\phi_n(x) = f_n(x)$, we could have

$$\begin{aligned} \phi_{n+1}(x) = g_{n+1}(x) - K &> f_{n+1}(x) \\ &\ge f_n(x) = \phi_n(x). \end{aligned}$$

In each case, $\phi_n(x) \le \phi_{n+1}(x)$. Moreover, since

$$g_n(x) \nearrow g(x), \ f_n(x) \nearrow f(x) \text{ and } |f(x) - g(x)| \le K$$

for all x,

$$\lim_{n \to \infty} \phi_n(x) = \lim_{n \to \infty} f_n(x) = f(x).$$

Now we can reason as follows.

$$|T(f - g)| = \left| \lim_{n \to \infty} T(\phi_n) - \lim_{n \to \infty} T(g_n) \right|$$
$$= \lim_{n \to \infty} |T(\phi_n - g_n)| \leq MK$$

Thus even on $B[a, b]$ we have the extended T with $\|T\| = M$.

iv. Now we define $\alpha(t) = T(1_{[a,t)})$ for all $t \in [a, b)$, and $\alpha(b) = T(1_{[a,b]})$. We observe that $\alpha(a) = T(1_{[a,a)}) = T(0) = 0$. The reason for the slight difference in the way $\alpha(b)$ is defined will become clear in part v of this proof.

We claim that

$$V_a^b(\alpha) \leq \|T\| = M.$$

To prove this, we let $P = \{x_0, x_1, \ldots, x_n\}$ be any partition of $[a, b]$, and we form

$$P(\alpha) = \sum_{k=1}^{n} |\alpha(x_k) - \alpha(x_{k-1})|.$$

Observe that for all k we can select a number $\epsilon_k \in \{\pm 1\}$ such that

$$|\alpha(x_k) - \alpha(x_{k-1})| = \epsilon_k [\alpha(x_k) - \alpha(x_{k-1})]$$
$$= T \left(\epsilon_k [1_{[a,x_k)} - 1_{[a,x_{k-1})}] \right).$$

Thus we can write

$$P(\alpha) = T \left\{ \sum_{k=1}^{n} \epsilon_k \left[1_{[a,x_k)} - 1_{[a,x_{k-1})} \right] \right\}$$
$$= T(f),$$

where we denote by f the argument of T. Note that $|f(x)| \leq 1$ for all x, so that

$$|T(f)| \leq M \|f\|_{\sup} = M.$$

Hence

$$V_a^b(\alpha) \leq \|T\|.$$

If we can show that $T = T_\alpha$, then we will know from Theorem 7.3.3 that the opposite inequality holds as well, and then we will know that $\|T\| = V_a^b(\alpha)$.

v. We claim that $T = T_\alpha$. To prove this, we let $P = \{x_0, x_1, \ldots, x_n\}$ be any partition of $[a, b]$. Let $f \in \mathcal{C}[a, b]$. We need to show that $T(f) = T_\alpha(f)$. Consider

$$P(f, \alpha, x) = \sum_{k=1}^{n} f(x_k)[\alpha(x_k) - \alpha(x_{k-1})] = \sum_{k=1}^{n} f(x_k) \Delta \alpha_k \qquad (7.4)$$

This is a Riemann–Stieltjes sum for $\int_a^b f\, d\alpha$, with μ_k selected to be x_k for all $k = 1, \ldots, n$. Thus the sum in Equation (7.4) converges to $\int_a^b f\, d\alpha$ as $\|P\| \to 0$.

On the other hand, $P(f, \alpha, x) = T(\phi_P)$, where

$$\phi_P(x) = f(x_1)1_{[a,x_1)} + f(x_2)1_{[x_1,x_2)} + \cdots$$
$$+ f(x_{n-1})1_{[x_{n-2},x_{n-1})} + f(x_n)1_{[x_{n-1},b]}.$$

However, note that f is *uniformly continuous* on [a,b]. Thus for each $\epsilon > 0$ there exists $\delta > 0$ such that $\|P\| < \delta$ implies $\|f - \phi_P\|_{\sup} < \epsilon$. (It is to enable this last inequality that we defined $\alpha(b) = T(1_{[a,b]})$, with the interval closed at both ends.) Since T is bounded on $B[a, b]$,

$$P(f, \alpha, x) = T(\phi_P) \to T(f)$$

as $\|P\| \to 0$. Hence $T(f) = \int_a^b f\, d\alpha$. ∎

We have observed already that if $\alpha \in \mathcal{BV}[a, b]$, then

$$T_{\alpha+c} = T_\alpha \in \mathcal{C}'[a, b]$$

for all $c \in \mathbb{R}$. This enables us to *restrict our attention, without loss of generality, to those $\alpha \in \mathcal{BV}[a, b]$ for which $\alpha(a) = 0$, which we assume henceforth.* Also, changing the value of α at just one point, or finitely many points, in (a, b) has no effect on T_α. We must ask ourselves under what circumstances in general T_α and T_β will be the same linear functionals on $\mathcal{C}[a, b]$, where α and $\beta \in \mathcal{BV}[a, b]$ with $\alpha(a) = 0 = \beta(a)$.

First, suppose $p \in (a, b)$ is a point at which $\alpha \in \mathcal{BV}[a, b]$ happens to be continuous. Consider the continuous function f_n that is identically 1 on $[a, p]$, 0 on $\left[p + \frac{1}{n}, b\right]$, and linear on $\left[p, p + \frac{1}{n}\right]$. We see that

$$T(f_n) = T_\alpha f_n = \int_a^p f_n\, d\alpha + \int_p^{p+\frac{1}{n}} f_n\, d\alpha + \int_{p+\frac{1}{n}}^b f_n\, d\alpha.$$

Now the third integral is 0 by definition of f_n. The middle integral is bounded in absolute value by

$$\|f_n\|_{\sup} V_p^{p+\frac{1}{n}}(\alpha) \to 0$$

as $n \to \infty$ since

$$V_p^{p+\frac{1}{n}}(\alpha) = V_a^{p+\frac{1}{n}}(\alpha) - V_a^p(\alpha),$$

which approaches 0 since $V_a^x(\alpha)$ is continuous function of x at each point of continuity of α (Exercise 7.12). But the first integral is just $\alpha(p) - \alpha(a) = \alpha(p)$. Thus we see that $\alpha(p)$ is determined uniquely by the action of the original linear functional T at each point p of continuity of α. Namely,

$$\alpha(p) = \lim_{n \to \infty} T(f_n).$$

Now suppose $p \in (a, b)$ is a point of *discontinuity* of α. We know from Exercise 2.25 that a monotone function has at most countably many discontinuities. Since α is a difference of two monotone functions, the same is true for α, because the union of two countable sets is again countable. Moreover,

$$\lim_{x \to p+} \alpha(x) = \alpha(p+)$$

and

$$\lim_{x \to p-} \alpha(x) = \alpha(p-)$$

both exist since this is true for monotone functions. Also, every interval $(p, p + \delta)$ or $(p - \delta, p)$ of positive length contains *uncountably* many points, and hence contains points of continuity of α. Thus $\alpha(p+)$ and $\alpha(p-)$ are uniquely determined by α at points of continuity, which means these one-sided limits are determined uniquely by the initial linear functional T that is being represented as T_α.

All that remains is to ask what is the value of $\alpha(p)$ itself. But changing the value of α at p has no effect on T_α since if $\gamma(x) = 1$ at $x = p$ and 0 everywhere else, $T_\gamma(f) = 0$ for all $f \in \mathcal{C}[a, b]$. But if we require α to be right-continuous on (a, b), then $\alpha(x)$ is uniquely determined by T. (See Exercise 7.39.)

Definition 7.4.1 *We define*

$$\mathcal{BV}_0[a, b] = \left\{ \alpha \in \mathcal{BV}[a, b] \,\middle|\, \alpha(a) = 0, \ \alpha \ right\text{-}continuous \ on \ (a, b) \right\}.$$

Remark 7.4.1 By virtue of the reasoning above, for all $T \in \mathcal{C}'[a, b]$ there exists a *unique* $\alpha \in \mathcal{BV}_0[a, b]$ such that $T = T_\alpha$. Note that we do not require right-continuity at $x = a$ since we have agreed to let $\alpha(a) = 0$, and we may need a jump discontinuity on the right at a–for example to provide for the functional $T(f) = f(a)$. And right-continuity would be a vacuous requirement at $x = b$, in addition to which we are forced to let $\alpha(b) = T\left(1_{[a,b]}\right)$ if $\alpha(a) = 0$. Moreover,

$$\|T\| = V_a^b(\alpha),$$

and $\alpha \to T_\alpha$ is a linear map of the vector space $\mathcal{BV}_0[a, b]$ injective (meaning one-to-one) surjective (meaning onto) $\mathcal{C}'[a, b]$. (A map that is both injective and surjective is called a *bijection*.) Since $\mathcal{C}'[a, b]$ is a Banach space it follows that $\mathcal{BV}_0[a, b]$ is also a Banach space with the so-called *total variation norm* $\|\alpha\|_{\text{tv}} = V_a^b(\alpha)$. Moreover, $\mathcal{C}'[a, b]$ and $\mathcal{BV}_0[a, b]$ are *isomorphic* as Banach spaces. Some of the details of the Banach space properties are in Exercises 7.39 and 7.40 below.

EXERCISES

7.33   Let $t \in [a, b]$. Find a sequence $f_k \in \mathcal{C}[a, b]$ such that $f_k(x) \nearrow 1_{[a,t)}(x)$ for all $x \in [a, b]$.

7.34   Prove: $f_1, f_2 \in B^+[a, b]$ implies $f_1 + f_2 \in B^+[a, b]$, where $B^+[a, b]$ is as defined in part (1) of the proof of Theorem 7.4.1. Show also that if $c \geq 0$ then $cf_1 \in B^+[a, b]$.

7.35   Prove: $B[a, b]$ is a *vector space* of functions, where $B[a, b]$ is defined in part (2) of the proof of Theorem 7.4.1.

7.36   If

$$f = f_1 - f_2 = g_1 - g_2$$

are two representations of f, where $f_1, f_2, g_1, g_2 \in B^+[a, b]$, then

$$T(f_1) - T(f_2) = T(g_1) - T(g_2).$$

Prove that T is linear on $B[a, b]$.

7.37   Prove that the functions ϕ_n defined in Equation (7.3) are continuous.

7.38

    a) Prove that $1_{[a,t]} \notin B^+[a, b]$ if $a \leq t < b$, but $-1_{[a,t]} \in B^+[a, b]$ so that $1_{[a,t]} \in B[a, b]$.

    b) Let $a < c < d \leq b$. Prove that $1_{[c,d)} \notin B^+[a, b]$, but $1_{[c,d)} \in B[a, b]$.

7.39   Let $\beta \in \mathcal{BV}[a, b]$ and $\{p_k \mid k \in \mathbb{N}\}$ be the set of points in (a, b) at which β is discontinuous. Let β_n be the same as β except that $\beta_n(p_k) = \beta(p_k+)$ for all $k = 1, \ldots, n$. And let β' be the same as β except that $\beta'(p_k) = \beta(p_k+)$ for all $k \in \mathbb{N}$. Show that $\beta' \in \mathcal{BV}[a, b]$ and that

$$V_a^b(\beta_n - \beta') \to 0$$

as $n \to \infty$. Prove that $T_{\beta'} = T_\beta$.

7.40   Let α_n be a Cauchy sequence in $\mathcal{BV}_0[a, b]$, in the sense of the total variation norm: For all $\epsilon > 0$ there exists $N \in \mathbb{N}$ such that n and $m \geq N$ implies

$$\|\alpha_n - \alpha_m\|_{\text{tv}} = V_a^b(\alpha_n - \alpha_m) < \epsilon.$$

Prove that there exists $\alpha \in \mathcal{BV}_0[a, b]$ such that

$$\|\alpha_n - \alpha\|_{\text{tv}} \to 0$$

as $n \to \infty$. That is, prove that $\mathcal{BV}_0[a, b]$ is complete. (Hint: Consider the sequence

$$\{T_{\alpha_n}\}_{n=1}^\infty \subset \mathcal{C}'[a, b],$$

which is already known to be complete, and use the uniqueness of the correspondence between α and T_α in Remark 7.4.1.)

7.5 TEST YOURSELF

EXERCISES

7.41 Let

$$f(x) = \begin{cases} x^2 \left(\sin \left(\frac{\pi}{x^2} \right) + \sin \left(\frac{\pi}{x} \right) \right) & \text{if } x \neq 0, \\ 0 & \text{if } x = 0. \end{cases}$$

True or False: $f \in \mathcal{BV}[0,1]$. Explain.

7.42 Give an example of an integrable function $f(x)$ that is not of bounded variation on $[0,1]$.

7.43 Find $\int_0^3 x \, d\lceil x \rceil$ (Note: The *ceiling function* $\lceil x \rceil$ denotes the least integer that is no less than x.)

7.44 True or False: The Riemann–Stieltjes integral $\int_{-1}^1 \lfloor x \rfloor \, d \operatorname{sgn} x$ exists.

7.45 True or False: The Riemann–Stieltjes integral

$$\int_{-1}^1 x^2 \sin \frac{\pi}{x} \, d \operatorname{sgn} x$$

exists. (We interpret the integrand function at $x = 0$ as having the value 0.)

7.46 Let $g(x) = 1_{\{1\}}$. Find $\int_0^2 \tan^{-1} x \, dg$.

7.47 Let $T : C[0,2] \to \mathbf{R}$ by $T(f) = 2f(1)$. Find a function $g \in BV[0,2]$ such that

$$T(f) = \int_0^2 f \, dg,$$

for all $f \in C[0,2]$.

7.48 Suppose that $f(x) = e^x$ for all x. Let

$$g(x) = \begin{cases} 1 & \text{if } 0 \leq x < 1, \\ 7 & \text{if } x = 1, \\ 4 & \text{if } 1 < x \leq 2. \end{cases}$$

Evaluate $\int_0^2 f \, dg$.

7.49 True or Give a Counterexample: If $\int_a^b f \, dg$ and $\int_b^c f \, dg$ both exist, then $\int_a^c f \, dg$ exists and

$$\int_a^c f \, dg = \int_a^b f \, dg + \int_b^c f \, dg.$$

PART III

ADVANCED CALCULUS IN SEVERAL VARIABLES

CHAPTER 8

EUCLIDEAN SPACE

8.1 EUCLIDEAN SPACE AS A COMPLETE NORMED VECTOR SPACE

Throughout pure and applied mathematics, it is necessary to consider functions of more than one variable. If a function depends on n real variables, $x_1, x_2, \ldots, x_n$, it is possible to combine these n real variables into one *vector variable*

$$\mathbf{x} = (x_1, x_2, \ldots, x_n).$$

In this notation, $\mathbf{x}$ is not a real number, but rather an n-tuple of real numbers. In addition, it is often necessary to consider functions that have vector values instead of real values. Up to this point, our course has focused on real-valued functions of one real variable. In this chapter we will begin the rigorous study of vector-valued functions of vector variables.

In the study of real-valued functions of a single real variable, we saw the advantages of considering normed vector spaces of more than one dimension. For example, the reader has seen that the vector space $\mathcal{C}[a, b]$ has infinitely many dimensions (Exercise 2.57). In the present chapter, we will focus on finite-dimensional vector spaces equipped with what is called the *Euclidean* norm. However, we begin with a more general context.

Advanced Calculus: An Introduction to Linear Analysis. By Leonard F. Richardson
Copyright © 2008 John Wiley & Sons, Inc.

Definition 8.1.1 *In any (real) vector space V (Tabel 2.1, p. 59), we call a function*

$$\langle \cdot, \cdot \rangle : V \times V \to \mathbb{R}$$

a scalar product *if and only if it has the following three properties.*

   i. $\langle a\mathbf{x} + \mathbf{y}, \mathbf{z} \rangle = a\langle \mathbf{x}, \mathbf{z} \rangle + \langle \mathbf{y}, \mathbf{z} \rangle$ *for all $a \in \mathbb{R}$ and for all $\mathbf{x}$ and $\mathbf{y}$ in V.*

   ii. $\langle \mathbf{x}, \mathbf{y} \rangle = \langle \mathbf{y}, \mathbf{x} \rangle$ *for all $\mathbf{x}$ and $\mathbf{y}$ in V.*

   iii. $\langle \mathbf{x}, \mathbf{x} \rangle \geq 0$ *for all $\mathbf{x} \in V$ and $\langle \mathbf{x}, \mathbf{x} \rangle = 0 \Leftrightarrow \mathbf{x} = \mathbf{0} \in V$.*

■ **EXAMPLE 8.1**

In the vector space $\mathbb{R}^n$, define the *Euclidean* scalar product by

$$\langle \mathbf{x}, \mathbf{y} \rangle = \sum_{i=1}^{n} x_i y_i. \tag{8.1}$$

The reader should verify that this product satisfies all three conditions to be called a scalar product.

Theorem 8.1.1 *In any (real) vector space V equipped with a scalar product as defined in Definition 8.1.1, we define*

$$\|\mathbf{x}\| = \sqrt{\langle \mathbf{x}, \mathbf{x} \rangle} \tag{8.2}$$

for all $\mathbf{x} \in V$. The function $\| \cdot \|$ as defined in Equation (8.2) is a norm, as in Definition 2.4.4, and the Cauchy–Schwarz *Inequality is satisfied:*

$$|\langle \mathbf{x}, \mathbf{y} \rangle| \leq \|\mathbf{x}\|\|\mathbf{y}\|.$$

Proof: To prove the Cauchy–Schwarz inequality, we fix $\mathbf{x}$ and $\mathbf{y}$, and we proceed as follows. For all $t \in \mathbb{R}$, define a polynomial

$$p(t) = \langle t\mathbf{x} + \mathbf{y}, t\mathbf{x} + \mathbf{y} \rangle.$$

Observe that $p(t) \geq 0$ for all t. By linearity of $\langle \cdot, \cdot \rangle$ in each variable we see that

$$p(t) = \|\mathbf{x}\|^2 t^2 + 2\langle \mathbf{x}, \mathbf{y} \rangle t + \|\mathbf{y}\|^2 = at^2 + bt + c,$$

where $a = \|\mathbf{x}\|^2$, $b = 2\langle \mathbf{x}, \mathbf{y} \rangle$, and $c = \|\mathbf{y}\|^2$. But the quadratic polynomial $p(t) \geq 0$ for all $t \in \mathbb{R}$ if and only if

$$b^2 - 4ac \leq 0,$$

which is equivalent to $b^2 \leq 4ac$. Hence

$$|\langle \mathbf{x}, \mathbf{y} \rangle|^2 \leq \|\mathbf{x}\|^2 \|\mathbf{y}\|^2.$$

The first two conditions of Definition 2.4.4 are easily verified for $\|\cdot\|$. The third condition, the triangle inequality, is left for Exercise 8.1.   ■

Definition 8.1.2 *If a vector space V is equipped with a norm $\|\cdot\|$, we define the* open ball *of radius $r \geq 0$ about $\mathbf{p} \in V$ by*

$$B_r(\mathbf{p}) = \left\{\mathbf{v} \in V \mid \|\mathbf{v} - \mathbf{p}\| < r\right\}.$$

Similarly, we define the closed ball *of radius $r \geq 0$ about $\mathbf{p} \in V$ by*

$$\bar{B}_r(\mathbf{p}) = \left\{\mathbf{v} \in V \mid \|\mathbf{v} - \mathbf{p}\| \leq r\right\}.$$

The *bar* above the symbol B_r indicates that the set $\bar{B}_r$ is the *closed* ball, meaning that it includes the spherical boundary surface.

■ **EXAMPLE 8.2**

In the familiar Cartesian plane of Euclidean geometry, with the norm of a vector being its geometrical length, we have

$$\langle \mathbf{x}, \mathbf{y} \rangle = x_1 y_1 + x_2 y_2 = \|\mathbf{x}\| \|\mathbf{y}\| \cos\theta,$$

where θ is the angle between $\mathbf{x}$ and $\mathbf{y}$. Thus the Cauchy–Schwarz inequality in the plane follows from the fact that $|\cos\theta| \leq 1$ for all θ. (The Cauchy–Schwarz inequality follows alternatively from the argument given in Theorem 8.1.1.) In the plane of Euclidean geometry, $B_1(\mathbf{0})$ is the region strictly inside the circle of radius 1 centered at the origin. The reader should check that $B_0(\mathbf{p}) = \emptyset$, the empty set, for all points $\mathbf{p}$.

Definition 8.1.3 *If $\mathbf{x} = (x_1, x_2, \ldots, x_n)$ and $\mathbf{y} = (y_1, y_2, \ldots, y_n)$ are in $\mathbb{R}^n$, the set of all n-tuples of real numbers, we define the* Euclidean scalar product *of $\mathbf{x}$ and $\mathbf{y}$ as in Equation (8.1) and the* Euclidean norm *of $\mathbf{x}$ as in Equation (8.2). The* Euclidean space $\mathbb{E}^n$ *of n dimensions is defined to be the vector space $\mathbb{R}^n$ equipped with the Euclidean scalar product and the Euclidean norm.*

The reader should have checked that the Euclidean scalar product as defined above does satisfy the three properties required to be a scalar product, and that the Euclidean norm is in fact a norm.

We remark that it is very common when dealing with $\mathbb{E}^n$ to call it $\mathbb{R}^n$ informally. This means that we should suppose that the Euclidean inner product and norm are in use unless stated explicitly to the contrary. (See Exercise 8.4.)

Theorem 8.1.2 *The Euclidean normed vector space $\mathbb{E}^n$ has the following two properties.*

i. *The Euclidean space $\mathbb{E}^n$ is complete in the sense of Definition 2.5.3.* (in the cauchy sense)

ii. *For a sequence of vectors $\mathbf{x}^{(j)}$ in $\mathbb{E}^n$, the sequence*

$$\mathbf{x}^{(j)} \to \mathbf{x} \in \mathbb{E}^n \iff x_k^{(j)} \to x_k$$

as $j \to \infty$, for each $k = 1, 2, \ldots, n$.

Remark 8.1.1 The second part of the theorem tells us that convergence of a sequence in $\mathbb{E}^n$ is equivalent to convergence in each separate coordinate sequence.

Proof: The reader will recall that the vectors $\mathbf{x} \in \mathbb{R}^n$, which are the vectors of $\mathbb{E}^n$, comprise a vector space using the familiar operations under which

$$c\mathbf{x} = (cx_1, \ldots, cx_n)$$

and

$$\mathbf{x} + \mathbf{y} = (x_1 + y_1, \ldots, x_n + y_n).$$

We begin by proving claim (i). To see that every convergent sequence in any normed vector space is Cauchy in the sense of Definition 2.5.3, see Exercise 8.8. We will prove that every Cauchy sequence $\mathbf{x}^{(k)}$ in V converges. By hypothesis, For each $\epsilon > 0$ there exists K such that j and $k \geq K$ implies $\left\| \mathbf{x}^{(j)} - \mathbf{x}^{(k)} \right\| < \epsilon$. However, for each $l = 1, \ldots, n$, we have

$$\left| x_l^{(j)} - x_l^{(k)} \right| \leq \sqrt{\sum_{m=1}^{n} \left(x_m^{(j)} - x_m^{(k)} \right)^2} = \left\| \mathbf{x}^{(j)} - \mathbf{x}^{(k)} \right\|.$$

Thus the sequence $x_l^{(j)}$ is Cauchy in $\mathbb{R}$, and so it has a limit: $x_l^{(j)} \to x_l$ as $j \to \infty$, for each $l = 1, \ldots, n$. Now we denote $\mathbf{x} = (x_1, \ldots, x_n)$. Then we conclude that

$$\to \quad \left\| \mathbf{x}^{(j)} - \mathbf{x} \right\| = \sqrt{\sum_{l=1}^{n} \left(x_l^{(j)} - x_l \right)^2} \to 0$$

as $j \to \infty$, by the theorems governing limits of sequences. Hence the original Cauchy sequence does converge in the sense of the norm of $\mathbb{E}^n$. Note that we have also proven that if each sequence of coordinates converges, then the sequence of vectors in $\mathbb{E}^n$ converges as well. We leave the second claim to Exercise 8.9.

■ **EXAMPLE 8.3**

Here are two examples of convergence and *divergence*, which means *failure of convergence*, for sequences of vectors.

(i) Let $\mathbf{x}^{(j)} = \left(\frac{j^2+1}{2j^2-3}, j \sin \left(\frac{1}{j} \right) \right)$. Then $x^{(j)} \to \left(\frac{1}{2}, 1 \right) \in \mathbb{E}^2$ as $j \to \infty$, because of the convergence of the sequence in each coordinate separately.

ii. In $\mathbb{E}^2$, the sequence $x^{(j)} = \left(\frac{1}{j}, j \right)$ diverges as $j \to \infty$.

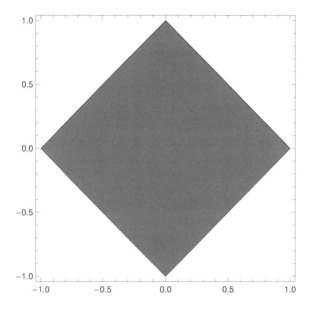

Figure 8.1 The unit ball $B_1(\mathbf{0})$ in the taxicab norm.

■ **EXAMPLE 8.4**

Here is an example of a *non-Euclidean* norm that can be placed on the vector space $\mathbb{R}^2$. Define the *taxicab* norm on $\mathbb{R}^2$ by

$$\|\mathbf{x}\|_t = |x_1| + |x_2|.$$

In Exercise 8.4 the reader will show that the taxicab norm satisfies the requirements of Definition 2.4.4 to be a norm. To understand the name *taxicab norm* consider a city in which all roads run parallel to the x-axis or the y-axis. The length of an honest taxicab ride would be determined using the taxicab norm. Fig. 8.1 shows the unit ball around the origin in the taxicab norm.

EXERCISES

8.1 † Complete the proof of Theorem 8.1.1 by proving the triangle inequality:

$$\|\mathbf{x} + \mathbf{y}\| \le \|\mathbf{x}\| + \|\mathbf{y}\|$$

for all $\mathbf{x}$ and $\mathbf{y}$ in V. That is, prove that the norm defined in terms of a given scalar product does satisfy the triangle inequality. (Hint: Apply the Cauchy–Schwarz Inequality.)

8.2 In $\mathbb{E}^n$, $\|\mathbf{x} - \mathbf{y}\|$ is interpreted geometrically as being the distance between $\mathbf{x}$ and $\mathbf{y}$. Prove the *geometrical version* of the triangle inequality in $\mathbb{E}^n$:

$$\|\mathbf{x} - \mathbf{z}\| \le \|\mathbf{x} - \mathbf{y}\| + \|\mathbf{y} - \mathbf{z}\|$$

for all $\mathbf{x}, \mathbf{y}$, and $\mathbf{z}$ in $\mathbb{E}^n$. Interpret this inequality in terms of a geometrical triangle in $\mathbb{E}^n$.

8.3   Two vectors $\mathbf{x}$ and $\mathbf{y}$ in $\mathbb{E}^n$ are called *orthogonal* if and only if

$$\langle \mathbf{x}, \mathbf{y} \rangle = 0.$$

We will prove the Pythagorean Theorem in $\mathbb{E}^n$.

     a) Prove: The vectors $\mathbf{x}$ and $\mathbf{y}$ are orthogonal if and only if

$$\|\mathbf{x}\|^2 + \|\mathbf{y}\|^2 = \|\mathbf{x} + \mathbf{y}\|^2.$$

     b) Interpret part (a) as a theorem about triangles in $\mathbb{E}^n$ with a vertex at the origin.

     c) Let $\mathbf{x}, \mathbf{y}$, and $\mathbf{z}$ be in $\mathbb{E}^n$. Prove that the triangle with vertices at $\mathbf{x}, \mathbf{y}$ and $\mathbf{z}$ is a right triangle with right angle at $\mathbf{x}$ if and only if

$$\|\mathbf{x} - \mathbf{y}\|^2 + \|\mathbf{x} - \mathbf{z}\|^2 = \|\mathbf{y} - \mathbf{z}\|^2.$$

8.4   † Define the *taxicab norm* as in Example 8.4. Show that $\| \cdot \|_t$ satisfies the requirements of Definition 2.4.4 to be a norm.

8.5   Find necessary and sufficient conditions on the real numbers a_1 and a_2 to assure that

$$\langle \mathbf{x}, \mathbf{y} \rangle_{\mathbf{a}} = a_1 x_1 y_1 + a_2 x_2 y_2$$

satisfies Definition 8.1.1 and is thus a bona fide scalar product on the two-dimensional vector space $\mathbb{R}^2$. Then sketch the unit ball around the origin in $\mathbb{R}^2$ using the norm $\| \cdot \|_{\mathbf{a}}$ that is determined by this scalar product.

8.6   In $\mathbb{E}^1$, what is the open ball $B_1(\mathbf{0})$? Answer the same question in $\mathbb{E}^3$.

8.7   Find the $\lim_{j \to \infty} \mathbf{x}^{(j)}$ in $\mathbb{E}^2$, or state that it does not exist.

     a) $\mathbf{x}^{(j)} = \left(\frac{1}{j}, \left(1 + \frac{1}{j}\right)^j \right)$.

     b) $\mathbf{x}^{(j)} = \left(\frac{1}{j}, \frac{e^j}{j^{100}} \right)$.

8.8   † Prove that every convergent sequence $\mathbf{x}^{(k)}$ in a normed vector space V must be Cauchy, as claimed in Theorem 8.1.2.

8.9   † Show that $\mathbf{x}^{(j)} \to \mathbf{x}$ in the sense of the norm of $\mathbb{E}^n$ if and only if the sequence $x_l^{(j)} \to x_l$ for each $l = 1, \dots, n$, completing the proof of Theorem 8.1.2.

8.10   Suppose $t_k \to t \in \mathbb{R}$ and $\mathbf{x}^{(k)} \to \mathbf{x} \in \mathbb{E}^n$ as $k \to \infty$. Prove: $t_k \mathbf{x}^{(k)} \to t\mathbf{x}$ as $k \to \infty$.

8.11   A sequence $\mathbf{x}^{(j)}$ in $\mathbb{E}^n$ is called *bounded* if there exists a number $M > 0$ such that $\|\mathbf{x}^{(j)}\| \le M$ for all $j \in \mathbb{N}$.

     a) Prove that every convergent sequence in $\mathbb{E}^n$ is bounded.

b) Give an example of a bounded sequence in $\mathbb{E}^2$ that is not convergent.

c) † *(The Bolzano–Weierstrass Theorem)* Prove that every bounded sequence in $\mathbb{E}^n$ has a convergent subsequence. (Hint: One approach is to give a proof by induction on the dimension n. Or one can work informally as follows, applying the Bolzano–Weierstrass Theorem for $\mathbb{R}$ a total of n times. Here is a notational suggestion: If one wishes to denote a subsequence of a subsequence, use two *strictly increasing* functions ϕ_1 and ϕ_2 mapping $\mathbb{N}$ into itself, and $\mathbf{x}^{(\phi_1 \circ \phi_2(j))}$ can be used as in Definition 1.5.1 to denote the jth term of a subsequence of a subsequence of a sequence $\mathbf{x}^j$.)

8.12 Prove: A sequence $\mathbf{x}_n \to \mathbf{L} \in \mathbb{E}^k$ if and only if *every* subsequence $\mathbf{x}_{n_i}$ possesses a sub-subsequence $\mathbf{x}_{n_{i_j}}$ that converges to $\mathbf{L}$ as $j \to \infty$. (Hint: To prove the *only if* part, suppose false and write out the logical negation of convergence of $\mathbf{x}_n$ to $\mathbf{L}$.)

8.13 *Prove or give a counterexample*: A sequence $\mathbf{x}_n \in \mathbb{E}^k$ converges if and only if *every* subsequence $\mathbf{x}_{n_i}$ possesses a sub-subsequence $\mathbf{x}_{n_{i_j}}$ that converges as $j \to \infty$.

8.14

a) Suppose a vector space V is equipped with an inner product $\langle \cdot, \cdot \rangle$, and suppose we define a corresponding norm by $\|\mathbf{x}\|^2 = \langle \mathbf{x}, \mathbf{x} \rangle$. Prove the *Parallelogram Law:*

$$\|\mathbf{x} + \mathbf{y}\|^2 + \|\mathbf{x} - \mathbf{y}\|^2 = 2\|\mathbf{x}\|^2 + 2\|\mathbf{y}\|^2.$$

b) Prove that the taxicab norm defined in Exercise 8.4 does not correspond as in part (a) above to any inner product on $\mathbb{R}^2$.

c) Under the hypotheses of part (a) above, prove the identity

$$\langle \mathbf{x}, \mathbf{y} \rangle = \frac{1}{4} \left(\|\mathbf{x} + \mathbf{y}\|^2 - \|\mathbf{x} - \mathbf{y}\|^2 \right).$$

d) *(For this part, you might be able to negotiate extra credit from your teacher.)* Suppose only that V has a norm. Define what is *hoped* to be a scalar product on V by the formula in the preceding part. Prove that this defines a legitimate scalar product on V provided the norm satisfies the Parallelogram Law of part. (Hints: Positivity and symmetry are easy to verify. To show additivity of the product in the first variable, express $4\langle \mathbf{x} + \mathbf{y}, \mathbf{z} \rangle$ in terms of the identity in given in the previous part. Then add and subtract $\|\mathbf{x} - \mathbf{y} + \mathbf{z}\|^2$ to show that

$$4\langle \mathbf{x} + \mathbf{y}, \mathbf{z} \rangle = 2(\|\mathbf{x} + \mathbf{z}\|^2 + \|\mathbf{y}\|^2 - \|\mathbf{x}\|^2 - \|\mathbf{y} - \mathbf{z}\|^2).$$

Then let $\mathbf{x}$ and $\mathbf{y}$ change roles and add the results. When this is done, show that $\langle \alpha \mathbf{x}, \mathbf{y} \rangle = \alpha \langle \mathbf{x}, \mathbf{y} \rangle$ for all $\alpha \in \mathbb{N}$. Extend this to $\alpha \in \mathbb{Q}$. Then justify that the inner product is a continuous function of $\mathbf{x}$ and extend to all $\alpha \in \mathbb{R}$.)

8.15   Prove that Euclidean space is *self-dual*. That is, prove that $T : \mathbb{E}^n \to \mathbb{R}$ is a bounded linear functional if and only if there exists $\mathbf{y} \in \mathbb{E}^n$ such that $T(\mathbf{x}) = \langle \mathbf{x}, \mathbf{y} \rangle$ for each $\mathbf{x} \in \mathbb{E}^n$.

8.2   OPEN SETS AND CLOSED SETS

In the calculus of one variable, it is possible to accomplish much with functions defined on an interval. Intervals are quite simple sets to consider as domains on which functions may be defined. Many useful theorems concerning the continuity, differentiability, or integrability of functions can be proven based on whether a domain is a closed or an open interval, or a finite or an infinite interval. In two or more variables, however, the domain of definition of a function can be much more complicated than an interval. For example, the real-valued function

$$f(x, y) = \frac{1}{x\sqrt{1 - 2x^2 - 3y^2}}$$

is defined on the domain that is the region lying *inside* but *not on* an ellipse, *excluding* those points inside the ellipse that are on the y-axis. It is easy to construct more intricate examples than this. Hence a rigorous study of the calculus of functions defined on a Euclidean space $\mathbb{E}^n$ necessitates a careful study of those *topological* properties of sets in $\mathbb{E}^n$ that are needed to prove important theorems regarding continuity, differentiability, and integrability of such functions. Most of the definitions and theorems work for all vector spaces V equipped with a norm, so we state them in this context.

Definition 8.2.1 *In a normed vector space V a subset $S \subseteq V$ is called an* open set *if and only if for each $\mathbf{x} \in S$ there exists a number $r > 0$ corresponding to $\mathbf{x}$ such that the open ball $B_r(\mathbf{x}) \subseteq S$.* S is open $\Leftrightarrow \forall \vec{x} \in S \ \exists r > 0 \ s.th \ B_r(\vec{x}) \in S$

Expressed informally, a set S is said to be open if each point $\mathbf{p}$ of S has at least some small open ball, or *buffer zone* around itself that remains entirely inside S. The radius of the buffer zone may need to be smaller for some points of S than for others.

■ **EXAMPLE 8.5**

The first example is the open ball $B_r(\mathbf{x})$, which we claim is an open set. (The language *open ball* would be a poor choice if the claim just made were false, but a name alone does not establish that a definition is satisfied.)

Proof: To show that the definition of openness is satisfied, we must show that if $\mathbf{y} \in B_r(\mathbf{x})$ then there exists $d > 0$ such that $B_d(\mathbf{y}) \subseteq B_r(\mathbf{x})$. We claim that $d = r - \|\mathbf{y} - \mathbf{x}\| > 0$ works. In fact, if $\mathbf{z} \in B_d(\mathbf{y})$, then

$$\|\mathbf{z} - \mathbf{x}\| \leq \|\mathbf{z} - \mathbf{y}\| + \|\mathbf{y} - \mathbf{x}\|$$
$$< d + \|\mathbf{y} - \mathbf{x}\| = r,$$

proving that $\mathbf{z} \in B_r(\mathbf{x})$ as claimed.   ∎

Theorem 8.2.1 *Suppose A is an arbitrary set. Suppose for each $a \in A$ there is an open set $O_a \subseteq V$ indexed by a. Then $\mathcal{O} = \bigcup_{a \in A} O_a$ is an open set.* See p. 28

Remark 8.2.1 In this theorem, the *index set* A may be finite or infinite, countable or uncountable.

Proof: In words, this theorem says that the union of a family of open sets must always be open. Let $\mathbf{x} \in \mathcal{O}$. We must show there exists $r > 0$ such that $B_r(\mathbf{x}) \subseteq \mathcal{O}$. But there exists $a \in A$ such that $\mathbf{x} \in O_a$. Since O_a is open, there exists $r > 0$ such that $B_r(\mathbf{x}) \subseteq O_a \subseteq \mathcal{O}$.   ∎

It is natural at this point to ask whether the intersection of an arbitrary family of open sets must be open. This is false, as shown in Exercise 8.21.

Theorem 8.2.2 *If $O_1, \ldots, O_k$ is a family of finitely many open sets, then $\mathcal{O} = \bigcap_{j=1}^{k} O_j$ is open.*

Proof: In words, this says the intersection of finitely many open sets must be open. For each $\mathbf{x} \in \mathcal{O}$, $\mathbf{x} \in O_j$ for each j. Hence there exists for each j a number $r_j > 0$ such that $B_{r_j}(\mathbf{x}) \subseteq O_j$. Let $r = \inf\{r_1, \ldots, r_k\}$, so that $r > 0$. Then $B_r(\mathbf{x}) \subseteq \mathcal{O}$.   ∎

Definition 8.2.2 *In a normed vector space V, a set $S \subseteq V$ is called a* closed set *if and only if its complement $S^c = V \setminus S$ is open.*

We will use the following definition to prove an alternative and equivalent form of the concept of *closed* set.

Definition 8.2.3 *A point $\mathbf{p}$ is called a* cluster point *of a subset S of a normed vector space V if and only if there exists a sequence of vectors $\mathbf{x}^{(j)} \in S \setminus \{\mathbf{p}\}$ such that $\mathbf{x}^{(j)} \to \mathbf{p}$ as $j \to \infty$. If $\mathbf{p} \in S$ but $\mathbf{p}$ is not a cluster point of S, then $\mathbf{p}$ is called an* isolated point *of S.*

Note that a cluster point of S may or may not be an element of the set S. An isolated point of S must, however, be an element of S.

Theorem 8.2.3 *A subset S of a normed vector space V is closed if and only if every cluster point $\mathbf{p}$ of S is an element of S.*   *ie, if it contains all its cluster points*

Proof: Suppose first that S is closed. Let $\mathbf{p}$ be any cluster point of S. We need to prove that $\mathbf{p} \in S$. By hypothesis, there is a sequence $\mathbf{x}^{(j)} \in S \setminus \{\mathbf{p}\}$ such that $\mathbf{x}^{(j)} \to \mathbf{p}$ as $j \to \infty$. Thus, for every open ball $B_r(\mathbf{p})$ there exists $J \in \mathbb{N}$ such that $j \geq J \implies \mathbf{x}^{(j)} \in B_r(\mathbf{p})$. Thus $\mathbf{p} \notin S^c = V \setminus S$ since the latter set is open. Hence $\mathbf{p} \in S$.

Next, we suppose every cluster point of S is in S. We need to prove that the complement of S is open. Let $\mathbf{x} \in S^c = V \setminus S$. We claim there exists $r > 0$ such that $B_r(\mathbf{x}) \subseteq S^c = V \setminus S$. Suppose this claim were false. Then for each $j \in \mathbb{N}$

there exists $\mathbf{x}^{(j)} \in S \setminus \{\mathbf{x}\}$ such that $\mathbf{x}^{(j)} \in B_{\frac{1}{j}}(\mathbf{x})$. But then $\mathbf{x}^{(j)} \to \mathbf{x}$. Thus $\mathbf{x}$ is a cluster point of S, which forces $\mathbf{x} \in S$. This is a contradiction, which proves the claim. ∎

■ EXAMPLE 8.6

Every closed ball $\bar{B}_r(\mathbf{x})$ is a closed set.

Proof: We will show that the complement of the closed ball is open. If $\mathbf{y} \in V \setminus \bar{B}_r(\mathbf{x})$, let $\delta = \|\mathbf{y} - \mathbf{x}\| - r$. We claim that $B_\delta(\mathbf{y})$ is contained in the complement of the closed ball. So let $\mathbf{z} \in B_\delta(\mathbf{y})$. Then

$$\|\mathbf{z} - \mathbf{x}\| + \|\mathbf{z} - \mathbf{y}\| \geq \|\mathbf{x} - \mathbf{y}\|$$

by the triangle inequality, so

$$\|\mathbf{z} - \mathbf{x}\| \geq \|\mathbf{x} - \mathbf{y}\| - \|\mathbf{z} - \mathbf{y}\|$$
$$> \|\mathbf{x} - \mathbf{y}\| - \delta = r.$$

Thus $\mathbf{z}$ lies outside the closed ball, as required. ∎

■ EXAMPLE 8.7

It follows from the preceding example that for each $\mathbf{p} \in \mathbb{E}^n$, the singleton set $\{\mathbf{p}\} = \bar{B}_0(\mathbf{p})$ is a closed set. The reader should show that this singleton set $\{\mathbf{p}\}$ is not an open set. Note that this does *not* follow from some linguistic relationship between the words *open* and *closed* in every day parlance. Instead, it is necessary to show that if $r > 0$ then $B_r(\mathbf{p})$ is not a subset of $\{\mathbf{p}\}$.

Theorem 8.2.4 *Suppose A is an arbitrary set. (A may be finite or infinite, countable or uncountable.) Suppose for each $a \in A$ there is a closed set $F_a \subseteq \mathbb{E}^n$ indexed by a. Then $\mathcal{F} = \bigcap_{a \in A} F_a$ is a closed set.*

Proof: It suffices to observe the

$$\mathcal{F} = \left(\bigcup_{a \in A} F_a^c \right)^c.$$

Since each set F_a^c is open being the complement of the closed set F_a, the union is open and its complement, $\mathcal{F}$, is closed. ∎

EXERCISES

8.16   Prove that Theorem 8.2.2 implies that the *empty set* $\emptyset$ is open in any normed vector space V, and that the whole space V is closed.

8.17   For each of the following sets, state whether it is open, closed, both, or neither.

    a) $A = \{\mathbf{x} \in \mathbb{E}^2 \mid x_1 \neq 0\}$.

    b) $B = \{\mathbf{x} \in \mathbb{E}^2 \mid x_1 = 0\}$.

    c) $C = \{\mathbf{x} \in \mathbb{E}^2 \mid x_1 \geq 0 \text{ and } x_2 > 0\}$.

    d) $E = \{\mathbf{x} \in \mathbb{E}^2 \mid x_1 + x_2 > 0\}$.

    e) $F = \{\mathbf{x} \in \mathbb{E}^2 \mid x_1 + x_2 \leq 0\}$.

    f) $D = A \cup B$.

    g) $G = \{\mathbf{x} \in \mathbb{E}^2 \mid 1 < x_1^2 + x_2^2 < 4\}$.

8.18   Prove that a point $\mathbf{p}$ is a cluster point of a set $D \subseteq \mathbb{E}^n$ if and only if for each $\delta > 0$ there exists $\mathbf{x} \in D$ such that $0 < \|\mathbf{x} - \mathbf{p}\| < \delta$.

8.19   Prove that if $\mathbf{p}$ is an isolated point of D, once $\delta > 0$ is sufficiently small the set $\{\mathbf{x} \in D \mid 0 < \|\mathbf{x} - \mathbf{p}\| < \delta\} = \emptyset$, the empty set.

8.20   Let $\mathcal{O}$ denote an *arbitrary open* subset of $\mathbb{E}^n$. Prove that $\mathcal{O}$ is the union of a (possibly infinite) family of open *balls*.

8.21   † Let $\mathbf{p} \in \mathbb{E}^n$. Show that the singleton set $\{\mathbf{p}\}$ can be written as the intersection of an infinite sequence of open balls, thereby showing that intersections of open sets need not be open. (Hint: Define a sequence $r_n > 0$ for which $\inf\{r_n \mid n \in \mathbb{N}\} = 0$.)

8.22   Prove or give a counterexample: No set $S \subseteq \mathbb{E}^n$ can be *both* open and closed.

8.23   Prove or give a counterexample: Every set $S \subseteq \mathbb{E}^n$ must be either open or closed.

8.24   Prove that if $F_1, \ldots, F_k$ is a family of finitely many closed sets, then $\mathcal{F} = \bigcup_{j=1}^{k} F_j$ is closed.

8.25   Use Example 8.7 to prove that every set $S \subseteq \mathbb{E}^n$ is the union of a family of closed sets.

8.26   † If $S \subset \mathbb{E}^n$, define the *interior* of S, denoted by S^o, to be

$$S^o = \{\mathbf{x} \in S \mid \exists\, r > 0 \text{ such that } B_r(\mathbf{x}) \subseteq S\}.$$

Prove that $S \subseteq \mathbb{E}^n$ is an open set if and only if $S = S^o$.

8.27   Prove that $\left(\bar{B}_r(\mathbf{x})\right)^o = B_r(\mathbf{x})$ in the space $\mathbb{E}^n$.

8.28   Define the *closure* $\bar{S}$ of $S \subseteq \mathbb{E}^n$ to be the intersection of all closed sets that contain S. Prove that $\bar{S}$ is a closed set and that $\bar{S} = S \cup C$, where C is the set of all cluster points of S.

8.29   Prove: The closed ball $\bar{B}_r(\mathbf{x})$, defined in Definition 8.1.2, is actually the closure of the open ball $B_r(\mathbf{x})$.

8.30   Define the *boundary* of $S \subseteq \mathbb{E}^n$, denoted by ∂S, to be $\partial S = \bar{S} \setminus S^o$. Find the boundary of $B_r(\mathbf{x})$ in $\mathbb{E}^n$.

8.31   Define a set $S \subseteq \mathbb{E}^n$ to be dense in $\mathbb{E}^n$ if and only if $\bar{S} = \mathbb{E}^n$. Denote by $\mathbb{Q}^n$ the subset of $\mathbb{E}^n$ consisting of vectors with exclusively rational coordinates. Prove that $\mathbb{Q}^n$ is dense in $\mathbb{E}^n$.

8.32   Prove or give a counterexample: Every open set in $\mathbb{E}^n$ can be expressed as the union of either a finite or a countable family of open balls.

8.33   Find $(\mathbb{Q}^n)^o$ and $\partial \mathbb{Q}^n$. Justify your answers. (Hint: It will help to consider the density of the set of irrational numbers in the set $\mathbb{R}$.)

8.3   COMPACT SETS

In Chapters 1 through 3, we saw that closed, finite intervals $[a, b]$ have very useful properties as domains of continuous functions. Such intervals were very important for establishing an Extreme Value Theorem (Theorem 2.4.2) and a uniform continuity theorem (Theorem 2.3.2), for example. In this section we seek to describe the class of subsets of $\mathbb{E}^n$ that have similar properties as domains for functions of vector variables. We take our cue from the Heine–Borel Theorem (Theorem 1.7.2) for functions defined on intervals $[a, b]$.

Definition 8.3.1   *If $E \subseteq \mathbb{E}^n$, we call a family*

$$\mathcal{O} = \{O_a \mid a \in A\}$$

of open sets O_a an open cover *of E if and only if*

$$E \subseteq \bigcup_{a \in A} O_a.$$

If there exists a finite subset F of the index set A such that

$$E \subseteq \bigcup_{a \in F} O_a,$$

then we call $\mathcal{O}_F = \{O_a \mid a \in F\}$ a finite subcover of E. We call the set E compact *if and only if every open cover of E has a finite subcover. A set E is called* bounded *if and only if $\sup \{\|\mathbf{x}\| \mid \mathbf{x} \in E\} < \infty$.*

⋆   The reader should take care not to confuse the concept of an open cover with the union of its elements. ~~The open cover is the set of sets~~

■ **EXAMPLE 8.8**

Let
$$S = \{(m, n) \mid m \in \mathbb{Z}, \, n \in \mathbb{Z}\} = \mathbb{Z}^2 = \mathbb{Z} \times \mathbb{Z}.$$

We claim that the subset S of $\mathbb{E}^2$ is not compact. For a proof, we observe that

$$S \subset \bigcup_{(m,n) \in S} B_1((m, n)).$$

However, the open cover $\{B_1((m, n)) \mid m \in \mathbb{Z},\, n \in \mathbb{Z}\}$ of S has no finite subcover. Do you see why not?

■ EXAMPLE 8.9

The set $B_1(\mathbf{0})$ is not a compact subset of $\mathbb{E}^2$. For a proof, show that

$$\mathcal{O} = \{B_r(\mathbf{0}) \mid 0 < r < 1\}$$

is an open cover of $B_1(\mathbf{0})$, yet there is no finite subcover. The reader should show why this is true.

The following theorem provides a powerful generalization of the phenomena exhibited by the two preceding examples.

Theorem 8.3.1 (Heine–Borel Theorem for $\mathbb{E}^n$) *A set $E \subseteq \mathbb{E}^n$ is compact if and only if E is* both *closed and bounded.*

Proof: We leave the proof that every compact set must be both bounded and closed to Exercises 8.38 and 8.39, respectively. Here we suppose that E is both closed and bounded, and we will prove that E is necessarily compact. We suppose that this were *false*, and deduce a contradiction.

Thus we assume there exists an open cover $\mathcal{O} = \{O_a \mid a \in A\}$ of E for which there exists *no* finite subcover. Since E is bounded, there exists $R > 0$ such that $E \subseteq B_R(\mathbf{0})$, which is, in turn, contained in the *hyper*cube $[-R, R]^{\times n}$, an n-fold Cartesian product of the interval $[-R, R]$. It will be convenient in this proof to denote the hypercube by a pair of diagonally opposite corner-vectors $\mathbf{a}_1 = (-R, -R, \ldots, -R)$ and $\mathbf{b}_1 = (R, R, \ldots, R)$. Thus we will denote the cube

$$[-R, R]^{\times n} = [\mathbf{a}_1, \mathbf{b}_1].$$

Next we subdivide the cube $[\mathbf{a}_1, \mathbf{b}_1]$ into 2^n congruent subcubes as follows. Simply bisect the interval $[-R, R]$ on each of the n coordinate axes, and form all 2^n possible Cartesian products of a half-interval from each of the axes. Each such subcube can be denoted $[\mathbf{a}, \mathbf{b}]$, where the coordinates of $\mathbf{a} = (a_1, a_2, \ldots, a_n)$ denote the left-hand endpoints of the n chosen subintervals on the axes, and with a similar convention is employed for $\mathbf{b}$ using right-hand endpoints. Observe that for each such subcube $[\mathbf{a}, \mathbf{b}] \subset [\mathbf{a}_1, \mathbf{b}_1]$, the set $E \cap [\mathbf{a}, \mathbf{b}]$ is covered by $\mathcal{O}$. Among the 2^n subcubes formed, there must exist at least one subcube $[\mathbf{a}_2, \mathbf{b}_2]$ having the property that $E \cap [\mathbf{a}_2, \mathbf{b}_2]$ has no finite subcover from $\mathcal{O}$. (Otherwise, there would exist a finite subcover for E itself.) Now we repeat the process by subdividing $[\mathbf{a}_2, \mathbf{b}_2]$ into 2^n subcubes and we select a subcube $[\mathbf{a}_3, \mathbf{b}_3]$ in the same manner as before. In this way, we obtain a decreasing nest

$$[\mathbf{a}_1, \mathbf{b}_1] \supset [\mathbf{a}_2, \mathbf{b}_2] \supset \ldots \supset [\mathbf{a}_k, \mathbf{b}_k] \supset \ldots$$

of subcubes having the property that for each $k \in \mathbb{N}$ the set $E_k = E \cap [\mathbf{a}_k, \mathbf{b}_k]$ is covered by $\mathcal{O}$ but has no finite subcover.

Select a point $\mathbf{p}_k \in E_k$ for each $k \in \mathbb{N}$. If j and $k \geq N$ then $\mathbf{p}_j$ and $\mathbf{p}_k \in E_N$. Hence

$$\|\mathbf{p}_j - \mathbf{p}_k\| \leq \frac{R\sqrt{n}}{2^{N-2}} \to 0$$

as $N \to \infty$. Thus the sequence $\mathbf{p}_k$ is a Cauchy sequence, and $\mathbf{p}_k \to \mathbf{p} \in E_N$ for each $N \in \mathbb{N}$. (Here we use the fact that E_N is a closed set.) Since $\mathbf{p} \in E$, there exists $a \in A$ such that $\mathbf{p} \in O_a$. Since O_a is open, there exists $r > 0$ such that $B_r(\mathbf{p}) \subseteq O_a$. Now select $N \in \mathbb{N}$ such that $k \geq N$ implies

$$\frac{R\sqrt{n}}{2^{k-2}} < r,$$

so that $\mathbf{p} \in E_N \subset B_r(\mathbf{p}) \subseteq O_a$. Thus we have covered E_N with a *single* set O_a from $\mathcal{O}$, contradicting the claim that E_N has no finite subcover from $\mathcal{O}$. ∎

EXERCISES

8.34   If E is a compact subset of $\mathbb{E}^n$, prove (*without using the Heine–Borel Theorem*) that every closed subset of E is compact.

8.35   Use *only* the definition of compactness to show that every finite subset of $\mathbb{E}^n$ is compact.

8.36

    a) Let $S = \{\mathbf{x}^{(j)} \mid j \in \mathbb{N}\} \subset \mathbb{E}^n$ be any *convergent* sequence. Show that S is a compact set if and only if $\lim_{j \to \infty} \mathbf{x}^{(j)} \in S$.

    b) Let $S = \{\mathbf{x}_n = \left(1 + \frac{n}{2^n}, \frac{n}{2^n}\right) \mid n = 0, 1, 2, 3, \ldots\} \subset \mathbb{E}^2$. True or False: S is compact.

8.37   Let $S = \{\mathbf{x}^{(j)} \mid j \in \mathbb{N}\} \subset \mathbb{E}^n$ be *any* sequence. Prove or give a counterexample: The set S is a compact set if and only if the sequence $\mathbf{x}^{(j)}$ is convergent to an element of S.

8.38   † Prove part of the Heine–Borel theorem by showing that if $E \subseteq \mathbb{E}^n$ is *compact*, then E is *bounded*. (Hint: Show how to cover E with suitable open balls centered at $\mathbf{0}$.)

8.39   † Prove part of the Heine–Borel theorem by showing that if $E \subseteq \mathbb{E}^n$ is *compact*, then E is *closed*. (Hint: Suppose false, and let $\mathbf{p} \notin E$. Show how to cover E with the complements of suitable closed balls centered at $\mathbf{p}$. Conclude that E^c is *open*.)

8.40   † Let f be a real-valued function defined on a closed finite interval $[a, b]$ in $\mathbb{E}^1$. Define the *graph* $G_f = \{(x_1, x_2) \in \mathbb{E}^2 \mid x_2 = f(x_1), x_1 \in [a, b]\}$.

    a) Prove: If $f \in C[a, b]$, then G_f is a compact subset of $\mathbb{E}^2$. (Hint: Use the Heine–Borel Theorem.)

b) Let

$$g(x_1) = \begin{cases} \sin \frac{\pi}{x_1} & \text{if } x_1 \in (0, 1], \\ 0 & \text{if } x_1 = 0. \end{cases}$$

Is the graph G_g a compact subset of $\mathbb{E}^2$? Prove your conclusion.

8.41 Let $E_1 \supseteq E_2 \supseteq \ldots \supseteq E_k \supseteq \ldots$ be a decreasing nest of *nonempty closed* subsets of $\mathbb{E}^n$.

a) Give an example to show it is possible for $\bigcap_{k=1}^{\infty} E_k$ to be empty.

b) If E_1 is compact, show that $\bigcap_{k=1}^{\infty} E_k \neq \emptyset$. (Hint: Select a point $\mathbf{x}_k \in E_k$ for each $k \in \mathbb{N}$. Apply Exercise 8.11.c.)

8.42 For each of the following subsets of $\mathbb{E}^n$, determine whether or not it is compact and justify your conclusion.

a) $B_r(\mathbf{x})$, with $r > 0$.

b) $\bar{B}_r(\mathbf{x})$, with $r > 0$.

c) $S^{n-1} = \bar{B}_r(\mathbf{x}) \setminus B_r(\mathbf{x})$, with $r > 0$.

8.43 Let $K \subset \mathbb{E}^n$ be compact. Suppose f and g are in $\mathcal{C}(K)$ and suppose that $f(\mathbf{x}) = 0$ for each $\mathbf{x} \in K$ such that $g(\mathbf{x}) = 0$. Prove: If $\epsilon > 0$, there exists $M \geq 0$ such that

$$|f(\mathbf{x})| < M|g(\mathbf{x})| + \epsilon$$

for all $\mathbf{x} \in K$. (Hint: Use the Heine–Borel Theorem. For interesting applications of this exercise in approximation theory, see [13].)

8.4 CONNECTED SETS

In order to generalize the Intermediate Value Theorem (Theorem 2.3.1) to functions defined on Euclidean space, it will be necessary to identify a class of subsets of $\mathbb{E}^n$ with properties sufficiently similar to those of intervals. Let us begin by writing a formal definition of the concept of *interval*.

Definition 8.4.1 *An* interval *is any subset I of $\mathbb{R}$ such that for all a and b in I with $a \leq b$ the set $\{x \in \mathbb{R} \mid a \leq x \leq b\} \subseteq I$.*

This concept includes all open, closed, half-open and half-closed, finite or infinite intervals. (See Exercise 8.44.) The difficulty in generalizing this concept is that there is no natural *linear ordering*, or notion of *inequality* in $\mathbb{E}^n$ when $n > 1$. The concept we seek is that of being a *connected* set. In order to define this concept it is convenient to begin with its opposite: a set is called *disconnected*, in the sense that it can be *separated* into two parts, or components.

Definition 8.4.2 *We say that a set $E \subseteq V$, a normed vector space, is* disconnected *if there exists a pair of* disjoint open *subsets A and B of V for which $E \subseteq A \cup B$ and such that both*

$$E_1 = E \cap A \neq \emptyset \text{ and } E_2 = E \cap B \neq \emptyset.$$

In this case, we say that A and B separate *E. If E is* not *disconnected, then E is called* connected.

 ■ **EXAMPLE 8.10**

Let $E = \mathbb{Q}$, the set of all rational numbers. We claim that E is a *disconnected* subset of $\mathbb{E}^1$. In fact, let

$$A = \left\{ x \in \mathbb{E}^1 \mid x < \sqrt{2} \right\} \quad \text{and} \quad B = \left\{ x \in \mathbb{E}^1 \mid x > \sqrt{2} \right\}.$$

Then A and B are disjoint open sets that separate E. The key to this construction is that there exists $\sqrt{2} \in \mathbb{E}^1 \setminus \mathbb{Q}$.

In Exercise 8.45 the reader will prove that a subset of $\mathbb{E}^1$ is connected if and only if it is an interval. The set $\mathbb{Q}$, which is shown above to be disconnected, is not an interval. We show below that the concept of a set E being disconnected can be expressed without reference to open sets A and B as in the definition. This will give us a more visually intuitive concept of the meaning of connectivity.

Theorem 8.4.1 *A set $E \subseteq V$, a normed vector space, is disconnected if and only if we can decompose E as $E = E_1 \cup E_2$, both sets nonempty, where $E_1 \cap E_2 = \emptyset$ and where neither E_1 nor E_2 contains any cluster points of the other set.*

Proof: First we suppose that E is separated. Thus A and B exist as in the definition. If $\mathbf{e} \in E_1$, then there exists $r > 0$ such that $B_r(\mathbf{e}) \subseteq A$. Since E_2 is disjoint from A, $\mathbf{e}$ is not a cluster point of E_2. Similarly, E_2 has no cluster point of E_1.

Next we suppose that $E = E_1 \cup E_2$, a disjoint union, where neither E_1 nor E_2 has any cluster point of the other, and each set is nonempty. Let $\mathbf{e} \in E$. Since $\mathbf{e}$ is in one of the two sets E_1 or E_2 without being a a cluster point of the other, there exists $r_{\mathbf{e}} > 0$ such that $B_{r_{\mathbf{e}}}(\mathbf{e})$ is disjoint from the set to which $\mathbf{e}$ does not belong. Let

$$A = \bigcup_{\mathbf{e} \in E_1} B_{\frac{r_{\mathbf{e}}}{2}}(\mathbf{e}).$$

Similarly, we let

$$B = \bigcup_{\mathbf{e} \in E_2} B_{\frac{r_{\mathbf{e}}}{2}}(\mathbf{e}).$$

It follows that A and B are open sets and that $A \cap E = E_1$ and $B \cap E = E_2$. We need prove only that $A \cap B = \emptyset$. Suppose this claim were false. Then there exists $\mathbf{p} \in A \cap B$. Hence there exist $\mathbf{e}_1 \in E_1$ and $\mathbf{e}_2 \in E_2$ such that $\|\mathbf{e}_1 - \mathbf{p}\| < \frac{r_{\mathbf{e}_1}}{2}$ and $\|\mathbf{e}_2 - \mathbf{p}\| < \frac{r_{\mathbf{e}_2}}{2}$. By the triangle inequality for norms, it follows that

$$\|\mathbf{e}_1 - \mathbf{e}_2\| < \frac{r_{\mathbf{e}_1} + r_{\mathbf{e}_2}}{2} \leq \max\{r_{\mathbf{e}_1}, r_{\mathbf{e}_2}\}.$$

This is a contradiction.   ■

Theorem 8.4.2 *Let f be any continuous real-valued function defined on an interval I. Let G_f denote the graph of f as defined in Exercise 8.40. We claim that G_f is a connected subset of $\mathbb{E}^2$.*

Proof: We suppose the claim were false, and we will deduce a contradiction. If G_f were disconnected, then $G_f = E_1 \cup E_2$, a disjoint union of two nonempty sets, neither one containing any cluster point of the other. So there exist $(x_i, f(x_i)) \in E_i$, for $i = 1, 2$. Suppose without loss of generality that $x_1 < x_2$. (We know $x_1 \neq x_2$, since G_f is the graph of a function.) Define

$$c = \sup \left\{ t \in [x_1, x_2] \,\middle|\, (q, f(q)) \in E_1, \forall q \in [x_1, t] \right\}.$$

Thus $x_1 \leq c \leq x_2$. If $(c, f(c)) \in E_1$ then $c < x_2$. Hence for $n \in \mathbb{N}$ there exists

$$c_n \in \left(c, c + \frac{1}{n} \right) \cap [x_1, x_2]$$

for which $(c_n, f(c_n)) \in E_2$. This would imply that $(c_n, f(c_n)) \to (c, f(c))$, by continuity of f. Hence $(c, f(c))$ is a cluster point of E_2, which is a contradiction. On the other hand, if $(c, f(c)) \in E_2$, then $c > x_1$, $(c, f(c))$ is a cluster point of E_1. This yields a similar contradiction. ∎

Corollary 8.4.1 *The Euclidean space $\mathbb{E}^n$ is connected, for each $n \in \mathbb{N}$.*

Proof: Suppose the corollary were false. Then $\mathbb{E}^n = E_1 \cup E_2$, a disjoint union of two sets, neither of which contains any of the other's cluster points. Let $\mathbf{p} \in E_1$ and $\mathbf{q} \in E_2$. Let

$$\phi(t) = \mathbf{p} + t(\mathbf{q} - \mathbf{p})$$

for each $t \in [0, 1]$. Without loss of generality, suppose $\phi(0) \in E_1$ and $\phi(1) \in E_2$. Define

$$c = \sup \left\{ t \in [0, 1] \,\middle|\, (q, \phi(q)) \in E_1, \forall q \in [0, t] \right\}.$$

Now complete the proof just as in Theorem 8.4.2. ∎

Theorem 8.4.3 *Suppose E and F are* nondisjoint *connected subsets of a normed vector space V. Then $G = E \cup F$ is connected.*

Proof: Suppose false. Then there exist nonempty disjoint open sets A and B such that $A \cap G \neq \emptyset \neq B \cap G$. Since E cannot be separated, E is contained entirely in one open set or the other. Without loss of generality, suppose $E \subseteq A$. Similarly, F is contained entirely in one of the open sets. Since $B \cap G \neq \emptyset$, $F \subseteq B$. But this is impossible since $E \cap F \neq \emptyset$ and $A \cap B = \emptyset$. ∎

EXERCISES

8.44 † Prove that if $I \subseteq \mathbb{R}$ is an interval, as defined in Definition 8.4.1, then I must have one of the following forms: $[a, b], (a, b), [a, b)$ or $(a, b]$. In this notation $-\infty \leq a \leq b \leq \infty$, but closed endpoints must be finite. (Hint: Consider $\sup(I)$ and $\inf(I)$.)

8.45 †

    a) Prove that every connected subset $S \subseteq \mathbb{E}^1$ is an interval.

    b) Prove that every interval I is a connected subset of $\mathbb{E}^1$. In particular, this includes the claim that the real line $\mathbb{E}^1$ is itself a connected set. (Hint: Consider Theorem 8.4.2 and the function $f(x) \equiv 0$. Explain how connectedness in E^2 implies connectedness in E^1.)

8.46   Prove that the union of the x_1-axis and the x_2-axis in $\mathbb{E}^2$ is a connected set.

8.47   Prove that $S^1 = \{ \mathbf{x} \in \mathbb{E}^2 \mid \|\mathbf{x}\| = 1 \}$ is a connected set.

8.48   Suppose $E \subseteq \mathbb{E}^2$ such that for all $\mathbf{x} \in E$, $x_2 = 0$. Suppose E is a connected subset of $\mathbb{E}^1$, which we identify with the x_1-axis of $\mathbb{E}^2$. Prove that E is also connected as a subset of $\mathbb{E}^2$.

8.49   Let $E = \{ \mathbf{x} \in \mathbb{E}^2 \mid x_1 x_2 > 0 \}$. Is E connected? Prove your conclusion.

8.50   Let $E = \{ \mathbf{x} \in \mathbb{E}^2 \mid x_1 x_2 \geq 0 \}$. Is E connected? Prove your conclusion.

8.51   Let $E = \{ \mathbf{x} \in \mathbb{E}^2 \mid x_1^2 = 1 \}$. Is E connected? Prove your conclusion.

8.52   In E^2, let $E = \{ \mathbf{x} \mid x_1 = 0 \} \cup \{ \mathbf{x} \mid x_1 x_2 = 1, x_1 > 0 \}$. Is E connected? Prove your conclusion.

8.53   Let

$$g(x) = \begin{cases} x \sin \frac{\pi}{x} & \text{if } x > 0, \\ 0 & \text{if } x = 0. \end{cases}$$

Prove that the graph of g is a connected subset of $\mathbb{E}^2$.

8.54   Let

$$f(x) = \begin{cases} \sin \frac{\pi}{x} & \text{if } x > 0, \\ 0 & \text{if } x = 0. \end{cases}$$

Prove that the graph of f is a connected subset of $\mathbb{E}^2$. (Hint: Try writing $G_f \subseteq A \cup B$ with A and B open and disjoint. Prove that G_f must be contained entirely within one of the two open sets.)

8.55   Let $E \subseteq \mathbb{E}^n$ be any connected set and let $\mathbf{p}$ be a cluster point of E. *Prove or disprove*: $E \cup \{\mathbf{p}\}$ is connected.

8.56   Let $E \subseteq \mathbb{E}^n$ be any connected set and let $\bar{E}$ be the *closure* of E. *Prove or disprove*: $\bar{E}$ is connected.

8.57   Let $f(x) = e^{-x^2}$ for all $x \in \mathbb{R}$. Denote the graph of f by G_f and the x-axis in $\mathbb{E}^2$ by G_0. *Prove or disprove*: $S = G_f \cup G_0$ is a connected subset of $\mathbb{E}^2$. (Hint: Apply Theorem 8.4.1.)

8.58   Prove or give a counterexample: Every open set $\mathcal{O} \subseteq \mathbb{E}^n$ can be expressed as the union of a finite or countable family of *disjoint* open balls. (Hint: See Exercise 8.32.)

8.5 TEST YOURSELF

EXERCISES

8.59   Let $\mathbf{x} = (5, -2, 3) \in \mathbb{E}^3$. Give an example of a non-$\mathbf{0}$ vector $\mathbf{y} \in \mathbb{E}^3$ for which $\|\mathbf{x}\|^2 + \|\mathbf{y}\|^2 = \|\mathbf{x} + \mathbf{y}\|^2$.

8.60   Let $S \subseteq \mathbb{E}^n$ be a set of vectors for which each sequence

$$\left\{ \mathbf{x}^{(k)} \mid k \in \mathbb{N} \right\} \subseteq S$$

has a convergent *sub*sequence. True or False: S must be bounded.

8.61   The *taxicab* norm in $\mathbb{R}^2$ is defined by $\|\mathbf{x}\|_t = |x_1| + |x_2|$. Draw a sketch of the unit ball $B_1(\mathbf{0})$ in $\mathbb{R}^2$ using the taxicab norm. Label your axes.

8.62   Give an example of a set $D \subseteq \mathbb{E}^2$ and a point $\mathbf{p}$ which is an *isolated* point of D.

8.63   Give an example of a set $S \subseteq \mathbb{E}^2$ for which S is dense $\mathbb{E}^2$ but has empty interior.

8.64   True or False: The interior of the closure of $S \subseteq \mathbb{E}^n$ need *not* be S itself.

8.65   Let $f(x) = \sin \frac{\pi}{x}$ for all $x \neq 0$. True or False: The graph G_f is compact.

8.66   Let $E_1 \supseteq E_2 \supseteq \ldots \supseteq E_k \supseteq \ldots$ be a decreasing nest of closed subsets of $\mathbb{E}^n$. True or False: $\bigcap_{k=1}^{\infty} E_k$ must be non-$\emptyset$.

8.67   Let $S = \left\{ \mathbf{x}_n = \left(1 + \frac{n}{2^n}, \frac{n}{2^n}\right) \mid n = 0, 1, 2, 3, \ldots \right\} \subset \mathbb{E}^2$. True or False: S is compact.

8.68   Let

$$g(x) = \begin{cases} x^2 \cos \frac{1}{x} & \text{if } x \neq 0, \\ 0 & \text{if } x = 0. \end{cases}$$

True or False: The graph G_g is connected.

8.69   For each one of the following two sets, decide whether it is *connected* or *not connected*.

    a) $S = \left\{ \mathbf{x} \in \mathbb{E}^2 \mid |x_2| = |x_1| \right\}$.

    b) $T = \left\{ \mathbf{x} \in \mathbb{E}^2 \mid |x_2| < |x_1| \right\}$.

8.70   True or False: If $S \neq \emptyset$ is open in $\mathbb{E}^n$, then $S = \cup_{a \in A} B_a$, a union of nonempty *disjoint* open balls of finite radius.

CHAPTER 9

CONTINUOUS FUNCTIONS ON EUCLIDEAN SPACE

9.1 LIMITS OF FUNCTIONS

We will consider vector-valued functions $\mathbf{f} : D \to \mathbb{E}^m$, where the *domain* of definition of $\mathbf{f}$ is a set $D \subseteq V$, a normed vector space. (Most often, V will be $\mathbb{E}^n$.) Since for each $\mathbf{x} \in D$ we have $\mathbf{f}(\mathbf{x}) \in \mathbb{E}^m$, we can write

$$\mathbf{f}(\mathbf{x}) = (f_1(\mathbf{x}), \ldots, f_m(\mathbf{x})),$$

where $f_i(\mathbf{x}) \in \mathbb{R}$ for all $\mathbf{x} \in D$. In elementary courses in the calculus of several variables, V is always $\mathbb{E}^n$ and it is common to write the real-valued function $f_i(\mathbf{x})$ in the form $f_i(x_1, \ldots, x_n)$.

In order to define the concept $\lim_{\mathbf{x} \to \mathbf{a}} \mathbf{f}(\mathbf{x})$, we need to restrict ourselves to the case in which $\mathbf{a}$ is a cluster point of D. Recall that a cluster point of D need not be an element of D. A cluster point $\mathbf{a}$ of a set D has the property that it is always possible to find points $\mathbf{x} \in D$ for which $\mathbf{x} \neq \mathbf{a}$ and yet $\mathbf{x}$ is as close to $\mathbf{a}$ as we like. If $\mathbf{a}$ is not a cluster point, it will be impossible to define $\lim_{\mathbf{x} \to \mathbf{a}} \mathbf{f}(\mathbf{x})$ because it is impossible for $\mathbf{x}$ to approach $\mathbf{a}$. For example, if $\mathbf{a}$ were an **isolated** point of D, then for all sufficiently small $\delta > 0$ we obtain the set

$$\{\mathbf{x} \mid 0 < \|\mathbf{x} - \mathbf{a}\| < \delta, \ \mathbf{x} \in D\} = \emptyset,$$

Advanced Calculus: An Introduction to Linear Analysis. By Leonard F. Richardson
Copyright © 2008 John Wiley & Sons, Inc.

the empty set. Every statement one might make about each element $\mathbf{x}$ of the empty set, such as the statement $\|\mathbf{f}(\mathbf{x}) - \mathbf{L}\| < \epsilon$, is vacuously true.

Definition 9.1.1 *Let* $\mathbf{a}$ *be any cluster point of the domain* $D_{\mathbf{f}}$ *of a function* $\mathbf{f}$. *Then* $\lim_{\mathbf{x} \to \mathbf{a}} \mathbf{f}(\mathbf{x}) = \mathbf{L} \in \mathbb{E}^m$ *if and only if for all* $\epsilon > 0$ *there exists* $\delta > 0$ *such that* $\mathbf{x} \in D_{\mathbf{f}}$ *and* $0 < \|\mathbf{x} - \mathbf{a}\| < \delta$ *implies* $\|\mathbf{f}(\mathbf{x}) - \mathbf{L}\| < \epsilon$. $\forall \epsilon > 0 \ \exists \delta > 0$ s.th. $\mathbf{x} \in D_f$ and $0 < \|\vec{x} - $ $\Rightarrow \|\vec{f}(\vec{x}) - \vec{L}$

This definition is a natural adaptation of Definition 2.1.2 to vector-valued functions defined on Euclidean space. We have the following theorem that permits more such adaptations to be developed.

Theorem 9.1.1 *Suppose* $\mathbf{a}$ *is a cluster point of the domain* $D_{\mathbf{f}}$ *of* $\mathbf{f}$. *Then the following two statements are equivalent.*

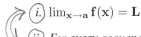

(i.) $\lim_{\mathbf{x} \to \mathbf{a}} \mathbf{f}(\mathbf{x}) = \mathbf{L}$

(ii.) *For every sequence of points* $\mathbf{x}_n \in D_{\mathbf{f}} \setminus \{\mathbf{a}\}$ *such that* $\mathbf{x}_n \to \mathbf{a}$, *we have the sequence* $\mathbf{f}(\mathbf{x}_n) \to \mathbf{L}$.

Proof: Let us prove first that (i) implies (ii). So suppose (i) and consider any sequence of points $\mathbf{x}_n \in D_{\mathbf{f}} \setminus \{\mathbf{a}\}$ such that $\mathbf{x}_n \to \mathbf{a}$. We know for each $\epsilon > 0$ there exists $\delta > 0$ such that $\mathbf{x} \in D_f$ and $0 < \|\mathbf{x} - \mathbf{a}\| < \delta$ implies $\|\mathbf{f}(\mathbf{x}) - \mathbf{L}\| < \epsilon$. But since $\mathbf{x}_n \to \mathbf{a}$, there exists N such that $n \geq N$ implies $0 < \|\mathbf{x}_n - \mathbf{a}\| < \delta$. This implies $\|\mathbf{f}(\mathbf{x}_n) - \mathbf{L}\| < \epsilon$, so that $\mathbf{f}(\mathbf{x}_n) \to \mathbf{L}$.

Next we prove (ii) implies (i). So suppose (ii) is true. We need to show $\lim_{\mathbf{x} \to \mathbf{a}} \mathbf{f}(\mathbf{x}) = \mathbf{L}$. We will show that if this conclusion *were false* then a self-contradiction would result. But if it were false that $\lim_{\mathbf{x} \to \mathbf{a}} \mathbf{f}(\mathbf{x}) = \mathbf{L}$, then there exists $\epsilon > 0$ such that for all $\delta > 0$ there exists $\mathbf{x} \in D_f$ such that $0 < \|\mathbf{x} - \mathbf{a}\| < \delta$ and yet $\|\mathbf{f}(\mathbf{x}) - \mathbf{L}\| \geq \epsilon$. In particular, if we let $\delta_n = \frac{1}{n}$ then we get $\mathbf{x}_n$ such that $0 < \|\mathbf{x}_n - \mathbf{a}\| < \frac{1}{n}$ and yet $\|\mathbf{f}(\mathbf{x}_n) - \mathbf{L}\| \geq \epsilon$. Now, $\mathbf{x}_n \in D_{\mathbf{f}} \setminus \{\mathbf{a}\}$ and $\mathbf{x}_n \to \mathbf{a}$, yet $\mathbf{f}(\mathbf{x}_n) \not\to \mathbf{L}$. This contradicts (ii), which was assumed to be true. ∎

The theorem above combined with Theorem 8.1.2 and Theorem 2.1.1 yields the following immediate corollary.

Corollary 9.1.1 *Let* $\mathbf{a}$ *be any cluster point of the domain* $D_{\mathbf{f}}$. *Then* $\lim_{\mathbf{x} \to \mathbf{a}} \mathbf{f}(\mathbf{x})$ *exists and equals* $\mathbf{L} = (l_1, \ldots, l_m)$ *if and only if* $\lim_{\mathbf{x} \to \mathbf{a}} f_i(\mathbf{x})$ *exists and equals* l_i *for all* $i = 1, \ldots, m$.

This corollary enables us to understand the concept of convergence of vector-valued functions $\mathbf{f}$ in terms of convergence of each real-valued component function f_i.

Remark 9.1.1 If two functions $\mathbf{f}$ and $\mathbf{g}$ are defined on subsets of the *same space* $\mathbb{E}^n$, with values in the *same space* $\mathbb{E}^m$, then we can form the sum or difference of these functions with the domain of this combination being $D_{\mathbf{f}} \cap D_{\mathbf{g}}$. If, furthermore, $\mathbf{f} = f$ and $\mathbf{g} = g$ are *real-valued* functions of vector variables, then we can also multiply and divide the two functions. The domain $D_{\frac{f}{g}}$ of the quotient will be only those points of $D_f \cap D_g$ for which $g(\mathbf{x}) \neq 0$. (Multiplication and division are *not* defined, however, if $\mathbf{f}$ and $\mathbf{g}$ map $\mathbb{E}^n$ to $\mathbb{E}^m$ with $m > 1$.)

Corollary 9.1.2 *Let $\mathbf{f}$ and $\mathbf{g}$ be defined on domains in V, a normed vector space, with values in $\mathbb{E}^m$, as in the* Remark *above. Suppose $\mathbf{a}$ is a cluster point of the intersection $D_{\mathbf{f}} \cap D_{\mathbf{g}}$ of the domains of $\mathbf{f}$ and $\mathbf{g}$, $\lim_{\mathbf{x} \to \mathbf{a}} \mathbf{f}(\mathbf{x}) = \mathbf{L}$ and $\lim_{\mathbf{x} \to \mathbf{a}} \mathbf{g}(\mathbf{x}) = \mathbf{M}$. Then*

(i.) $\lim_{\mathbf{x} \to \mathbf{a}} (\mathbf{f} \pm \mathbf{g})(\mathbf{x}) = \mathbf{L} \pm \mathbf{M}$.

(ii.) *If $m = 1$, then $\lim_{\mathbf{x} \to \mathbf{a}} (fg)(x) = LM$.* , ie if $\vec{f} : D_f \to \mathbb{R}$ and $\vec{g} : D_g \to \mathbb{R}$

(iii.) *If $m = 1$, then $\lim_{\mathbf{x} \to \mathbf{a}} \frac{f}{g}(\mathbf{x}) = \frac{L}{M}$, provided that $M \neq 0$ and that $\mathbf{a}$ is a cluster point of $D_{\frac{f}{g}}$.*

Proof: The proofs are very similar to those of Corollary 2.1.1. Here we will treat only conclusion (i). Consider a sequence of points $\mathbf{x}_n \in D_{\mathbf{f}} \cap D_{\mathbf{g}} \setminus \{\mathbf{a}\}$ such that $\mathbf{x}_n \to \mathbf{a}$. Then $(\mathbf{f} \pm \mathbf{g})(\mathbf{x}_n) = \mathbf{f}(\mathbf{x}_n) \pm \mathbf{g}(\mathbf{x}_n) \to \mathbf{L} \pm \mathbf{M}$, by the corresponding theorem for limits of sequences. The other proofs are similar. ∎

This corollary suggests that limits of functions on $\mathbb{E}^n$ are very similar in behavior to limits of real-valued functions of one real variable, subject only to certain essential hypotheses to insure the requisite operations are defined for the objects in question. However, the reader must be very careful: it is much harder for the limit of even a real-valued function of two or more variables to exist than it is for functions of only one variable. See Exercise 9.4.

Definition 9.1.2 *Let $\mathbf{x} \in \mathbb{E}^n$ and let $\mathbf{k} = (k_1, \ldots, k_n)$ be any n-tuple of nonnegative integers. We define*

$$\mathbf{x}^{\mathbf{k}} = x_1^{k_1} x_2^{k_2} \cdots x_n^{k_n},$$

a real-valued monomial on $\mathbb{E}^n$. A real-valued polynomial on $\mathbb{E}^n$ is any function of the form $p(\mathbf{x}) = \sum_{j=1}^q c_j \mathbf{x}^{\mathbf{k}_j}$, where each $c_j \in \mathbb{R}$. A rational function is any ratio of two real-valued polynomials. A rational function is defined wherever the denominator is nonzero.

In the exercises below, the reader will show that the behavior of polynomials and rational functions on $\mathbb{E}^n$ with respect to limits is very similar to the behavior of such functions on $\mathbb{R}$.

Definition 9.1.3 *Let $\mathbf{f} : D \to \mathbb{E}^m$. Suppose $D \subseteq \mathbb{E}^n$, where $n \geq 2$.*

(i.) *If the domain $D_{\mathbf{f}}$ is not a bounded subset of $\mathbb{E}^n$, we say $\lim_{\mathbf{x} \to \infty} \mathbf{f}(\mathbf{x}) = \mathbf{L}$ provided for all sequences $\mathbf{x}_n \in D_{\mathbf{f}}$ such that $\|\mathbf{x}_n\| \to \infty$ we have $\mathbf{f}(\mathbf{x}_n) \to \mathbf{L}$.*

(ii.) *If $\mathbf{a}$ is a cluster point of $D_{\mathbf{f}}$ we say $\lim_{\mathbf{x} \to \mathbf{a}} \mathbf{f}(\mathbf{x}) = \infty$ provided for all sequences $\mathbf{x}_n \in D_{\mathbf{f}}$ such that $\mathbf{x}_n \to \mathbf{a}$ we have $\|\mathbf{f}(\mathbf{x}_n)\| \to \infty$.*

Remark 9.1.2 In case (i) we say that $\mathbf{f}(\mathbf{x})$ *converges* to $\mathbf{L}$ as $\mathbf{x}$ *diverges* to infinity. In case (ii) we say that $\mathbf{f}(\mathbf{x})$ *diverges* to ∞ as $\mathbf{x}$ *converges* to $\mathbf{a}$. The use of the word *converges* versus *diverges* hinges on whether or not the limit exists as a point of $\mathbb{E}^n$.

Remark 9.1.3 The restriction to $n \geq 2$ in Definition 9.1.3 is employed in order to avoid conflict with the normal terminology in $\mathbb{E}^1$. For functions of a single real variable, it is customary to distinguish between $x \to \infty$ and $x \to -\infty$ and to distinguish both of these concepts from $|x| \to \infty$. The rationale for this is that in $\mathbb{E}^1$ there are only two directions in which to move away from 0 as far as one pleases. But in $\mathbb{E}^n$ for $n \geq 2$ there are infinitely many directions along which one might wander as far from $\mathbf{0}$ as one pleases. Thus distinguishing among directions in the concept of divergence to infinity becomes hopeless.

EXERCISES

9.1   Write a detailed proof for Corollary 9.1.1.

9.2   Let p be any real-valued polynomial on $\mathbb{E}^n$, as in Definition 9.1.2. In the parts below, prove that $\lim_{\mathbf{x} \to \mathbf{a}} p(\mathbf{x})$ exists and equals $p(\mathbf{a})$.
   a) If $m(\mathbf{x}) = x_i$, $1 \leq i \leq n$, prove that $\lim_{\mathbf{x} \to \mathbf{a}} m(\mathbf{x}) = m(\mathbf{a})$.
   b) If $m(\mathbf{x}) = \mathbf{x}^k$, prove that $\lim_{\mathbf{x} \to \mathbf{a}} m(\mathbf{x}) = m(\mathbf{a})$.
   c) Let p be any polynomial on $\mathbb{E}^n$. Prove that $\lim_{\mathbf{x} \to \mathbf{a}} p(\mathbf{x})$ exists and equals $p(\mathbf{a})$. That is, prove that a polynomial is continuous everywhere. (See Definition 9.2.1.)

9.3   Let $Q(\mathbf{x})$ be any real-valued rational function on $\mathbb{E}^n$, as in Definition 9.1.2. Prove that $\lim_{\mathbf{x} \to \mathbf{a}} Q(\mathbf{x})$ exists and equals $Q(\mathbf{a})$, provided that $\mathbf{a} \in D_Q$. That is, prove that a rational function is continuous wherever it is defined. (See Definition 9.2.1.)

9.4   † Let $f : \mathbb{E}^2 \to \mathbb{R}$ by the formula

$$f(\mathbf{x}) = \begin{cases} \frac{x_1^2 x_2}{x_1^4 + x_2^2} & \text{if } \mathbf{x} \neq \mathbf{0}, \\ 0 & \text{if } \mathbf{x} = \mathbf{0}. \end{cases}$$

(See Figs. 9.1 and 9.2)
   a) Show that if $\mathbf{x} \to \mathbf{0}$ along either the x_1- or the x_2-coordinate axis, $f(\mathbf{x}) \to 0$.
   b) Show that if $\mathbf{x} \to \mathbf{0}$ along any straight line $x_2 = kx_1$ through the origin, $f(\mathbf{x}) \to 0$.
   c) Show that $\lim_{\mathbf{x} \to \mathbf{0}} f(\mathbf{x})$ *does not exist*. Hint: Show there exist points $\mathbf{x}$ arbitrarily close to $\mathbf{0}$ at which $f(\mathbf{x}) = \frac{1}{2}$.
   d) Does $\lim_{\mathbf{x} \to \infty} f(\mathbf{x})$ exist? Prove your conclusion.

9.5   Let $f : \mathbb{E}^2 \to \mathbb{R}$ by the formula

$$f(\mathbf{x}) = \begin{cases} \frac{x_1 x_2^2}{x_1^4 + x_2^2} & \text{if } \mathbf{x} \neq \mathbf{0}, \\ 0 & \text{if } \mathbf{x} = \mathbf{0}. \end{cases}$$

Prove that $\lim_{\mathbf{x} \to \mathbf{0}} f(\mathbf{x}) = 0$. (Hint: Let $\epsilon > 0$. Find $\delta > 0$ such that $\|\mathbf{x}\| < \delta$ implies $|f(\mathbf{x})| < \epsilon$. Compare with Exercise 9.4.)

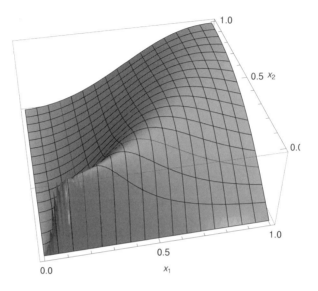

Figure 9.1 $f(x_1, x_2) = \frac{x_1^2 x_2}{x_1^4 + x_2^2}$ in the first quadrant.

9.6   For $\mathbf{x} \in \mathbb{E}^2$, evaluate each of the following limits.
   a) $\lim_{\mathbf{x} \to 0} \frac{1}{\|\mathbf{x}\|}$.
   b) $\lim_{\mathbf{x} \to \infty} \frac{1}{\|\mathbf{x}\|}$.

9.7   Let $f(x) = x \sin \frac{\pi}{x}$, $f : \mathbb{E}^1 \setminus \{0\} \to \mathbb{R}$. Prove each statement True or False:
   a) $\lim_{x \to 0} f(x) = 0$.
   b) $\lim_{x \to 0} \frac{1}{f(x)} = \infty$.
   c) $\lim_{x \to 0} \frac{1}{|f(x)|} = \infty$.

9.8   Let $\mathbf{a}$ be a cluster point of the domain $D_{\mathbf{f}}$ of a *nonvanishing* (that is *nowhere-0*) function $\mathbf{f}(\mathbf{x})$. Prove that

$$\lim_{\mathbf{x} \to \mathbf{a}} \mathbf{f}(\mathbf{x}) = \mathbf{0} \Leftrightarrow \lim_{\mathbf{x} \to \mathbf{a}} \frac{1}{\|\mathbf{f}(\mathbf{x})\|} = \infty.$$

9.9   Suppose the domain $D \subseteq \mathbb{E}^n$ of a function $\mathbf{f} : D \to \mathbb{E}^m$ is not bounded. Prove: $\lim_{\mathbf{x} \to \infty} \mathbf{f}(\mathbf{x}) = \mathbf{L} \in \mathbb{E}^m$ if, and only if, for each $\epsilon > 0$ there exists $M > 0$ such that $\mathbf{x} \in D \setminus B_M(\mathbf{0})$ implies $\|\mathbf{f}(\mathbf{x}) - \mathbf{L}\| < \epsilon$.

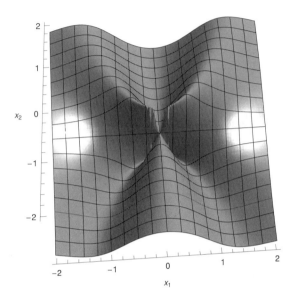

Figure 9.2 $f(x_1, x_2) = \frac{x_1^2 x_2}{x_1^4 + x_2^2}$ in all four quadrants.

9.10   Suppose that $\mathbf{a}$ is a cluster point of the domain $D \subseteq \mathbb{E}^n$ of a function $\mathbf{f} : D \to \mathbb{E}^m$. Prove: $\lim_{\mathbf{x} \to \mathbf{a}} \mathbf{f}(\mathbf{x}) = \infty$ if, and only if, for each $M > 0$ there exists $\delta > 0$ such that $\mathbf{x} \in (D \setminus \{\mathbf{a}\}) \cap B_\delta(\mathbf{a})$ implies $\|\mathbf{f}(\mathbf{x})\| > M$.

9.11   Prove the following *Cauchy Criterion* for the existence of the limit of a function. Let $\mathbf{f} : D \to \mathbb{E}^m$ and let $\mathbf{a}$ be a cluster point of $D \subseteq \mathbb{E}^n$. Then $\lim_{\mathbf{x} \to \mathbf{a}} \mathbf{f}(\mathbf{x})$ exists if and only if for each $\epsilon > 0$ there exists $\delta > 0$ such that $\mathbf{x}$ and $\mathbf{x}' \in D \cap (B_\delta(\mathbf{a}) \setminus \{\mathbf{a}\})$ implies $\|\mathbf{f}(\mathbf{x}) - \mathbf{f}(\mathbf{x}')\| < \epsilon$.

9.2   CONTINUOUS FUNCTIONS

The following definition is analogous to Definition 2.2.1.

Definition 9.2.1 *Let V be any normed vector space. A function $\mathbf{f} : D \to \mathbb{E}^m$ is called* continuous *at a point $\mathbf{a} \in D \subseteq V$ provided that for each $\epsilon > 0$ there exists a corresponding $\delta > 0$ such that $\mathbf{x} \in D \cap B_\delta(\mathbf{a})$ implies*

$$\|\mathbf{f}(\mathbf{x}) - \mathbf{f}(\mathbf{a})\| < \epsilon.$$

If $\mathbf{f}$ is continuous at every point $\mathbf{a} \in D$, we say $\mathbf{f} \in \mathcal{C}(D)$, the family of all continuous functions on D with values in $\mathbb{E}^m$. In this notation, the value of m is fixed. If there is *more than one possible range-space* $\mathbb{E}^m$ *under discussion, we will denote the family of continuous functions instead as* $\mathcal{C}(D, \mathbb{E}^m)$.

If $D_{\mathbf{f}}$ has any isolated points $\mathbf{a}$, then $\mathbf{f}$ is automatically continuous at $\mathbf{a}$. The more interesting case, however, is that of a nonisolated point $\mathbf{a} \in D_{\mathbf{f}}$.

Theorem 9.2.1 *A function* $\mathbf{f}$ *is continuous at a cluster point* $\mathbf{a} \in D$ *if and only if* $\lim_{\mathbf{x} \to \mathbf{a}} \mathbf{f}(\mathbf{x})$ *exists and equals* $\mathbf{f}(\mathbf{a})$.

Proof: First suppose $\mathbf{f}$ is continuous at $\mathbf{a} \in D$. Then for each $\epsilon > 0$ there exists $\delta > 0$ such that $\|\mathbf{x} - \mathbf{a}\| < \delta$ and $\mathbf{x} \in D$ implies $\|\mathbf{f}(\mathbf{x}) - \mathbf{f}(\mathbf{a})\| < \epsilon$. Hence for all $\mathbf{x} \in D$ such that $0 < \|\mathbf{x} - \mathbf{a}\| < \delta$ we have $\|\mathbf{f}(\mathbf{x}) - \mathbf{f}(\mathbf{a})\| < \epsilon$ which implies $\lim_{\mathbf{x} \to \mathbf{a}} \mathbf{f}(\mathbf{x})$ exists and equals $\mathbf{f}(\mathbf{a})$.

Now suppose $\lim_{\mathbf{x} \to \mathbf{a}} \mathbf{f}(\mathbf{x})$ exists and is $\mathbf{f}(\mathbf{a})$. Then for all $\epsilon > 0$ there exists $\delta > 0$ such that $0 < \|\mathbf{x} - \mathbf{a}\| < \delta$ and $\mathbf{x} \in D$ implies $\|\mathbf{f}(\mathbf{x}) - \mathbf{f}(\mathbf{a})\| < \epsilon$. This implies for all $\mathbf{x} \in D \cap \mathcal{B}_\delta(\mathbf{a})$ we have $\|\mathbf{f}(\mathbf{x}) - \mathbf{f}(\mathbf{a})\| < \epsilon$ so that $\mathbf{f}$ is continuous at $\mathbf{a}$. ∎

This theorem enables us to prove some properties of combinations of continuous functions that are analogous to the results in Theorem 2.2.2 to the extent possible.

Theorem 9.2.2 *Let* $\mathbf{f}$ *and* $\mathbf{g}$ *be defined on domains in a normed vector space* V *with values in* $\mathbb{E}^m$. *Suppose* $\mathbf{f}$ *and* $\mathbf{g}$ *are both continuous at* $\mathbf{a}$. *Then*

i. $\mathbf{f} \pm \mathbf{g}$ *is continuous at* $\mathbf{a}$.

ii. *If* $m = 1$, *then the function* fg *is continuous at* $\mathbf{a}$.

iii. *If* $m = 1$ *and if* $\mathbf{g}(\mathbf{a}) \neq 0$, *then* $\frac{f}{g}$ *is continuous at* $\mathbf{a}$.

iv. *If we write* $\mathbf{f} = (f_1, \ldots, f_m)$ *in terms of its scalar-valued components, then* $\mathbf{f}$ *is continuous at* $\mathbf{a}$ *if and only if* $f_1, \ldots, f_m$ *are all continuous at* $\mathbf{a}$.

Proof: This is a simple application of Corollary 9.1.2 and Corollary 9.1.1 and it is left to the Exercises. Note that the proof is trivial if $\mathbf{a}$ is not a cluster point of the domain of the appropriate combination of $\mathbf{f}$ and $\mathbf{g}$. ∎

The following definition is useful for the study of continuous functions in Euclidean space.

Definition 9.2.2 *A set* E *is said to be* relatively open *in* $D \subseteq V$, *a normed vector space, provided there exists an open set* $U \subseteq V$ *such that* $E = U \cap D$. *Similarly, a set* E *is said to be* relatively closed *in* $D \subseteq V$ *provided there exists a closed set* $K \subseteq V$ *such that* $E = K \cap D$.

■ **EXAMPLE 9.1**

In $\mathbb{E}^1$, $[0, 1)$ is *relatively open* in $[0, 2]$, since $[0, 1) = (-1, 1) \cap [0, 2]$. Similarly, $(0, 1]$ is *relatively closed* in $(0, 2]$, since $(0, 1] = [-1, 1] \cap (0, 2]$

Theorem 9.2.3 *A subset $E \subseteq D$ is relatively open in D if and only if for each $\mathbf{x} \in E$ there exists $\delta_\mathbf{x} > 0$ such that $B_{\delta_\mathbf{x}}(\mathbf{x}) \cap D \subseteq E$.*

Proof: First suppose that E is relatively open in D. Then there exists an open set U such that $U \cap D = E$. If $\mathbf{x} \in E \subseteq U$, it follows that there exists a corresponding $\delta_\mathbf{x} > 0$ such that $B_{\delta_\mathbf{x}}(\mathbf{x}) \subseteq U$. Thus $B_{\delta_\mathbf{x}}(\mathbf{x}) \cap D \subseteq U \cap D = E$. This completes the proof in one direction. For the other direction see Exercise 9.27.   ∎

Corollary 9.2.1 *A subset $E \subseteq D$ is relatively open in D if and only if its complement in D, $D \setminus E$, is relatively closed in D.*

The proof of this corollary is left to the Exercises.

It is a remarkable and very useful fact that the continuity of a function $\mathbf{f}$ on a domain D is closely connected to the way in which open sets and closed sets are transformed by the *inverse* of $\mathbf{f}$. In this terminology we make no supposition that $\mathbf{f}$ is invertible, as we see below.

Definition 9.2.3 *Let $\mathbf{f} : D \to \mathbb{E}^m$, where $D \subseteq V$, a normed vector space. We define the inverse image of $E \subseteq \mathbb{E}^m$ by $\mathbf{f}^{-1}(E) = \{\mathbf{x} \in D \mid \mathbf{f}(\mathbf{x}) \in E\}$.*

Theorem 9.2.4 *Let $\mathbf{f}$ be defined on a domain $D \subseteq V$, a normed vector space, where $\mathbf{f} : D \to \mathbb{E}^m$. Then $\mathbf{f} \in \mathcal{C}(D)$ if and only if $\mathbf{f}^{-1}(U)$ is relatively open in D for each open set $U \subseteq \mathbb{E}^m$.*

Proof: First we suppose $\mathbf{f} \in \mathcal{C}(D)$. Let $U \subseteq \mathbb{E}^m$ be open. We need to show $\mathbf{f}^{-1}(U)$ is relatively open in D. If $\mathbf{f}^{-1}(U) = \emptyset$ then there is nothing to prove. Without loss of generality, let $\mathbf{x} \in \mathbf{f}^{-1}(U)$, so that $\mathbf{f}(\mathbf{x}) \in U$. Since U is open, there exists $r > 0$ such that $B_r(\mathbf{f}(\mathbf{x})) \subseteq U$. Because $\mathbf{f}$ is continuous at $\mathbf{x}$, there exists $\delta_\mathbf{x} > 0$ such that

$$f : B_{\delta_\mathbf{x}}(\mathbf{x}) \cap D \to B_r(\mathbf{f}(\mathbf{x})).$$

Thus $B_{\delta_\mathbf{x}}(\mathbf{x}) \cap D \subseteq \mathbf{f}^{-1}(U)$. By Theorem 9.2.3 $\mathbf{f}^{-1}(U)$ relatively open in D.

Next we suppose $\mathbf{f}^{-1}(U)$ is relatively open in D for each open set $U \subseteq \mathbb{E}^m$, and we must prove that for each $\mathbf{x} \in D$, $\mathbf{f}$ is continuous at $\mathbf{x}$. Let $\epsilon > 0$. We need to show that there exists $\delta > 0$ such that $\mathbf{f} : B_\delta(\mathbf{x}) \cap D \to B_\epsilon(\mathbf{f}(\mathbf{x}))$. Since $\mathbf{f}^{-1}(B_\epsilon(\mathbf{f}(\mathbf{x})))$ is relatively open in D, there exists $\delta > 0$ such that

$$B_\delta(\mathbf{x}) \cap D \subseteq \mathbf{f}^{-1}(B_\epsilon(\mathbf{f}(\mathbf{x}))).$$

(See Theorem 9.2.3.) Thus $\mathbf{f}$ is continuous at $\mathbf{x}$.   ∎

EXERCISES

9.12   Define $\pi_i : \mathbb{E}^n \to \mathbb{R}$ by $\pi_i(\mathbf{x}) = x_i, i = 1, \dots, n$. Prove that π_i is continuous.

9.13   † If V is any normed vector space, define $N : V \to \mathbb{R}$ by $N(\mathbf{x}) = \|\mathbf{x}\|$. Prove that N is continuous. Hint: See Exercise 2.20. Show that

$$\big| \|\mathbf{x}\| - \|\mathbf{y}\| \big| \leq \|\mathbf{x} - \mathbf{y}\|.$$

9.14   Let $\mathbf{f} : \mathbb{E}^2 \to \mathbb{E}^2$ by the formula

$$
\mathbf{f}(\mathbf{x}) = \begin{cases} \left(x_1 x_2, \frac{x_1^2 x_2}{x_1^4 + x_2^2} \right) & \text{if } \mathbf{x} \neq \mathbf{0}, \\ (0, 0) & \text{if } \mathbf{x} = \mathbf{0}. \end{cases}
$$

Determine whether or not $\mathbf{f} \in \mathcal{C} \left(\mathbb{E}^2 \right)$ and prove your conclusion.

9.15   Let $\mathbf{f} : \mathbb{E}^2 \to \mathbb{E}^2$ by the formula

$$
\mathbf{f}(\mathbf{x}) = \begin{cases} \left(x_1 x_2, \sin\left(\frac{x_1^2 x_2}{x_1^2 + x_2^2} \right) \right) & \text{if } \mathbf{x} \neq \mathbf{0}, \\ (0, 0) & \text{if } \mathbf{x} = \mathbf{0}. \end{cases}
$$

a) Determine whether or not $\mathbf{f} \in \mathcal{C} \left(\mathbb{E}^2 \right)$ and prove your conclusion.

b) Let $S = \{\mathbf{x} \mid \mathbf{f}(\mathbf{x}) = \mathbf{0}\}$. Is the complement of S an open set? Prove your conclusion.

9.16   Suppose $f : \mathbb{E}^2 \to \mathbb{R}$ has the property that for *each* fixed value $x_2 = b$ the function $g_b(x_1) = f(x_1, b)$ is a continuous function of x_1. Suppose also that for *each* fixed value $x_1 = a$ the function $h_a(x_2) = f(a, x_2)$ is a continuous function of x_2. Does it follow that $f \in \mathcal{C} \left(\mathbb{E}^2, \mathbb{R} \right)$? If *yes*, prove it. If *no*, give a counterexample and *prove* that it is a counterexample. (Remark: This question is often expressed verbally as follows. Does the continuity of $f(\mathbf{x})$ in each variable x_1, x_2 *separately* imply *joint continuity*, meaning continuity as a function of the vector variable $\mathbf{x}$? See Problem 9.14.)

9.17   Prove all four parts of Theorem 9.2.2.

9.18   Prove Corollary 9.2.1.

9.19   Give an example of a *continuous* function $\mathbf{f} : \mathbb{E}^1 \to \mathbb{E}^2$ that maps the *open* set $(-2\pi, 2\pi)$ onto the circle $x_1^2 + x_2^2 = 1$, which is *not* an open set.

9.20   Prove that if $D \subseteq \mathbb{E}^n$ is open then a relatively open subset $E \subseteq D$ is open.

9.21   Prove that $\mathcal{C}(D, \mathbb{E}^m)$ is a vector space.

9.22   Let $D = \{\mathbf{x} \mid 1 < \|\mathbf{x}\| \leq 3\} \subset \mathbb{E}^2$. Show that the set

$$
E = \{\mathbf{x} \mid 1 < \|\mathbf{x}\| \leq 2\}
$$

is relatively closed in D, and that the set $F = \{\mathbf{x} \mid 2 < \|\mathbf{x}\| \leq 3\}$ is relatively open in D.

9.23   Identify the relatively open subsets of the set $\mathbb{Z}$ contained in $\mathbb{E}^1$. Identify the relatively closed subsets as well.

9.24   Give an example of a function $f \in \mathcal{C}([0, 1), \mathbb{R})$ and an open subset $U \subseteq \mathbb{R}$ such that $f^{-1}(U)$ is *not* open.

9.25   Let $\mathbf{f}$ be defined on a domain $D \subseteq \mathbb{E}^n$, where $\mathbf{f} : D \to \mathbb{E}^m$. Prove that $\mathbf{f} \in \mathcal{C}(D)$ if and only if $\mathbf{f}^{-1}(K)$ is relatively closed in D for each closed set $K \subseteq \mathbb{E}^m$.

9.26   Is the set $\{\mathbf{x} \in \mathbb{E}^4 \mid x_1 x_4 - x_2 x_3 = 1\}$ open, closed, neither, or both?

9.27   † Complete the proof of Theorem 9.2.3 by showing that $E \subseteq D$ is relatively open in D if for each $\mathbf{x} \in E$ there exists $\delta_{\mathbf{x}} > 0$ such that $B_{\delta_{\mathbf{x}}}(\mathbf{x}) \cap D \subseteq E$.

9.28   Let $\mathbf{f} : D \to \mathbb{E}^m$ be continuous on its domain $D \subseteq \mathbb{E}^n$, and suppose $\mathbf{g} : \mathbf{f}(D) \to \mathbb{E}^p$ is continuous on $\mathbf{f}(D)$. Prove: The composition $\mathbf{g} \circ \mathbf{f} \in \mathcal{C}(D, \mathbb{E}^p)$, where $\mathbf{g} \circ \mathbf{f}(\mathbf{x}) = \mathbf{g}(\mathbf{f}(\mathbf{x})) \in \mathbb{E}^p$, for each $\mathbf{x} \in D$. (Hint: Use Theorem 9.2.4.)

9.3   CONTINUOUS IMAGE OF A COMPACT SET

In the Extreme Value Theorem (Theorem 2.4.2) we saw that a continuous *real-valued* function on a closed finite interval $[a, b]$ must achieve both a maximum and a minimum value. The type of set analogous to $[a, b]$ that enables us to prove an Extreme Value Theorem for *real-valued* functions on $\mathbb{E}^n$ is a compact set. (If $\mathbf{f}$ were not real-valued but rather $\mathbb{E}^m$-valued with $m > 1$, then it would not make sense to speak of an extreme value for $\mathbf{f}$ since there is no natural order relation among the vectors of $\mathbb{E}^m$.) We begin with a more general theorem that does not require real values, however.

Theorem 9.3.1 *Let $D \subset \mathbb{E}^n$ be compact, and suppose $\mathbf{f} : D \to \mathbb{E}^m$ is continuous. Then the set $\mathbf{f}(D) = \{\mathbf{f}(\mathbf{x}) \mid \mathbf{x} \in D\}$ is a compact subset of $\mathbb{E}^m$.*

Proof: In words, this theorem states that *the continuous image of a compact set is compact.* We begin by letting $O = \{O_a \mid a \in A\}$ be an arbitrary open cover of $\mathbf{f}(D)$, where A is an index set and each O_a is an open set in $\mathbb{E}^m$.

It will suffice to prove that there exists a finite subcover of $\mathbf{f}(D)$ selected from the given open cover O. Since $\mathbf{f}$ is continuous, each set $V_a = \mathbf{f}^{-1}(O_a)$ is relatively open in D. Thus there exists an open set U_a in $\mathbb{E}^n$ such that $V_a = U_a \cap D$. Since $\mathbf{f}(D)$ is covered by O, $U = \{U_a \mid a \in A\}$ is an open cover of D. Since D is compact, there exists a finite subcover $\{U_j \mid j = 1, 2, \ldots, p\}$ of D. Thus $D = \bigcup_{j=1}^{p} V_j$, and $\mathbf{f}(D) \subseteq \bigcup_{j=1}^{p} O_j$. This is a finite subcover of $\mathbf{f}(D)$, which must therefore be compact.   ∎

Now we can prove an Extreme Value Theorem for $\mathbb{E}^n$.

Theorem 9.3.2 (Extreme Value Theorem for $\mathbb{E}^n$) *If $D \subset \mathbb{E}^n$ is compact and if $f : D \to \mathbb{R}$ is continuous, then f achieves both a maximum value and a minimum value on D.*

Proof: Since $f(D)$ is a compact subset of $\mathbb{R}$, it follows that $f(D)$ is both closed and bounded. Let $M = \sup f(D)$ and $m = \inf f(D)$, both of which are necessarily real-valued (not infinite). It will suffice to prove that both M and m are elements of $f(D)$. We will see that this follows from the fact that $f(D)$ is closed. Since M is the least upper bound of $f(D)$, $M - \frac{1}{k}$ fails to be an upper bound of $f(D)$, and this is true for every $k \in \mathbb{N}$. Thus, for each $k \in \mathbb{N}$ we have a point

$$M - \frac{1}{k} < f(\mathbf{x}_k) \leq M.$$

If there exists k such that $f(\mathbf{x}_k) = M$, then $M \in f(D)$ and we are finished. Otherwise, the sequence $f(\mathbf{x}_k) \to M$, with none of the points of this sequence being M, so that M is a cluster point $f(D)$. Hence $M \in f(D)$ as claimed. The similar argument for m is an exercise. ∎

Here is a surprising theorem that can be quite useful.

Theorem 9.3.3 *If* $\mathbf{f} : D \to \mathbb{E}^m$ *is a continuous,* one-to-one *function defined on a compact set* $D \subset \mathbb{E}^n$, *Then* $\mathbf{f}^{-1}$ *is continuous.*

Proof: Since $\mathbf{f}$ is one-to-one (also called *injective*), it makes sense to define $\mathbf{f}^{-1}(\mathbf{y}) = \mathbf{x}$ if and only if $\mathbf{f}(\mathbf{x}) = \mathbf{y}$. An invertible continuous function that has a continuous inverse is called a *homeomorphism*. Thus in words the theorem says that a continuous injective map of a compact set is a homeomorphism. It suffices to prove the continuity of $\mathbf{f}^{-1}$ at any cluster point $\mathbf{y} \in \mathbf{f}(D)$. If

$$\mathbf{y}_j = \mathbf{f}(\mathbf{x}_j) \to \mathbf{y} = \mathbf{f}(\mathbf{x}),$$

we need to show that

$$\mathbf{x}_j = \mathbf{f}^{-1}(\mathbf{y}_j) \to \mathbf{x} = \mathbf{f}^{-1}(\mathbf{y}).$$

We claim that the entire sequence $\mathbf{x}_j$ must converge to $\mathbf{x}$. If this were false, then there exists $\epsilon > 0$ and a subsequence $\mathbf{x}_{j_i}$ for which $\|\mathbf{x}_{j_i} - \mathbf{x}\| \geq \epsilon$ for all i. In that case, the *bounded* sequence $\mathbf{x}_{j_i}$ must itself have a subsequence $\mathbf{x}_{j_{i_l}}$ that converges to some vector $\mathbf{x}' \neq \mathbf{x}$, since $\|\mathbf{x}' - \mathbf{x}\| \geq \epsilon$. But $\mathbf{f}(\mathbf{x}') = \mathbf{y} = \mathbf{f}(\mathbf{x})$. This is a contradiction since $\mathbf{f}$ is one-to-one. ∎

We can extend the concept of the sup-norm to vector-valued functions as follows.

Definition 9.3.1 *If* $\mathbf{f} : D \to \mathbb{E}^m$ *where* $D \subseteq \mathbb{E}^n$, *we define the* sup-norm *of* $\mathbf{f}$ *to be*

$$\|\mathbf{f}\|_{\mathrm{sup}} = \sup \left\{ \|\mathbf{f}(\mathbf{x})\| \,\big|\, \mathbf{x} \in D \right\}.$$

It is left to the Exercises to show that this is in fact a norm. Next we can extend the concept of uniform convergence as follows to vector-valued functions.

Definition 9.3.2 *If* $\mathbf{f}$ *and* $\mathbf{f}_j$ *are all defined on a domain* D *with values in* $\mathbb{E}^m$, *we say that* $\mathbf{f}_j \to \mathbf{f}$ uniformly on D *provided that* $\|\mathbf{f}_j - \mathbf{f}\|_{\mathrm{sup}} \to 0$ *as* $j \to \infty$.

The following theorem shows that uniform convergence behaves very well with respect to the continuity of functions.

Theorem 9.3.4 *Suppose* $\mathbf{f}_j \in \mathcal{C}(D, \mathbb{E}^m)$ *for all* j, *and suppose also that* $\mathbf{f}_j \to \mathbf{f}$ *uniformly on* D. *Then* $\mathbf{f} \in \mathcal{C}(D, \mathbb{E}^m)$.

The proof is left to the Exercises.

EXERCISES

9.29 Prove or give a counterexample: If a *continuous injective* function f mapping $(-1, 1)$ to $\mathbb{E}^1$ has closed range then its range must be bounded.

9.30 Prove or give a counterexample: If a *continuous injective* function f mapping $(-1, 1)$ to $\mathbb{E}^1$ has bounded range, then its range must be closed.

9.31 Give an example of a noncontinuous image of a compact set that is *not* compact because
    a) it is *closed but not bounded.*
    b) it is *bounded but not closed.*

9.32 The set $E = (0, 1]$ is a bounded subset of $\mathbb{E}^1$ and E is *relatively closed* in the set $F = (0, 2]$. Is E compact? Justify your answer.

9.33 Let $D = \left\{ \frac{1}{n} \mid n \in \mathbb{N} \right\} \cup \{0\} \subset \mathbb{E}^1$. Prove each of the following statements.
    a) D is compact.
    b) If $\mathbf{f} \in \mathcal{C}(D, \mathbb{E}^m)$, then function $\|\mathbf{f}\|$ must achieve both a maximum value and a minimum value on D.
    c) If $\mathbf{g} : D \to \mathbb{E}^m$, then $\mathbf{g} \in \mathcal{C}(D, \mathbb{E}^m)$ if and only if $\mathbf{g}\left(\frac{1}{n}\right) \to \mathbf{g}(0)$ as $n \to \infty$. (Caution: Be sure to prove that if the latter condition is satisfied and if $\{x_k \mid k \in \mathbb{N}\} \subseteq D$ is such that $x_k \to 0$, then $\mathbf{g}(x_k) \to \mathbf{g}(0)$.)

9.34 Complete the proof of Theorem 9.3.2 by proving that m is a cluster point of $f(D)$.

9.35 Let $\phi : [0, 1] \to \mathbb{E}^2$ be defined by $\phi(t) = (\cos[2\pi t], \sin[2\pi t])$. Let

$$\mathbf{f} = \phi|_{[0,1)}$$

be the *restriction* of ϕ to $[0, 1)$.
    a) Show that $\mathbf{f}$ maps the interval $[0, 1)$ one-to-one and continuously onto S^1, the circle of radius 1 in $\mathbb{E}^2$.
    b) Show that $\mathbf{f}^{-1}$ is *not* continuous. Why doesn't this contradict Theorem 9.3.3? Evaluate or prove nonexistent

$$\lim_{\substack{\mathbf{y} \to (1,0) \\ \mathbf{y} \in S^1}} \mathbf{f}^{-1}(\mathbf{y}).$$

    c) The function ϕ maps the compact set $[0, 1]$ continuously onto the circle S^1 of radius 1 in $\mathbb{E}^2$. Does ϕ have a continuous inverse? Does this contradict Theorem 9.3.3?

9.36 Let $\mathbf{f} : [0, \infty) \to \mathbb{E}^2$ be defined by

$$\mathbf{f}(t) = \left(\cos\left[4 \tan^{-1}\right], \sin\left[4 \tan^{-1} t\right] \right).$$

Show that $\mathbf{f}$ maps the interval $[0, \infty)$ one-to-one and continuously onto the circle of radius 1 in $\mathbb{E}^2$. Show that $\mathbf{f}^{-1}$ is *not* continuous.

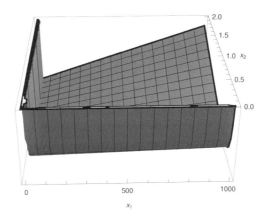

Figure 9.3 $f(x_1, x_2) = x_1 x_2 + \frac{96}{x_1} + \frac{96}{x_2}$.

9.37

  a) Suppose $S \subseteq \mathbb{E}^n, K \subset \mathbb{E}^m$, and K is compact. Suppose $\mathbf{f} : S \to K$ is one-to-one and onto K and that $\mathbf{f}^{-1}$ is continuous. Prove: S is compact.

  b) Give an example of a non-compact set $S \subseteq \mathbb{E}^n$, a compact set $K \subset \mathbb{E}^m$, $\mathbf{f} : S \to K$ is one-to-one, continuous and *into* K although $\mathbf{f}^{-1}$ is continuous as well.

9.38   Prove that the sup-norm introduced in Definition 9.3.1 satisfies Definition 2.4.4 and is thus a norm.

9.39   Prove Theorem 9.3.4. (Hint: Adapt the proof of Theorem 2.5.1.)

9.40   Let $D \subseteq \mathbb{E}^n$ and $\mathbf{f}_j : D \to \mathbb{E}^m$, for all $j \in \mathbb{N}$. Prove that in the *normed* vector space $\mathcal{C}(D, \mathbb{E}^m)$ equipped with the sup-norm, a sequence $\mathbf{f}_j$ converges if and only if it is a Cauchy sequence. Prove that $\mathcal{C}(D, \mathbb{E}^m)$ is a complete normed linear space. (Hint: Adapt the proof of Theorem 2.5.4.)

9.41   † Let $D \subseteq \mathbb{E}^n$ and $\mathbf{f} : D \to \mathbb{E}^m$. We call $\mathbf{f}$ *uniformly continuous* on D if and only if for each $\epsilon > 0$ there exists $\delta > 0$ such that $\|\mathbf{x} - \mathbf{x}'\| < \delta$ implies $\|\mathbf{f}(\mathbf{x}) - \mathbf{f}(\mathbf{x}')\| < \epsilon$, for all $\mathbf{x}$ and $\mathbf{x}' \in D$. If D is compact and if $\mathbf{f} \in \mathcal{C}(D, \mathbb{E}^m)$, prove that $\mathbf{f}$ is uniformly continuous on D. (Hint: Suppose false and use the Bolzano–Weierstrass Theorem to deduce a contradiction.)

9.42   It is commonly necessary to adapt Theorem 9.3.2 to special circumstances in which the domain D is not compact. Suppose we wish to construct an *open-topped* rectangular box of volume 48 with base measuring x_1 by x_2 in such a way as to achieve minimum total surface area

$$f(x_1, x_2) = x_1 x_2 + \frac{96}{x_1} + \frac{96}{x_2}$$

for the *five* faces of the box. Prove: The function f has a minimum value on the domain described by the conditions $x_1 > 0$ and $x_2 > 0$. See Fig. 9.3. (Hint: The idea is to reduce the problem to one on a compact domain. It may be helpful to consider the curves $x_1 x_2 = k$ for positive constants k, as well as the lines $x_1 = a$ and $x_2 = b$ for constants a and b.)

9.4   CONTINUOUS IMAGE OF A CONNECTED SET

We saw in Theorem 2.3.1 that every continuous function defined on an interval has the Intermediate Value Property, which is stated in Definition 2.3.1. In order to generalize the Intermediate Value Theorem to Euclidean Space, we will need to consider functions defined on a connected subset $D \subseteq \mathbb{E}^n$, and we will need to restrict our attention to real-valued functions so that the concept of k being *between* $f(\mathbf{a})$ and $f(\mathbf{b})$ will make sense. But first we prove a more general theorem.

Theorem 9.4.1 *Let $D \subseteq V$, a normed vector space, be a connected set, and suppose $\mathbf{f} \in \mathcal{C}(D, \mathbb{E}^m)$. Then the range $\mathbf{f}(D)$ is a connected subset of $\mathbb{E}^m$.*

Proof: Suppose the theorem were false: We will deduce a contradiction. By Theorem 8.4.1, the range $\mathbf{f}(D)$ can be separated as $\mathbf{f}(D) = A \cup B$, where $A \neq \emptyset \neq B$, $A \cap B = \emptyset$, and neither A nor B contains any cluster points of the other set. It follows that $D = \mathbf{f}^{-1}(A) \cup \mathbf{f}^{-1}(B)$, where the two non-$\emptyset$ sets $A' = \mathbf{f}^{-1}(A)$ and $B' = \mathbf{f}^{-1}(B)$ are disjoint. We claim that neither A' nor B' can have any cluster point of the other set. Suppose false: For example, suppose there exists $\mathbf{a}' \in A'$ that is also a cluster point of B'. Then there exists a sequence $\mathbf{b}'_j \in B'$ such that $\mathbf{b}'_j \to \mathbf{a}'$. By continuity, $\mathbf{b}_j = \mathbf{f}(\mathbf{b}'_j) \to \mathbf{a} = \mathbf{f}(\mathbf{a}')$. Hence A has a cluster point of B, which contradicts the hypothesis that D is connected. It follows that neither A' nor B' can have a cluster point of the other set, which contradicts the supposition that the theorem is false. Hence $\mathbf{f}(D)$ is connected.   ∎

Theorem 9.4.2 (Intermediate Value Theorem) *Let D be a connected subset of V, a normed vector space, and suppose $f \in \mathcal{C}(D, \mathbb{E}^1)$, so that f is real-valued. If $\mathbf{a}$ and $\mathbf{b} \in D$ and if $f(\mathbf{a}) < k < f(\mathbf{b})$, then there exists $\mathbf{c} \in D$ such that $f(\mathbf{c}) = k$.*

Proof: By the preceding theorem, we see that $f(D)$ is a connected subset of $\mathbb{E}^1$ and is thus an interval I. (See Exercise 8.45.) Since $f(\mathbf{a})$ and $f(\mathbf{b}) \in I$, it follows that $k \in I = f(D)$. Thus there exists $\mathbf{c} \in D$ such that $f(\mathbf{c}) = k$.   ∎

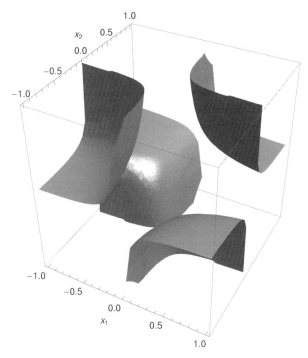

Figure 9.4 $\mathbf{f}(x_1, x_2) = \left(x_1, x_2, \frac{1}{x_1 x_2}\right)$.

EXERCISES

9.43 † A subset $D \subseteq \mathbb{E}^n$ is called *convex* if and only if for each pair of points **a** and $\mathbf{b} \in D$ the straight-line segment joining them is contained in D. That is,

$$S(\mathbf{a}, \mathbf{b}) = \left\{\phi(t) \mid \phi(t) = \mathbf{a} + t(\mathbf{b} - \mathbf{a}), \ t \in [0, 1]\right\} \subseteq D.$$

a) Prove that every convex subset of $\mathbb{E}^n$ is connected.

b) Prove that $\mathbb{E}^n$ is a connected set for each $n \in \mathbb{N}$.

9.44 The plane $\mathbb{E}^2 = \{\mathbf{x} \mid x_1 x_2 = 0\} \dot{\cup} \text{QI} \dot{\cup} \cdots \dot{\cup} \text{QIV}$. Here QI, for example, denotes the open first quadrant, where $x_1 > 0$ and $x_2 > 0$, and the dots denote disjointness of the union. Prove: Each of the five sets in the union is connected.

9.45 Let $\mathbf{f} : D \to \mathbb{E}^3$ by

$$\mathbf{f}(x_1, x_2) = \left(x_1, x_2, \frac{1}{x_1 x_2}\right),$$

where D is the Euclidean plane without either of the two coordinate axes. Prove: $\mathbf{f}(D)$ can be written as the union of 4 mutually disjoint sets, each set connected and each set maximal in the sense that it could not be enlarged with additional points from $\mathbf{f}(D)$ without losing connectivity. See Fig. 9.4.

9.46   Prove that each open ball $B_r(\mathbf{a})$ is a convex set in $\mathbb{E}^n$. Do the same thing for each closed ball $\bar{B}_r(\mathbf{a})$.

9.47   Consider the polynomial $p \in C\left(\mathbb{E}^2, \mathbb{E}^1\right)$ given by

$$p(\mathbf{x}) = x_1^2 x_2^3 - 2x_1^2 x_2 + x_1 x_2 - 3.$$

Show that there exists $\mathbf{x} \in \mathbb{E}^2$ such that $p(\mathbf{x}) = 0$.

9.48   Consider the function $\mathbf{f} \in C\left(\mathbb{E}^2, \mathbb{E}^2\right)$ given by $\mathbf{f}(\mathbf{x}) = e^{x_1}(\cos x_2, \sin x_2)$. Note that $\mathbf{f}(1,0) = (e,0)$ and $\mathbf{f}(1,\pi) = (-e,0)$. Is there a point $\mathbf{x} \in \mathbb{E}^2$ such that $\mathbf{f}(\mathbf{x}) = (0,0)$? Justify your claim.

9.49   Suppose $D \subset \mathbb{E}^n$ is compact and $\mathbf{f} \in C\left(D, \mathbb{E}^m\right)$ is one-to-one. Prove: If $\mathbf{f}(D)$ is connected, then D is connected.

9.50   Show how to use Theorem 9.4.1 to give an alternative proof of Theorem 8.4.2.

9.51   Show how to use Theorem 9.4.1 to give an alternative proof that $\mathbb{E}^n$ is connected. (Hint: If false, show that the continuous image of an interval into $\mathbb{E}^n$ could fail to be connected, which is a contradiction.)

9.5   TEST YOURSELF

EXERCISES

9.52   True or Give a Counterexample: If a set $S = A \dot{\cup} B \subset \mathbb{E}^n$ is the disjoint union of two non-$\emptyset$ sets, neither A nor B containing a cluster point of the other, then there is a positive distance $\delta > 0$ such that for all $\mathbf{a} \in A$ and $\mathbf{b} \in B$ we have $\|\mathbf{a} - \mathbf{b}\| \geq \delta$.

9.53   Let

$$f(x) = \begin{cases} \sin \frac{\pi}{x} & \text{if } x > 0, \\ 1 & \text{if } x = 0. \end{cases}$$

True or False: The graph of f is a *connected* set.

9.54   Give an example of a function $f : \mathbb{E}^2 \to \mathbb{R}$ for which

$$\lim_{x_1 \to 0} f(x_1, kx_1) = 0$$

for all $k \in \mathbb{R}$, yet $\lim_{\mathbf{x} \to \mathbf{0}} f(\mathbf{x})$ does *not* exist.

9.55   *State* the *Cauchy Criterion* for the existence of $\lim_{\mathbf{x} \to \mathbf{a}} \mathbf{f}(\mathbf{x})$ at a cluster point $\mathbf{a}$ of D_f.

9.56   True or False: If $f : \mathbb{E}^n \to \mathbb{R}$ is defined by

$$f(\mathbf{x}) = \frac{\sum_{i=1}^n (-1)^i x_i^2}{1 + \|\mathbf{x}\|^2},$$

then the set $S = \{\mathbf{x} \mid f(\mathbf{x}) = 0\}$ is a *closed* set.

9.57   True or False: If $\mathbf{f} \in \mathcal{C}\left(D, \mathbb{E}^m\right)$ with $D \subset \mathbb{E}^n$ and if $\mathcal{O}$ is open in $\mathbb{E}^m$, then $\mathbf{f}^{-1}(\mathcal{O})$ is an *open* set in $\mathbb{E}^n$.

9.58   Let

$$f(x_1, x_2) = \begin{cases} \frac{x_1 x_2}{x_1^2 + x_2^2} & \text{if } \mathbf{x} \neq \mathbf{0}, \\ 0 & \text{if } \mathbf{x} = \mathbf{0}. \end{cases}$$

Either find $\lim_{\mathbf{x} \to \mathbf{0}} f(\mathbf{x})$ or else state that it does not exist.

9.59   Let $D = \{0\} \cup \left\{ \frac{1}{n} \mid n \in \mathbb{N} \right\} \subset \mathbb{E}^1$. True or False: If $\mathbf{g} : D \to \mathbb{E}^m$, then $\mathbf{g} \in \mathcal{C}(D, \mathbb{E}^m)$ if and only if $\mathbf{g}\left(\frac{1}{n}\right) \to \mathbf{g}(0)$ as $n \to \infty$.

9.60   True or Give a Counterexample: If S is a *relatively closed* subset of $B_1(\mathbf{0}) \subset \mathbb{E}^n$, then S must be compact.

9.61   True or False: $B_1(0, 0) \cup B_1(1, 1)$ is a *connected* subset of $\mathbb{E}^2$.

9.62   Let $\mathbf{f} \in \mathcal{C}(\mathbb{E}^1, \mathbb{E}^2)$. Suppose $\mathbf{f}(0) = (1, 0)$ and $\mathbf{f}(1) = (-1, 0)$. True or False: there exists a number x between 0 and 1 such that $\mathbf{f}(x) = (0, 0)$.

9.63   Let $D = \{\mathbf{0}\} \cup \left\{ \left(x_1, \sin \frac{\pi}{x_1}\right) \mid x_1 \neq 0 \right\}$. True or False: If $\mathbf{f} \in \mathcal{C}\left(D, \mathbb{E}^2\right)$, then $\mathbf{f}(D)$ is connected.

CHAPTER 10

THE DERIVATIVE IN EUCLIDEAN SPACE

10.1 LINEAR TRANSFORMATIONS AND NORMS

We saw in Theorem 4.1.1 that a differentiable function $f : D \to \mathbb{R}$ defined on a domain $D \subseteq \mathbb{R}$ is a function for which the increments are given by

$$\Delta f = f(x + h) - f(x) = L(h) + \epsilon(h)h, \qquad (10.1)$$

where L is linear and $\epsilon(h) \to 0$ as $h \to 0$. We saw that if a function is differentiable, then the derivative $f'(x)$ exists and we have

$$L(h) = f'(x)h.$$

What the linear function L actually does is assign to the increment h in the variable x the corresponding rise of the *tangent line* to the graph of f at $(x, f(x))$.

The problem in generalizing this concept to $\mathbf{f} : \mathbb{E}^n \to \mathbb{E}^m$ is that in this context $f, L, h,$ and ϵ will all need to be replaced by *vectors* and by *vector-valued* functions $\mathbf{f}, \mathbf{L}, \mathbf{h},$ and $\epsilon(\mathbf{h})$. Unfortunately, it would be meaningless to multiply the vector values of $\epsilon(\mathbf{h})$ and $\mathbf{h}$. Thus, for purposes of generalization to $\mathbf{f} : \mathbb{E}^n \to \mathbb{E}^m$ it will be convenient to reformulate Equation (10.1) in the following equivalent form. A

Advanced Calculus: An Introduction to Linear Analysis. By Leonard F. Richardson
Copyright © 2008 John Wiley & Sons, Inc.

differentiable function $f : D \to \mathbb{R}$ is a function for which the increments are given by

$$\Delta f = f(x + h) - f(x) = L(h) + \tilde{\epsilon}(h), \tag{10.2}$$

where L is linear and

$$\frac{|\tilde{\epsilon}(h)|}{|h|} \to 0$$

as $h \to 0$. In this reformulation, $\tilde{\epsilon}(h) = \epsilon(h)h$. When we pass to functions $\mathbf{f} : \mathbb{E}^n \to \mathbb{E}^m$, we cannot divide by a vector $\mathbf{h}$. However, we can divide by its *norm*, which means the same thing as the length of the vector increment.

Since for functions $f : \mathbb{R} \to \mathbb{R}$ the derivative $f'(x)$ is the single coefficient of a 1×1 matrix representing the linear map L, it is easy to overlook the important distinction between the number $f'(x)$ and the linear transformation L for which this number is its sole matrix coefficient. But for $\mathbf{f} : \mathbb{E}^n \to \mathbb{E}^m$ the linear transformation L will map $\mathbb{E}^n$ to $\mathbb{E}^m$, and it will be represented by an $m \times n$ matrix. In the present section we begin by reviewing some basic information about linear transformations and their matrices. We will need to introduce the concept of the *norm* of a matrix, which may be new to the reader.

Let us denote by $\mathcal{L}(\mathbb{E}^n, \mathbb{E}^m)$ the set of all linear transformations $L : \mathbb{E}^n \to \mathbb{E}^m$. (In the special case in which $m = n$, we will write $\mathcal{L}(\mathbb{E}^n)$ in place of $\mathcal{L}(\mathbb{E}^n, \mathbb{E}^n)$.) The reader will recall [10] that if L and $L' \in \mathcal{L}(\mathbb{E}^n, \mathbb{E}^m)$ and if $c \in \mathbb{R}$, then $cL + L' \in \mathcal{L}(\mathbb{E}^n, \mathbb{E}^m)$, so that $\mathcal{L}(\mathbb{E}^n, \mathbb{E}^m)$ is a *vector space*.

Let $\mathbf{x} \in \mathbb{E}^n$ and $L \in \mathcal{L}(\mathbb{E}^n, \mathbb{E}^m)$. We can decompose

$$\mathbf{x} = \sum_{j=1}^{n} x_j \mathbf{e}_j,$$

where $\mathbf{e}_j = (0, \ldots, 0, 1, 0, \ldots, 0)$, where all entries are 0 except for the jth entry, which is 1. The family $\{\mathbf{e}_j \mid j = 1, \ldots, n\}$ is called the *standard basis* of $\mathbb{E}^n$. For the range space $\mathbb{E}^m$ we will denote the standard basis $\{\mathbf{f}_1, \ldots, \mathbf{f}_m\}$ to avoid confusion. Now we can write $L(\mathbf{x}) = \sum_{j=1}^{n} L(\mathbf{e}_j) x_j$. We can introduce an $m \times n$ matrix

$$[L] = [\ \mathbf{C}_1 \quad \mathbf{C}_2 \quad \ldots \quad \mathbf{C}_n\ ],$$

where each column vector

$$\mathbf{C}_j = \begin{bmatrix} l_{1j} \\ l_{2j} \\ \vdots \\ l_{mj} \end{bmatrix}$$

and $L(\mathbf{e}_j) = \sum_{i=1}^{m} l_{ij} \mathbf{f}_i$. Then $L(\mathbf{x}) = [L]\mathbf{x}$, in which the $m \times n$ matrix $[L]$ multiplies the column vector $\mathbf{x}$ in the following customary way. We express the matrix $[L]$ in terms of its m row vectors:

$$[L] = \begin{bmatrix} \mathbf{R}_1 \\ \mathbf{R}_2 \\ \vdots \\ \mathbf{R}_m \end{bmatrix}.$$

Thus $L(\mathbf{x})$ can be represented by the column vector

$$[L]\begin{bmatrix} x_1 \\ x_2 \\ \vdots \\ x_n \end{bmatrix} = \begin{bmatrix} \mathbf{R}_1 \cdot \mathbf{x} \\ \mathbf{R}_2 \cdot \mathbf{x} \\ \vdots \\ \mathbf{R}_m \cdot \mathbf{x} \end{bmatrix}.$$

Hence $L(\mathbf{x}) = \sum_{i=1}^{m}(\mathbf{R}_i \cdot \mathbf{x})\mathbf{f}_i$.

For the purposes of this book, we will need to introduce a *norm* on the vector space $\mathcal{L}(\mathbb{E}^n, \mathbb{E}^m)$. We begin with the following theorem.

Theorem 10.1.1 *Let $L \in \mathcal{L}(\mathbb{E}^n, \mathbb{E}^m)$ with matrix $[L] = [l_{ij}]_{m \times n}$. Let*

$$M = \sqrt{\sum_{i,j} l_{ij}^2}.$$

Then for each $\mathbf{x} \in \mathbb{E}^n$ we have $\|L(\mathbf{x})\| \leq M\|\mathbf{x}\|$.

Proof: In the notation introduced just above the statement of the theorem, we proceed as follows using the Cauchy–Schwarz inequality and the orthogonality of the basis vectors, each of which has norm 1.

$$\begin{aligned}
\|L(\mathbf{x})\|^2 &= \left\| \sum_{i=1}^{m}(\mathbf{R}_i \cdot \mathbf{x})\mathbf{f}_i \right\|^2 \\
&= \sum_{i=1}^{m}(\mathbf{R}_i \cdot \mathbf{x})^2 \\
&\leq \sum_{i=1}^{m}\|\mathbf{R}_i\|^2\|\mathbf{x}\|^2 \\
&= \|\mathbf{x}\|^2 \sum_{i=1}^{m}\|\mathbf{R}_i\|^2 \\
&= \|\mathbf{x}\|^2 \sum_{1 \leq i \leq m, 1 \leq j \leq n} l_{ij}^2 \\
&= M^2\|\mathbf{x}\|^2.
\end{aligned}$$

■

We are ready to introduce a norm on $\mathcal{L}(\mathbb{E}^n, \mathbb{E}^m)$ by generalizing a definition from Theorem 5.4.2 as follows.

Definition 10.1.1 *For each $L \in \mathcal{L}(\mathbb{E}^n, \mathbb{E}^m)$ we define a norm by*

$$\|L\| = \inf \left\{ K \mid \|L(\mathbf{x})\| \leq K\|\mathbf{x}\| \; \forall \mathbf{x} \in \mathbb{E}^n \right\}.$$

Remark 10.1.1 In Exercise 10.1, the reader will show that

$$\|L(\mathbf{x})\| \le \|L\| \cdot \|\mathbf{x}\|$$

for all $\mathbf{x} \in \mathbb{E}^n$. In order to show that the preceding definition satisfies Definition 2.4.4 of a norm on a vector space, the most significant step is the triangle inequality, which we leave to Exercise 10.4.

Remark 10.1.2 In Exercise 10.10 the reader will prove that $\mathcal{L}(\mathbb{E}^n, \mathbb{E}^m)$ is a *complete* normed linear space.

■ **EXAMPLE 10.1**

If $T \in \mathcal{L}(\mathbb{E}^n)$, the linear transformation $T : \mathbb{E}^n \to \mathbb{E}^n$ is called *invertible* if and only if T is both one-to-one and onto $\mathbb{E}^n$. This can be expressed also as requiring that T be both an *injection* and a *surjection*, which is called a *bijection*.

This condition is equivalent to the existence of a map $T^{-1} \in \mathcal{L}(\mathbb{E}^n)$ such that

$$T \circ T^{-1} = T^{-1} \circ T = I,$$

the *identity* transformation of $\mathbb{E}^n \to \mathbb{E}^n$. We will have a special interest in invertible linear transformations of $\mathbb{E}^n$. We define

$$\mathcal{GL}(n, \mathbb{R}) = \left\{ T \in \mathcal{L}(\mathbb{E}^n) \,\middle|\, T^{-1} \text{ exists} \right\}.$$

It will be useful to recall from linear algebra that T^{-1} exists if and only if the determinant is nonzero: $\det T \ne 0$.

In this connection it may help the reader to remember that the determinant of the *matrix* of a linear transformation is *independent* of the basis with respect to which the matrix is written. This is because if $\mathcal{B}$ is the standard basis and $\mathcal{B}'$ is any other basis, there will be a change of basis matrix A such that

$$[T]_{\mathcal{B}'} = A^{-1}[T]_{\mathcal{B}} A,$$

to which we apply the multiplicative property of the determinant function to see that

$$\det[T]_{\mathcal{B}'} = (\det A)^{-1} \det[T]_{\mathcal{B}} \det A = \det[T]_{\mathcal{B}}.$$

EXERCISES

10.1 † If $L \in \mathcal{L}(\mathbb{E}^n, \mathbb{E}^m)$, prove

$$\|L(\mathbf{x})\| \le \|L\| \|\mathbf{x}\|$$

for all $\mathbf{x} \in \mathbb{E}^n$. Hint: Express $\|L\|$ in terms of the set $\left\{ \frac{\|L(\mathbf{x})\|}{\|\mathbf{x}\|} \,\middle|\, \mathbf{x} \ne \mathbf{0} \right\}$.

10.2 If $L \in \mathcal{L}(\mathbb{E}^n, \mathbb{E}^m)$, prove $L = 0$, the zero linear transformation, if and only if $\|L\| = 0$.

10.3   If $L \in \mathcal{L}(\mathbb{E}^n, \mathbb{E}^m)$, prove $\|cL\| = |c|\|L\|$ for all $c \in \mathbb{R}$.

10.4   † If L and $L' \in \mathcal{L}(\mathbb{E}^n, \mathbb{E}^m)$, prove the *triangle inequality*

$$\|L + L'\| \leq \|L\| + \|L'\|.$$

10.5   Let A and B be in $\mathcal{L}(\mathbb{E}^2)$ with matrices in the standard basis

$$[A] = \begin{pmatrix} 1 & 0 \\ 0 & 0 \end{pmatrix} \text{ and } [B] = \begin{pmatrix} 0 & 0 \\ 0 & 1 \end{pmatrix}.$$

Find $\|A\|, \|B\|$, and $\|A + B\|$.

10.6   Give an example of $X \in \mathcal{L}(\mathbb{E}^2)$ for which $\|X + X\| < \|X\| + \|X\|$ or else prove impossible.

10.7   Let $L \in \mathcal{L}(\mathbb{E}^n, \mathbb{E}^m)$. Prove that L is a *uniformly continuous* function from $\mathbb{E}^n \to \mathbb{E}^m$, as defined in Exercise 9.41. (Hint: Use the result of Exercise 10.1.)

10.8   Let $L \in \mathcal{L}(\mathbb{E}^n, \mathbb{E}^m)$ and $T \in \mathcal{L}(\mathbb{E}^m, \mathbb{E}^k)$, so that $T \circ L \in \mathcal{L}(\mathbb{E}^n, \mathbb{E}^k)$.
   a) Prove: $\|T \circ L\| \leq \|T\|\|L\|$.
   b) Now let $k = m = n$. Denote $T^2 = T \circ T$ and $T^{j+1} = T^j \circ T$ for all $j \in \mathbb{N}$. Show that $\|T^j\| \leq \|T\|^j$.
   c) Give an example in which $\|T^2\| < \|T\|^2$. (Hint: Find a linear transformation $T \neq 0$ for which $T^2 = 0 \in \mathcal{L}(\mathbb{E}^n)$.)

10.9   † Let $T^{(k)} \in \mathcal{L}(\mathbb{E}^n, \mathbb{E}^m)$ for each $k \in \mathbb{N}$. Prove that the sequence $T^{(k)} \to T$ as $k \to \infty$ in the norm of $\mathcal{L}(\mathbb{E}^n, \mathbb{E}^m)$ if and only if for each i, j the sequence of matrix coefficients $t_{ij}^{(k)} \to t_{ij}$. Hint: Prove that for each i, j we have

$$|t_{ij}| \leq \|T\| \leq \sqrt{\sum_{1 \leq \kappa \leq m,\, 1 \leq \lambda \leq n} t_{\kappa\lambda}^2}.$$

10.10   Prove that the normed vector space $\mathcal{L}(\mathbb{E}^n, \mathbb{E}^m)$ is *complete*, meaning that each Cauchy sequence in the norm converges to an element of $\mathcal{L}(\mathbb{E}^n, \mathbb{E}^m)$.

10.11   Suppose $X \in \mathcal{L}(\mathbb{E}^n)$ is such that $\|X\| < 1$.
   a) Prove that

$$T_K = \sum_{k=0}^{K} X^k$$

   is Cauchy, so that T_K converges to some $T \in \mathcal{L}(\mathbb{E}^n)$ as $K \to \infty$.
   b) Prove that $(I - X)T_K \to I$ as $K \to \infty$.
   c) Prove that $(I - X)^{-1}$ exists and equals T.
       (Hints: Here the power $X^0 = I$, the *identity* transformation in $\mathcal{L}(\mathbb{E}^n)$. See Exercise 10.8.)

10.12   Suppose $A \in \mathcal{L}\left(\mathbb{E}^2\right)$ has the matrix

$$[A] = \begin{pmatrix} 2 & 1 \\ 3 & \sqrt{2} \end{pmatrix}$$

with respect to the standard basis of $\mathbb{E}^2$. Prove that $\sqrt{13} \leq \|A\| \leq 4$.

10.13   † Show that the set of linear transformations $\{T_{i,j}\}$ defined by $T_{i,j} : \mathbf{e}_i \to \mathbf{f}_j$ and $T_{i,j} : \mathbf{e}_l \to \mathbf{0}$ for all $l \neq i$ is a *basis* for the finite-dimensional vector space $\mathcal{L}(\mathbb{E}^n, \mathbb{E}^m)$, and show that the *dimension* is $\dim \mathcal{L}(\mathbb{E}^n, \mathbb{E}^m) = mn$.

10.14   Let the function $f : \mathcal{L}(\mathbb{E}^n, \mathbb{E}^m) \to \mathbb{R}$ defined by $f(T) = t_{i_0 j_0}$, where $[T] = [t_{ij}]_{m \times n}$, $1 \leq i \leq m$, and $1 \leq j \leq n$. Prove that f is *uniformly* continuous. (Hint: Show that f is a *linear* map from $\mathcal{L}(\mathbb{E}^n, \mathbb{E}^m)$ to $\mathbb{R}$ and use the hint from Exercise 10.9.)

10.15   Define $S \subset \mathcal{L}\left(\mathbb{E}^2\right)$ by letting $S = \left\{A \in \mathcal{L}\left(\mathbb{E}^2\right) \mid a_{11} \neq 0\right\}$, where the matrix $[A] = [a_{ij}]_{2 \times 2}$. Prove: S is an *open* but *not connected* subset of $\mathcal{L}\left(\mathbb{E}^2\right)$.

10.16   † Prove that the determinant function $\det : \mathcal{L}(\mathbb{E}^n) \to \mathbb{R}$ is continuous. (Hint: The determinant is a polynomial in the matrix coefficients. See [10].)

10.17   Prove that $\mathcal{GL}(n, \mathbb{R})$, defined in Example 10.1, is a *group* with the operation being composition of the linear transformations. That is, if S, T, and $U \in \mathcal{GL}(n, \mathbb{R})$ and if I is the identity transformation, then

   a) $S \circ T \in \mathcal{GL}(n, \mathbb{R})$.
   b) $S \circ (T \circ U) = (S \circ T) \circ U$.
   c) $I \in \mathcal{GL}(n, \mathbb{R})$.
   d) $S^{-1} \in \mathcal{GL}(n, \mathbb{R})$.

10.18   Show that the subset $\mathcal{GL}(n, \mathbb{R}) \subset \mathcal{L}(\mathbb{E}^n)$ is *not* a vector space.

10.19   Show that the group $\mathcal{GL}(n, \mathbb{R})$ is not commutative if $n > 1$. That is, show there exist S and $T \in \mathcal{GL}(n, \mathbb{R})$ such that $S \circ T \neq T \circ S$.

10.20   Prove that $\mathcal{GL}(n, \mathbb{R})$ is an *open* subset of the normed vector space $\mathcal{L}(\mathbb{E}^n)$. (Hint: Use the result of Exercise 10.16 together with the fact that T is invertible if and only if $\det(T) \neq 0$.)

10.21   Let $S \in \mathcal{GL}(n, \mathbb{R})$, and let $\sigma = \|S^{-1}\|$. Use the following steps to prove that the open ball $B_{\frac{1}{\sigma}}(S) \subset \mathcal{GL}(n, \mathbb{R})$.

   a) Show that $\sigma > 0$.
   b) Show for each $\mathbf{x} \in \mathbb{E}^n$ that $\|\mathbf{x}\| \leq \sigma \|S(\mathbf{x})\|$, so that $\|S(\mathbf{x})\| \geq \frac{1}{\sigma}\|\mathbf{x}\|$.
   c) Consider any $T \in B_{\frac{1}{\sigma}}(S) \subset \mathcal{L}(\mathbb{E}^n)$. Write

   $$\|T(\mathbf{x})\| = \|S(\mathbf{x}) - [S - T](\mathbf{x})\|.$$

   Prove that $\|T(\mathbf{x})\| \geq p\|\mathbf{x}\|$, where $p = \frac{1}{\sigma} - \|S - T\| > 0$. (Hint: See Exercise 9.13.)
   d) Show that the kernel $\ker(T) = \{\mathbf{x} \mid T(\mathbf{x}) = \mathbf{0}\} = \{\mathbf{0}\}$, so that T is both one-to-one and onto $\mathbb{E}^n$. Hence T is invertible.

10.22   Prove that $\mathcal{GL}(n, \mathbb{R})$ is not a connected subset of $\mathcal{L}(\mathbb{E}^n)$. (Hint: Consider the result of Exercise 10.16. Use the fact that T is invertible if and only if $\det(T) \neq 0$.)

10.23   If $T \in \mathcal{L}(\mathbb{E}^n)$, let $[T]_{\mathcal{B}} = (t_{ij})_{n \times n}$ and denote the *trace* of T by

$$tr(T) = \sum_{i=1}^{n} t_{ii}.$$

a) If $\mathcal{B}'$ is any other basis for $\mathbb{E}^n$, and if $[T]_{\mathcal{B}'} = (t'_{ij})_{n \times n}$, then prove that

$$\sum_{i=1}^{n} t_{ii} = \sum_{i=1}^{n} t'_{ii}.$$

Hint: Use the theorem from Linear Algebra that

$$tr(AB) = tr(BA).$$

b) Denote by $\mathfrak{sl}(\mathbb{E}^n) = \{T \in \mathcal{L}(\mathbb{E}^n) \mid tr(T) = 0\}$. Prove that $\mathfrak{sl}(\mathbb{E}^n)$ is a *closed* subset of $\mathcal{L}(\mathbb{E}^n)$.

10.2   DIFFERENTIABLE FUNCTIONS

Recall from the opening of the preceding section our intended meaning of the concept of differentiability of $\mathbf{f}$ at a point $\mathbf{x} \in D \subseteq \mathbb{E}^n$. We mean that the increments $\Delta\mathbf{f} = \mathbf{f}(\mathbf{x} + \mathbf{h}) - \mathbf{f}(\mathbf{x})$ should be locally approximated by a linear transformation $A \in \mathcal{L}(\mathbb{E}^n, \mathbb{E}^m)$ applied to the increment $\mathbf{h}$. The formal definition is as follows.

Definition 10.2.1 *Suppose $D \subseteq \mathbb{E}^n$ and $\mathbf{f} : D \to \mathbb{E}^m$. If $\mathbf{x} \in D$ is a cluster point of D we say that $\mathbf{f}$ is* differentiable *at $\mathbf{x}$ and that $\mathbf{f}'(\mathbf{x}) = A \in \mathcal{L}(\mathbb{E}^n, \mathbb{E}^m)$ if*

$$\lim_{\mathbf{h} \to 0} \frac{\|\mathbf{f}(\mathbf{x} + \mathbf{h}) - \mathbf{f}(\mathbf{x}) - A(\mathbf{h})\|}{\|\mathbf{h}\|} = 0.$$

Remark 10.2.1 The concept of derivative is most useful at points in the interior of the domain of definition. At such points, for *all* sufficiently small $\mathbf{h}$, we have $\mathbf{x} + \mathbf{h} \in D$. However, the definition still makes sense even for boundary points of D, at which only some $\mathbf{h}$ will be admissible, however short $\mathbf{h}$ may be. In that case, it is understood in this definition that $\mathbf{h} \to 0$ through values of $\mathbf{h}$ such that $\mathbf{x} + \mathbf{h} \in D$, the domain of $\mathbf{f}$. An equivalent form of the condition in this definition is the following statement.

Theorem 10.2.1 *Suppose $D \subseteq \mathbb{E}^n$ and $\mathbf{f} : D \to \mathbb{E}^m$. Let $\mathbf{x} \in D$ be a cluster point of D. Then $\mathbf{f}$ is* differentiable *at $\mathbf{x}$ and $\mathbf{f}'(\mathbf{x}) = A \in \mathcal{L}(\mathbb{E}^n, \mathbb{E}^m)$ if and only if*

$$\Delta\mathbf{f} = \mathbf{f}(\mathbf{x} + \mathbf{h}) - \mathbf{f}(\mathbf{x}) = A(\mathbf{h}) + \epsilon(\mathbf{h}),$$

where $\frac{\|\epsilon(\mathbf{h})\|}{\|\mathbf{h}\|} \to 0$ as $\mathbf{h} \to 0$.

The proof of this theorem is left to Exercise 10.33.

Theorem 10.2.2 *If* f *is differentiable at* $\mathbf{x}_0 \in D$, *then* f *is continuous at* $\mathbf{x}_0$.

Proof: For each $\mathbf{x} \in D$, denote $\mathbf{h} = \mathbf{x} - \mathbf{x}_0$. Note that A is continuous and $A(\mathbf{0}) = \mathbf{0}$. Then

$$\mathbf{f}(\mathbf{x}) - \mathbf{f}(\mathbf{x}_0) = A(\mathbf{h}) + \epsilon(\mathbf{h}) \rightarrow \mathbf{0} + \mathbf{0} = \mathbf{0}$$

as $\mathbf{x} \rightarrow \mathbf{x}_0$, so that $\mathbf{h} \rightarrow \mathbf{0}$. Thus $\mathbf{f}(\mathbf{x}) \rightarrow \mathbf{f}(\mathbf{x}_0)$ as $\mathbf{x} \rightarrow \mathbf{x}_0$.   ∎

If $f : \mathbb{R} \rightarrow \mathbb{R}$, we have seen that the differentiability of f at x is equivalent to the existence of

$$\lim_{h \to 0} \frac{f(x + h) - f(x)}{h}.$$

But if $\mathbf{f} : D \rightarrow \mathbb{E}^m$ where $D \subseteq \mathbb{E}^n$ with $n > 1$, then the relationship between differentiability and the existence of the so-called *partial derivatives* is significant but imperfect. (See Exercise 10.38.)

Definition 10.2.2 *Let* $\mathbf{f} : D \rightarrow \mathbb{E}^m$, *where* D *is a subset of* $\mathbb{E}^n$, *equipped with the standard orthonormal basis* $\mathcal{B}$. *Then the* partial derivatives *are defined by*

$$\frac{\partial f_i}{\partial x_j} = \lim_{t \to 0} \frac{f_i(\mathbf{x} + t\mathbf{e}_j) - f_i(\mathbf{x})}{t}$$

for all $1 \le i \le m$, $1 \le j \le n$, *provided these limits exist. Furthermore, the* directional derivatives *are defined by*

$$D_{\mathbf{v}}\mathbf{f}(\mathbf{x}) = \lim_{t \to 0} \frac{\mathbf{f}(\mathbf{x} + t\mathbf{v}) - \mathbf{f}(\mathbf{x})}{t},$$

provided that the limit exists.

Remark 10.2.2   In elementary courses about several variable calculus, it is customary to define the directional derivative for *real*-valued functions, though not for *vector*-valued functions, and to limit the directional derivative to directions specified by *unit* vectors $\mathbf{v}$. We do *not* require $\mathbf{v}$ to be a unit vector here, because if we did this we would lose the very satisfying property of $D_{\mathbf{v}}\mathbf{f}(\mathbf{x})$ presented in Exercise 10.42.

Theorem 10.2.3 *Let* $\mathbf{f} : D \rightarrow \mathbb{E}^m$ *be differentiable at* $\mathbf{x} \in D$, *where* D *is an* open *subset of* $\mathbb{E}^n$. *Then the* directional derivatives

$$D_{\mathbf{v}}\mathbf{f}(\mathbf{x}) = \lim_{t \to 0} \frac{\mathbf{f}(\mathbf{x} + t\mathbf{v}) - \mathbf{f}(\mathbf{x})}{t}$$

exist and equal $\mathbf{f}'(\mathbf{x})\mathbf{v}$ *for all* $\mathbf{v} \in \mathbb{E}^n$. *Moreover, the* partial derivatives

$$\frac{\partial f_i}{\partial x_j} = \left(D_{\mathbf{e}_j}\mathbf{f}(\mathbf{x})\right)_i = \lim_{t \to 0} \frac{f_i(\mathbf{x} + t\mathbf{e}_j) - f_i(\mathbf{x})}{t}$$

*exist for all $1 \leq i \leq m$, $1 \leq j \leq n$. With respect to the standard orthonormal basis
$\mathcal{B}$, the matrix*

$$[\mathbf{f}'(\mathbf{x})]_\mathcal{B} = \left[\frac{\partial f_i}{\partial x_j}\right]_{m \times n}.$$

Proof: In this theorem, the purpose of the assumption that D is open is to ensure
that for all $\mathbf{v} \in \mathbb{E}^n$ we will have $\mathbf{x} + t\mathbf{v} \in D$ for all sufficiently small t. Denote
$\mathbf{f}'(\mathbf{x}) = A \in \mathcal{L}(\mathbb{E}^n, \mathbb{E}^m)$.

For the directional derivatives we write

$$\Delta \mathbf{f} = \mathbf{f}(\mathbf{x} + t\mathbf{v}) - \mathbf{f}(\mathbf{x}) = A(t\mathbf{v}) + \epsilon(t\mathbf{v})$$

and we see that

$$\frac{\Delta \mathbf{f}}{t} = A\mathbf{v} + \frac{\epsilon(t\mathbf{v})}{\|t\mathbf{v}\|} \frac{|t|\|\mathbf{v}\|}{t} \longrightarrow A\mathbf{v} = D_\mathbf{v} f(\mathbf{x})$$

as $t \to 0$. Here we use the fact that $\frac{\epsilon(t\mathbf{v})}{\|t\mathbf{v}\|} \to \mathbf{0}$ as $t \to 0$ because its norm approaches
zero as a result of the hypothesis of differentiability. And we use the fact that $\frac{|t|\|\mathbf{v}\|}{t}$
is bounded.

It is shown in Exercise 10.34 that $\frac{\partial f_i}{\partial x_j}$ is the ith component of $D_{\mathbf{e}_j}\mathbf{f}$. Thus each
partial derivative $\frac{\partial f_i}{\partial x_j}$ exists because of the first part.

Denote the matrix $[\mathbf{f}'(x)]_\mathcal{B} = A = [a_{ij}]_{m \times n}$. We need to show that $\frac{\partial f_i}{\partial x_j} = a_{ij}$.
Let $\mathbf{C}_j$ denote the jth column vector of A, which is the matrix form of the vector
$D_{\mathbf{e}_j}\mathbf{f} = A\mathbf{e}_j$. This completes the proof that

$$[\mathbf{f}'(\mathbf{x})]_\mathcal{B} = \left[\frac{\partial f_i}{\partial x_j}\right]_{m \times n}.$$

∎

Corollary 10.2.1 *In Theorem 10.2.3, if $\mathbf{f}'(\mathbf{x})$ exists, then it is unique.*

■ **EXAMPLE 10.2**

Let $f : \mathbb{E}^2 \to \mathbb{E}^1$ be defined by $f(x_1, x_2) = x_1 x_2$, with $0 \leq x_1 \leq 1$ and
$0 \leq x_2 \leq 1$. (See Fig. 10.1.) Since the matrix is given by

$$[f'(x_1, x_2)] = (\begin{array}{cc} x_2 & x_1 \end{array}),$$

we calculate that

$$[f'(0.5, 0.5)] \left(\begin{array}{c} 0.1 \\ 0.1 \end{array}\right) = (\begin{array}{cc} 0.5 & 0.5 \end{array}) \left(\begin{array}{c} 0.1 \\ 0.1 \end{array}\right) = 0.05 + 0.05 = 0.1.$$

This is the rise of the *plane* that is *tangent* to the graph of this function at
$(0.5, 0.5, 0.25)$ as both of the first two coordinates are increased from 0.5 each
to 0.6 each. We compare this result with the actual rise of the function, which
is $f(0.6, 0.6) - f(0.5, 0.5) = 0.36 - 0.25 = 0.11$. The derivative gives an
error in the estimate of the 0.11 rise of the function of 0.01.

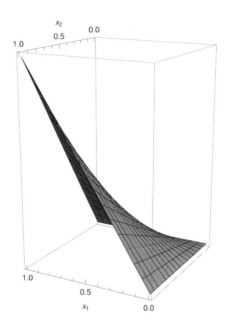

Figure 10.1 $x_3 = x_1 x_2$.

■ EXAMPLE 10.3

This example is a computational illustration of the meaning of the derivative as a linear approximation to the increments in **f**. Let $\mathbf{f} : \mathbb{R}^2 \to \mathbb{R}^2$ by $\mathbf{f}(r, \theta) = (r \cos \theta, r \sin \theta)$. A simple calculation using Theorem 10.2.3 shows that

$$\left[\mathbf{f}' \left(1, \frac{\pi}{4} \right) \right] \left(\begin{array}{c} \frac{1}{10} \\ \frac{1}{10} \end{array} \right) = \left(\begin{array}{cc} \frac{1}{\sqrt{2}} & -\frac{1}{\sqrt{2}} \\ \frac{1}{\sqrt{2}} & \frac{1}{\sqrt{2}} \end{array} \right) \left(\begin{array}{c} \frac{1}{10} \\ \frac{1}{10} \end{array} \right) = \left(\begin{array}{c} 0 \\ \frac{\sqrt{2}}{10} \end{array} \right).$$

The reader should use a calculator to compare this result with the numerical approximations to $f \left(\frac{11}{10}, \frac{\pi}{4} + \frac{1}{10} \right) - f \left(1, \frac{\pi}{4} \right) \approx (-0.011, 0.145)$.

Although Exercise 10.38 shows that the existence of all the partial derivatives $\frac{\partial f_i}{\partial x_j}$ is not sufficient to ensure that f is differentiable, we do have the following definition and theorem.

Definition 10.2.3 *Let $D \subseteq \mathbb{E}^n$ and $\mathbf{f} : D \to \mathbb{E}^m$. We say that $\mathbf{f}$ is* continuously differentiable, *denoted by* $\mathbf{f} \in \mathcal{C}^1(D, \mathbb{E}^m)$, *provided that $\mathbf{f}$ is differentiable at every* point of $\mathbf{a} \in D$ and that $\mathbf{f}' : D \to \mathcal{L}(\mathbb{E}^n, \mathbb{E}^m)$ is continuous. (That is, we require that $\|\mathbf{f}'(\mathbf{x}) - \mathbf{f}'(\mathbf{a})\| \to 0$ as $\mathbf{x} \to \mathbf{a}$, for each $\mathbf{a} \in D$.)

Theorem 10.2.4 Let $D \subseteq \mathbb{E}^n$ be an open set and $\mathbf{f} : D \to \mathbb{E}^m$. Then $\mathbf{f} \in \mathcal{C}^1(D, \mathbb{E}^m)$ if and only if every partial derivative $\frac{\partial f_i}{\partial x_j}$ is continuous on D.

Proof: First we suppose that $\mathbf{f} \in \mathcal{C}^1(D, \mathbb{E}^m)$. We note that $\frac{\partial f_i}{\partial x_j}\mathbf{e}_i$ is the ith component of $D_{\mathbf{e}_j}\mathbf{f}$. However,

$$\left| \frac{\partial f_i}{\partial x_j}(\mathbf{x}) - \frac{\partial f_i}{\partial x_j}(\mathbf{a}) \right| \leq \|D_{\mathbf{e}_j}\mathbf{f}(\mathbf{x}) - D_{\mathbf{e}_j}\mathbf{f}(\mathbf{a})\|$$
$$= \|(\mathbf{f}'(\mathbf{x}) - \mathbf{f}'(\mathbf{a}))\mathbf{e}_j\|$$
$$\leq \|\mathbf{f}'(\mathbf{x}) - \mathbf{f}'(\mathbf{a})\|\|\mathbf{e}_j\| \to 0$$

as $\mathbf{x} \to \mathbf{a}$ because $\mathbf{f}'$ is continuous. Thus $\frac{\partial f_i}{\partial x_j}$ is continuous at each $a \in D$.

Next, we assume that each of the partial derivatives $\frac{\partial f_i}{\partial x_j}$ is continuous on D, and we must prove that $\mathbf{f} \in \mathcal{C}^1(D, \mathbb{E}^m)$. We let A be the linear transformation that has the matrix

$$[A]_{m \times n} = \left[\frac{\partial f_i}{\partial x_j}(\mathbf{x}) \right]$$

with respect to the standard bases. We need to prove first that

$$\mathbf{f}(\mathbf{x} + \mathbf{h}) - \mathbf{f}(\mathbf{x}) = A(\mathbf{h}) + \epsilon(\mathbf{h})$$

where $\frac{\|\epsilon(\mathbf{h})\|}{\|\mathbf{h}\|} \to 0$ as $\mathbf{h} \to \mathbf{0}$. However, this is equivalent to proving for each $i = 1, \ldots, m$ that

$$f_i(\mathbf{x} + \mathbf{h}) - f_i(\mathbf{x}) = \mathbf{R}_i \cdot \mathbf{h} + \epsilon_i(\mathbf{h})$$
$$= \sum_{j=1}^{n} \frac{\partial f_i}{\partial x_j}(\mathbf{x})h_j + \epsilon_i(\mathbf{h}),$$

where

$$\frac{\epsilon_i(\mathbf{h})}{\|\mathbf{h}\|} \to 0$$

as $\mathbf{h} \to \mathbf{0}$ and where $\mathbf{R}_i$ denotes the ith row vector of the matrix A. Thus we can fix i arbitrarily and prove the latter condition for f_i, which is real-valued.

Let $\epsilon > 0$. There exists $\delta > 0$ such that $B_\delta(\mathbf{x}) \subseteq D$ and such that $\|\mathbf{h}\| < \delta$ implies

$$\left| \frac{\partial f_i}{\partial x_j}(\mathbf{x} + \mathbf{h}) - \frac{\partial f_i}{\partial x_j}(\mathbf{x}) \right| < \frac{\epsilon}{n}$$

for each $j = 1, \ldots, n$. Suppose that $\|\mathbf{h}\| < \delta$ and denote $\mathbf{h} = \sum_{j=1}^{n} h_j \mathbf{e}_j$. Denote $\mathbf{x}_0 = \mathbf{x}$ and

$$\mathbf{x}_k = \mathbf{x}_0 + \sum_{j=1}^{k} h_j \mathbf{e}_j \in B_\delta(\mathbf{x})$$

which is a convex set as defined in Exercise 9.43, for all $k = 1, \ldots, n$. Applying the Mean Value Theorem from one-variable calculus we find the existence of real numbers $\bar{t}_k$ to see that

$$
\begin{aligned}
f_i(\mathbf{x} + \mathbf{h}) - f_i(\mathbf{x}) &= \sum_{k=1}^{n} (f_i(\mathbf{x}_k) - f_i(\mathbf{x}_{k-1})) \\
&= \sum_{k=1}^{n} \frac{\partial f_i}{\partial x_k}(\mathbf{x}_{k-1} + \bar{t}_k \mathbf{e}_k) h_k \\
&= \sum_{k=1}^{n} \left(\frac{\partial f_i}{\partial x_k}(\mathbf{x}) + \eta_k \right) h_k,
\end{aligned}
$$

where $|\eta_k| < \frac{\epsilon}{n}$. It follows that

$$
\begin{aligned}
\|f_i(\mathbf{x} + \mathbf{h}) - f_i(\mathbf{x}) - \mathbf{R}_i \cdot \mathbf{h}\| &\leq \sum_{k=1}^{n} |\eta_k| |h_k| \\
&\leq \sum_{k=1}^{n} |\eta_k| \|\mathbf{h}\| \\
&< \epsilon \|\mathbf{h}\|.
\end{aligned}
$$

This proves that f_i is differentiable, and so is $\mathbf{f}$. To see that $\mathbf{f}'(\mathbf{x})$ is continuous we apply Theorem 10.1.1 to see that

$$
\|\mathbf{f}'(\mathbf{x}) - \mathbf{f}'(\mathbf{a})\| \leq \sqrt{ \sum_{i,j} \left(\frac{\partial f_i}{\partial x_j}(\mathbf{x}) - \frac{\partial f_i}{\partial x_j}(\mathbf{a}) \right)^2 } \to 0
$$

as $\mathbf{x} \to \mathbf{a}$ since the partial derivatives are continuous by hypothesis.   ∎

The following theorem will be useful when we study the Inverse Function Theorem.

Theorem 10.2.5 *Let $\mathbf{f}$ and $\mathbf{g} \in \mathcal{C}^1(D, \mathbb{E}^m)$, where D is an open subset of $\mathbb{E}^n$. Define $\phi : D \to \mathbb{R}$ by $\phi(\mathbf{x}) = \mathbf{f}(\mathbf{x}) \cdot \mathbf{g}(\mathbf{x})$, a scalar product in $\mathbb{E}^m$. Then $\phi \in \mathcal{C}^1(D, \mathbb{R})$ and*

$$
\phi'(\mathbf{x})\mathbf{h} = \mathbf{g}(\mathbf{x}) \cdot \mathbf{f}'(\mathbf{x})\mathbf{h} + \mathbf{f}(\mathbf{x}) \cdot \mathbf{g}'(\mathbf{x})\mathbf{h}
$$

for all $\mathbf{h} \in \mathbb{E}^n$.

Remark 10.2.3 In the equation that is the conclusion of this theorem, the dot products indicated on the right have meaning because both sides of the equation are applied to an arbitrary vector $\mathbf{h} \in \mathbb{E}^n$. Remembering that matrix multiplication is associative when it is defined, the conclusion can be expressed as follows in terms of matrices:

$$
\begin{aligned}
[\phi'(\mathbf{x})]_{1 \times n} [\mathbf{h}]_{n \times 1} &= [\mathbf{g}(\mathbf{x})]_{1 \times m} [\mathbf{f}'(\mathbf{x})]_{m \times n} [\mathbf{h}]_{n \times 1} \\
&+ [\mathbf{f}(\mathbf{x})]_{1 \times m} [\mathbf{g}'(\mathbf{x})]_{m \times n} [\mathbf{h}]_{n \times 1}
\end{aligned}
$$

or

$$[\phi'(\mathbf{x})]_{1\times n} = [\mathbf{g}(\mathbf{x})]_{1\times m}[\mathbf{f}'(\mathbf{x})]_{m\times n} + [\mathbf{f}(\mathbf{x})]_{1\times m}[\mathbf{g}'(\mathbf{x})]_{m\times n}.$$

Proof: We write

$$\phi(\mathbf{x} + \mathbf{h}) - \phi(\mathbf{x})$$

$$\begin{aligned}
&= (\mathbf{f}(\mathbf{x} + \mathbf{h}) - \mathbf{f}(\mathbf{x})) \cdot \mathbf{g}(\mathbf{x} + \mathbf{h}) \\
&\quad + (\mathbf{g}(\mathbf{x} + \mathbf{h}) - \mathbf{g}(\mathbf{x})) \cdot \mathbf{f}(\mathbf{x}) \\
&= (\mathbf{f}'(\mathbf{x})\mathbf{h} + \epsilon_1(\mathbf{h})) \cdot (\mathbf{g}(\mathbf{x}) + \mathbf{g}'(\mathbf{x})\mathbf{h} + \epsilon_2(\mathbf{h})) \\
&\quad + \mathbf{f}(\mathbf{x}) \cdot (\mathbf{g}'(\mathbf{x})\mathbf{h} + \epsilon_2(\mathbf{h})) \\
&= \mathbf{g}(\mathbf{x}) \cdot \mathbf{f}'(\mathbf{x})\mathbf{h} + \mathbf{f}(\mathbf{x}) \cdot \mathbf{g}'(\mathbf{x})\mathbf{h} + \epsilon(\mathbf{h}),
\end{aligned}$$

where $\epsilon(\mathbf{h})$ is a *real*-valued function defined by the latter equation. It remains only to prove that

$$\frac{|\epsilon(\mathbf{h})|}{\|\mathbf{h}\|} \to 0$$

as $\mathbf{h} \to \mathbf{0}$, which is left to Exercise 10.36. The continuity of the derivative follows from the continuity of its matrix coefficients.   ∎

EXERCISES

10.24 Follow the model of Example 10.3 to estimate $\mathbf{f}\left(\frac{11}{10}, \frac{\pi}{6} + \frac{1}{10}\right) - \mathbf{f}\left(1, \frac{\pi}{6}\right)$. Compare this linear estimate with the actual difference estimated numerically using a calculator.

10.25 In each of the following examples we have $\mathbf{f} : \mathbb{E}^1 \to \mathbb{E}^2$. Find all x for which $\mathbf{f}'(x)$ exists, and find the matrix $[\mathbf{f}'(x)]$. If $\mathbf{f}'(x)$ does not exist for all x, prove this also.
   a) $\mathbf{f}(x) = (\cos x, \sin x)$.
   b) $\mathbf{f}(x) = \left(x, \sqrt{x^2}\right)$.

10.26 For each of the following functions $\mathbf{f}$, Show that $\mathbf{f}'(\mathbf{x})$ exists, find the matrix $[\mathbf{f}'(\mathbf{x})]$ and calculate $\det \mathbf{f}'(\mathbf{x})$.
   a) $\mathbf{f} : \mathbb{E}^2 \to \mathbb{E}^2$ by $\mathbf{f}(\mathbf{x}) = (x_1 \cos x_2, x_1 \sin x_2)$.
   b) $\mathbf{f} : \mathbb{E}^3 \to \mathbb{E}^3$ by $\mathbf{f}(\mathbf{x}) = (x_1 \cos x_2, x_1 \sin x_2, x_3)$.
   c) $\mathbf{f} : \mathbb{E}^3 \to \mathbb{E}^3$ by $\mathbf{f}(\mathbf{x}) = (x_1 \sin x_3 \cos x_2, x_1 \sin x_3 \sin x_2, x_1 \cos x_3)$.

10.27 Let $\mathbf{f} : \mathbb{E}^2 \to \mathbb{E}^2$ be defined by $\mathbf{f}(\mathbf{x}) = (e^{x_1} \cos x_2, e^{x_1} \sin x_2)$.
   a) Prove that $\mathbf{f} \in C^1\left(\mathbb{E}^2, \mathbb{E}^2\right)$.
   b) Calculate the matrix $[\mathbf{f}'(\mathbf{x})]$, with respect to the standard basis for $\mathbb{E}^2$.
   c) Find $\det[\mathbf{f}'(\mathbf{x})]$.
   d) Calculate $D_\mathbf{v}\mathbf{f}(\mathbf{x})$, where $\mathbf{v} = (1, \sqrt{3})$.

10.28 Suppose $\mathbf{f} \in C^1\left(\mathbb{E}^2, \mathbb{E}^2\right)$ is such that the matrix $[\mathbf{f}'(1,2)] = \begin{pmatrix} 1 & 3 \\ -2 & 4 \end{pmatrix}$. Find the *directional derivative* $D_{(1,-2)}\mathbf{f}(1,2)$.

10.29   In elementary courses in the calculus of several variables, one learns about the *gradient*

$$\nabla f = \sum_{i=1}^{n} \frac{\partial f}{\partial x_i} \mathbf{e}_i$$

of a differentiable function $f : \mathbb{E}^n \to \mathbb{R}$. Determine the relationship between ∇f and $f'(\mathbf{x})$ as defined in this section.

10.30   † Let $T \in \mathcal{L}(\mathbb{E}^n, \mathbb{E}^m)$. Prove that T is differentiable at each point $\mathbf{x} \in \mathbb{E}^n$ and that $T' \equiv T$. That is, show that T is its own local linear approximation.

10.31   Let $\mathbf{f} \in C\left(D, \mathbb{E}^2\right)$, where D is an open subset of $\mathbb{E}^2$. Suppose also that $\mathbf{f}' \in \mathcal{C}\left(D, \mathcal{L}\left(\mathbb{E}^2, \mathbb{E}^3\right)\right)$. Prove or give a counterexample:

$$\mathbf{f}'(c\mathbf{x_1} + \mathbf{x_2}) = c\mathbf{f}'(\mathbf{x_1}) + \mathbf{f}'(\mathbf{x_2})$$

for all $\mathbf{x_1}$ and $\mathbf{x_2}$ in D and $c \in \mathbb{R}$.

10.32   † Suppose that $\mathbf{f}$ and $\mathbf{g} : D \to \mathbb{E}^m$ and that $\mathbf{x} \in D$ is a cluster point of $D \subseteq \mathbb{E}^n$. Let $c \in \mathbb{R}$. If both $\mathbf{f}$ and $\mathbf{g}$ are differentiable at $\mathbf{x}$, prove that $(c\mathbf{f} + \mathbf{g})'(\mathbf{x})$ exists and equals $c\mathbf{f}'(\mathbf{x}) + \mathbf{g}'(\mathbf{x})$.

10.33   † Prove Theorem 10.2.1.

10.34   Prove that if $\mathbf{f}$ is differentiable, then $\frac{\partial f_i}{\partial x_j}$ is the ith component of $D_{\mathbf{e}_j}\mathbf{f}$.

10.35   Prove Corollary 10.2.1.

10.36   † Complete the proof of Theorem 10.2.5 by proving the remaining limit.

10.37   Give an example of a function $\mathbf{f}$ that is continuous at $\mathbf{x}$ but not differentiable at $\mathbf{x}$. Prove that your example has the required properties.

10.38   † Define $f : \mathbb{E}^2 \to \mathbb{E}^1$ by

$$f(\mathbf{x}) = \begin{cases} \frac{x_1 x_2}{x_1^2 + x_2^2} & \text{if } \mathbf{x} \in \mathbb{E}^2 \setminus \{\mathbf{0}\}, \\ 0 & \text{if } \mathbf{x} = \mathbf{0}. \end{cases}$$

See Figs. 10.2 and 10.3.
    a) Prove: $\frac{\partial f}{\partial x_1}$ and $\frac{\partial f}{\partial x_2}$ exist at each $\mathbf{x} \in \mathbb{E}^2$.
    b) Prove: f is not differentiable at $\mathbf{0}$.

10.39   Define $f : \mathbb{E}^2 \to \mathbb{E}^1$ by

$$f(\mathbf{x}) = \begin{cases} \frac{x_1^2 x_2}{x_1^4 + x_2^2} & \text{if } \mathbf{x} \in \mathbb{E}^2 \setminus \{\mathbf{0}\}, \\ 0 & \text{if } \mathbf{x} = \mathbf{0}. \end{cases}$$

See Figs. 9.1 and 9.2.
    a) Prove: The directional derivative $D_{\mathbf{v}}f(\mathbf{x})$ exists at each $\mathbf{x} \in \mathbb{E}^2$ and for each $\mathbf{v} \in \mathbb{E}^2 \setminus \{\mathbf{0}\}$.

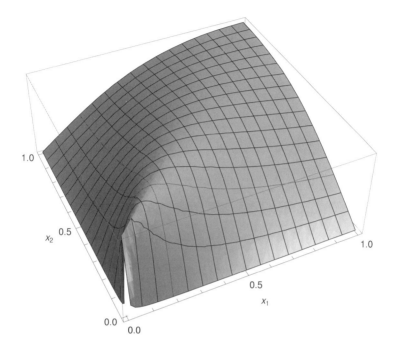

Figure 10.2 $\frac{x_1 x_2}{x_1^2 + x_2^2}$ in the first quadrant.

b) Prove: f is not differentiable at $\mathbf{0}$.

10.40   Prove that $\mathbf{f} \in \mathcal{C}^1(D, \mathbb{E}^m)$, where D is an open subset of $\mathbb{E}^n$, if and only if the directional derivative $D_{\mathbf{v}}\mathbf{f} \in \mathcal{C}(D, \mathbb{E}^m)$ for all $\mathbf{v} \in \mathbb{E}^n$.

10.41   Let the *open* set $\mathcal{U} \subset \mathbb{E}^n$. Suppose that $f : \mathcal{U} \to \mathbb{R}$ and that there is a number $M \in \mathbb{R}$ such that $\left| \frac{\partial f}{\partial x_j}(\mathbf{x}) \right| \le M$ for all $\mathbf{x} \in \mathcal{U}$ and for all $j = 1, \ldots, n$.

   a) Prove that $f \in C(\mathcal{U}, \mathbb{R})$. (Hint: Let $\mathbf{p} \in \mathcal{U}$ and take $r > 0$ such that $B_r(\mathbf{p}) \subseteq \mathcal{U}$. Prove that f is continuous at $\mathbf{p}$. Adapt the technique from the proof of Theorem 10.2.4.)

   b) Now suppose that $\mathbf{f} : \mathcal{U} \to \mathbb{E}^m$ and that there is a number $M \in \mathbb{R}$ such that $\left| \frac{\partial f_i}{\partial x_j}(\mathbf{x}) \right| \le M$ for all $\mathbf{x} \in \mathcal{U}$ and for all $j = 1, \ldots, n$ and all $i = 1, \ldots, m$. Prove that $\mathbf{f} \in C(\mathcal{U}, \mathbb{E}^m)$.

10.42   If $\mathbf{f} \in \mathcal{C}^1(\mathbb{E}^n, \mathbb{E}^m)$ and $\mathbf{v}$ and $\mathbf{w} \in \mathbb{E}^n$, express $D_{t\mathbf{v}+\mathbf{w}}\mathbf{f}(\mathbf{x})$ in terms of $D_{\mathbf{v}}\mathbf{f}(\mathbf{x})$, $D_{\mathbf{w}}\mathbf{f}(\mathbf{x})$ and t, for each $t \in \mathbb{R}$. Conclude that the mapping $\mathcal{D} :$

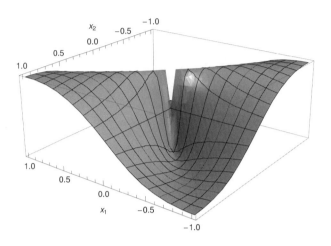

Figure 10.3 $\frac{x_1 x_2}{x_1^2 + x_2^2}$ in all four quadrants.

$\mathcal{C}^1(\mathbb{E}^n, \mathbb{E}^m) \to \mathbb{E}^m$ defined by $\mathcal{D}(\mathbf{v}) = D_\mathbf{v}\mathbf{f}(\mathbf{x})$ for fixed $\mathbf{f}$ and fixed $\mathbf{x}$ is a *linear* map.

10.43   Let $\mathbf{f}$ and $\mathbf{g} \in \mathcal{C}^1\left(\mathbb{E}^2, \mathbb{E}^2\right)$. Define $\phi : \mathbb{E}^2 \to \mathbb{R}$ by $\phi(\mathbf{x}) = \mathbf{f}(\mathbf{x}) \cdot \mathbf{g}(\mathbf{x})$, a scalar product in $\mathbb{E}^2$. Find the matrix $[\phi'(\mathbf{0})]$ given that

$$[\mathbf{f}(\mathbf{0})] = (1, 1), \qquad\qquad [\mathbf{g}(\mathbf{0})] = (-1, -1),$$

$$[\mathbf{f}'(\mathbf{0})] = \left(\begin{array}{cc} 1 & 2 \\ -2 & 1 \end{array} \right), \qquad\qquad [\mathbf{g}'(\mathbf{0})] = \left(\begin{array}{cc} 2 & 1 \\ -1 & 2 \end{array} \right).$$

10.3   THE CHAIN RULE IN EUCLIDEAN SPACE

Here we generalize Theorem 4.1.3 (the Chain Rule) to $\mathbb{E}^n$.

Theorem 10.3.1 (The Chain Rule) *Suppose $D \subseteq \mathbb{E}^n$ and $\mathbf{g} : D \to \mathbb{E}^m$ is differentiable at $\mathbf{x}_0 \in D$. Suppose $\mathbf{f} : \mathbf{g}(D) \to \mathbb{E}^p$ is differentiable at $\mathbf{g}(\mathbf{x}_0)$. Then $(\mathbf{f} \circ \mathbf{g})$ is differentiable at $\mathbf{x}_0$ and*

$$(\mathbf{f} \circ \mathbf{g})'(\mathbf{x}_0) = \mathbf{f}'(\mathbf{g}(\mathbf{x}_0))\mathbf{g}'(\mathbf{x}_0),$$

a composition of linear transformations.

Proof: To show the composition $\mathbf{f} \circ \mathbf{g}$ is differentiable at $\mathbf{x}_0$, we denote

$$\mathbf{k} = \mathbf{g}(\mathbf{x}_0 + \mathbf{h}) - \mathbf{g}(\mathbf{x}_0) \to \mathbf{0}$$

as $\mathbf{h} \to \mathbf{0}$ since $\mathbf{g}$ is continuous at $\mathbf{x}_0$ by Theorem 10.2.2. Also,

$$\mathbf{k} = \mathbf{g}'(\mathbf{x}_0)\mathbf{h} + \tilde{\epsilon}(\mathbf{h}),$$

where $\frac{\|\tilde{\epsilon}(\mathbf{h})\|}{\|\mathbf{h}\|} \to 0$ as $\mathbf{h} \to \mathbf{0}$. Next we observe that

$$\Delta \mathbf{f} = \mathbf{f}(\mathbf{g}(\mathbf{x}_0 + \mathbf{h})) - \mathbf{f}(\mathbf{g}(\mathbf{x}_0)) = \mathbf{f}'(\mathbf{g}(\mathbf{x}_0))\mathbf{k} + \epsilon(\mathbf{k}),$$

where $\frac{\|\epsilon(\mathbf{k})\|}{\|\mathbf{k}\|} \to 0$ as $\mathbf{k} \to \mathbf{0}$. Therefore

$$\Delta \mathbf{f} = \mathbf{f}'(\mathbf{g}(\mathbf{x}_0))\mathbf{g}'(\mathbf{x}_0)\mathbf{h} + \mathbf{f}'(\mathbf{g}(\mathbf{x}_0))\tilde{\epsilon}(\mathbf{h}) + \epsilon(\mathbf{g}'(\mathbf{x}_0)\mathbf{h} + \tilde{\epsilon}(\mathbf{h}))$$

Denoting $\tilde{\tilde{\epsilon}}(\mathbf{h}) = \mathbf{f}'(\mathbf{g}(\mathbf{x}_0))\tilde{\epsilon}(\mathbf{h}) + \epsilon(\mathbf{g}'(\mathbf{x}_0)\mathbf{h} + \tilde{\epsilon}(\mathbf{h}))$ it suffices to show that

$$
\begin{aligned}
\frac{\|\tilde{\tilde{\epsilon}}(\mathbf{h})\|}{\|\mathbf{h}\|} &= \frac{\|\mathbf{f}'(\mathbf{g}(\mathbf{x}_0))\tilde{\epsilon}(\mathbf{h}) + \epsilon(\mathbf{g}'(\mathbf{x}_0)\mathbf{h} + \tilde{\epsilon}(\mathbf{h}))\|}{\|\mathbf{h}\|} \\
&\leq \frac{\|\mathbf{f}'(\mathbf{g}(\mathbf{x}_0))\tilde{\epsilon}(\mathbf{h})\| + \|\epsilon(\mathbf{g}'(\mathbf{x}_0)\mathbf{h} + \tilde{\epsilon}(\mathbf{h}))\|}{\|\mathbf{h}\|} \\
&= \frac{\|\mathbf{f}'(\mathbf{g}(\mathbf{x}_0))\tilde{\epsilon}(\mathbf{h})\|}{\|\mathbf{h}\|} + \frac{\|\epsilon(\mathbf{g}'(\mathbf{x}_0)\mathbf{h} + \tilde{\epsilon}(\mathbf{h}))\|}{\|\mathbf{h}\|} \to 0
\end{aligned}
$$

as $\mathbf{h} \to \mathbf{0}$. Since

$$\frac{\|\mathbf{f}'(\mathbf{g}(\mathbf{x}_0))\tilde{\epsilon}(\mathbf{h})\|}{\|\mathbf{h}\|} \leq \frac{\|\mathbf{f}'(\mathbf{g}(\mathbf{x}_0))\|\|\tilde{\epsilon}(\mathbf{h})\|}{\|\mathbf{h}\|} \to 0$$

as $\mathbf{h} \to \mathbf{0}$, it suffices to show that

$$\frac{\|\epsilon(\mathbf{g}'(\mathbf{x}_0)\mathbf{h} + \tilde{\epsilon}(\mathbf{h}))\|}{\|\mathbf{h}\|} = \frac{\|\epsilon(\mathbf{k})\|}{\|\mathbf{h}\|} \to 0$$

as $\mathbf{h} \to \mathbf{0}$. Thus it suffices to show that if $\eta > 0$, then for all sufficiently small $\|\mathbf{h}\|$ we have

$$\|\epsilon(\mathbf{k})\| < \eta\|\mathbf{h}\|.$$

Since $1 > 0$ there exists $\delta_1 > 0$ such that $\|\mathbf{h}\| < \delta_1$ implies

$$\|\tilde{\epsilon}(\mathbf{h})\| \leq 1\|\mathbf{h}\|,$$

which implies in turn that

$$\|\mathbf{k}\| \leq (1 + \|\mathbf{g}'(\mathbf{x}_0)\|)\|\mathbf{h}\|.$$

Also, if

$$\eta_2 = \frac{\eta}{1 + \|\mathbf{g}'(\mathbf{x}_0)\|},$$

then there exists $\delta_2 > 0$ such that $\|\mathbf{k}\| < \delta_2$ implies

$$\|\epsilon(\mathbf{k})\| < \eta_2\|\mathbf{k}\| \leq \eta_2(1 + \|\mathbf{g}'(\mathbf{x}_0)\|)\|\mathbf{h}\| = \eta\|\mathbf{h}\|.$$

Thus it suffices to pick $\mathbf{h}$ such that

$$\|\mathbf{h}\| < \min\left\{\delta_1, \frac{\delta_2}{1 + \|\mathbf{g}'(\mathbf{x}_0)\|}\right\}$$

since then

$$\|\mathbf{k}\| \leq (1 + \|\mathbf{g}'(\mathbf{x}_0)\|)\|\mathbf{h}\| < \delta_2.$$

∎

10.3.1 The Mean Value Theorem

In one-variable calculus, the Mean Value Theorem (Theorem 4.2.3) plays a very important role. In Exercise 10.49 the reader will see that a direct adaptation of that theorem to vector-valued functions is not possible. However, the following version of the theorem is true and is useful.

Theorem 10.3.2 (Mean Value Theorem) *Suppose* $\mathbf{f} : D \to \mathbb{E}^m$ *is a differentiable function, where* D *is a* convex *subset of* $\mathbb{E}^n$, *as defined in Exercise 9.43. Suppose* $M = \sup_{\mathbf{x} \in D} \|\mathbf{f}'(\mathbf{x})\| < \infty$. *Then, for all* $\mathbf{a}$ *and* $\mathbf{b}$ *in* D,

$$\|\mathbf{f}(\mathbf{b}) - \mathbf{f}(\mathbf{a})\| \leq M\|\mathbf{b} - \mathbf{a}\|.$$

Proof: Observe that the straight-line segment

$$S = \{\mathbf{a} + t(\mathbf{b} - \mathbf{a}) \mid 0 \leq t \leq 1\} \subseteq D$$

because D is convex. Define a differentiable function $\phi : [0, 1] \to \mathbb{R}$ by

$$\phi(t) = (\mathbf{f}(\mathbf{b}) - \mathbf{f}(\mathbf{a})) \cdot \mathbf{f}(\mathbf{a} + t(\mathbf{b} - \mathbf{a}))$$

for all $t \in [0, 1]$. By the ordinary Mean Value Theorem from one-variable calculus together with Exercise 10.45.b, we obtain

$$
\begin{aligned}
\|\mathbf{f}(\mathbf{b}) - \mathbf{f}(\mathbf{a})\|^2 &= \phi(1) - \phi(0) \\
&= \phi'(\bar{t})(1 - 0) \\
&= (\mathbf{f}(\mathbf{b}) - \mathbf{f}(\mathbf{a})) \cdot (\mathbf{f}'\,(\mathbf{a} + \bar{t}(\mathbf{b} - \mathbf{a}))\,(\mathbf{b} - \mathbf{a})) \\
&\leq \|\mathbf{f}(\mathbf{b}) - \mathbf{f}(\mathbf{a})\|\|\mathbf{f}'(\mathbf{a} + \bar{t}(\mathbf{b} - \mathbf{a}))\|\|\mathbf{b} - \mathbf{a}\| \\
&\leq \|\mathbf{f}((\mathbf{b}) - \mathbf{f}(\mathbf{a})\|M\|\mathbf{b} - \mathbf{a}\|,
\end{aligned}
$$

where the inequality comes from the Cauchy–Schwarz inequality combined with the Chain Rule and the property of the norm of a linear transformation. Therefore

$$\|\mathbf{f}(\mathbf{b}) - \mathbf{f}(\mathbf{a})\| \leq M\|\mathbf{b} - \mathbf{a}\|$$

∎

The following corollary is an immediate consequence of the proof of the Mean Value Theorem.

Corollary 10.3.1 *Suppose* $\mathbf{f} : D \to \mathbb{E}^m$ *is a differentiable function. Suppose* $\mathbf{x}$ *and* $\mathbf{y} \in D$ *and suppose the straight-line segment* L *between* $\mathbf{x}$ *and* $\mathbf{y}$ *lies in* D. *Suppose* $M = \sup_{\mathbf{w} \in L} \|\mathbf{f}'(\mathbf{w})\| < \infty$. *Then*

$$\|\mathbf{f}(\mathbf{x}) - \mathbf{f}(\mathbf{y})\| \leq M \|\mathbf{x} - \mathbf{y}\|.$$

Remark 10.3.1 The corollary differs from the Mean Value Theorem in that D does not need to be convex. Instead, x and y are special points for which $L \subseteq D$. In the corollary, M is the supremum only over L, not over D.

If $\mathbf{f} : D \to \mathbb{E}^m$, with $D \subseteq \mathbb{E}^n$, it makes sense to speak of a local extreme point (either a local maximum or a local minimum) at a point $\mathbf{x} \in D$ if and only if $m = 1$. We say that $f : D \to \mathbb{R}$ has a local extreme point at $\mathbf{x} \in D$, provided that there is a number $r > 0$ such that $f(\mathbf{x})$ is either the largest or the smallest value achieved by f in the open ball $B_r(\mathbf{x})$.

Theorem 10.3.3 (Local Extreme Point) *Suppose that* $D \subseteq \mathbb{E}^n$ *and that* $f : D \to \mathbb{R}$ *has a local extreme point at* $\mathbf{x} \in D^\circ$, *the interior of* D. *If* f *is differentiable at* $\mathbf{x}$, *then*

$$f'(\mathbf{x}) = 0 \in \mathcal{L}\left(\mathbb{E}^n, \mathbb{E}^1\right).$$

Proof: Let $\mathbf{v} \in \mathbb{E}^n \setminus \{\mathbf{0}\}$. Define $\phi(t) = f(\mathbf{x} + t\mathbf{v})$. Thus $\phi : \mathbb{R} \to \mathbb{R}$ has a local extreme point at $t = 0$. By the first derivative test from one-variable calculus, combined with the multivariable chain rule, we see that $\phi'(0) = f'(\mathbf{x})\mathbf{v} = 0$ for all nonzero vectors $\mathbf{v}$. Thus $f' = 0 \in \mathcal{L}\left(\mathbb{E}^n, \mathbb{E}^1\right)$. Note that the latter condition is equivalent to the condition that $\frac{\partial f}{\partial x_i} = 0$ for all $i = 1, \ldots, n$. ∎

10.3.2 Taylor's Theorem

Next we will generalize Taylor's Theorem (Theorem 4.6.1). For this purpose we restrict our attention to functions $f : D \to \mathbb{R}$, where D is an open, convex subset of $\mathbb{E}^n$. Note that for such functions $\mathbf{f}'(\mathbf{x})$ is a $1 \times n$ matrix, and its one row vector is the *gradient* of f, denoted by

$$\nabla f = \sum_{i=1}^n \frac{\partial f}{\partial x_i} \mathbf{e}_i.$$

We denote by $\mathcal{C}^{N+1}(D, \mathbb{R})$ the set of all functions $f : D \to \mathbb{R}$ such that f and each of its derivatives of order up to N has in turn a continuous derivative. (See Exercise 10.53 for elaboration of this concept, including its equivalence to the continuity of all partial derivatives of order up to and including $N + 1$.) We will let $\mathbf{k}$ denote an arbitrary n-tuple of nonnegative integers, and we will denote by $|\mathbf{k}| = \sum_{i=1}^n k_i$. We will denote

$$\frac{\partial^{|\mathbf{k}|} f}{\partial \mathbf{x}^{\mathbf{k}}} = \frac{\partial^{k_1 + \cdots + k_n} f}{\partial x_1^{k_1} \ldots \partial x_n^{k_n}}$$

and

$$\mathbf{x}^{\mathbf{k}} = x_1^{k_1} \times \cdots \times x_n^{k_n}.$$

Theorem 10.3.4 (Taylor's Theorem in $\mathbb{E}^n$) *Let $f \in \mathcal{C}^{N+1}(D, \mathbb{R})$, where D is an open, convex subset of $\mathbb{E}^n$. Let $\mathbf{a}$ and $\mathbf{b} \in D$. Then there exists a point μ on the straight-line segment between $\mathbf{a}$ and $\mathbf{b}$ such that*

$$f(\mathbf{b}) = \sum_{k=0}^{N} \frac{1}{k!} \left(\sum_{j=1}^{n} (b_j - a_j) \frac{\partial}{\partial x_j} \right)^k f \bigg|_{\mathbf{x}=\mathbf{a}} + R_N(\mathbf{b}), \qquad (10.3)$$

where

$$R_N(\mathbf{b}) = \frac{1}{(N+1)!} \left(\sum_{j=1}^{n} (b_j - a_j) \frac{\partial}{\partial x_j} \right)^{N+1} f \bigg|_{\mathbf{x}=\mu}.$$

Equivalently, we can express these formulas in the form

$$f(\mathbf{b}) = \sum_{|\mathbf{k}| \leq N} \frac{\partial^{|\mathbf{k}|} f}{\partial \mathbf{x}^{\mathbf{k}}}(\mathbf{a}) \frac{(\mathbf{b} - \mathbf{a})^{\mathbf{k}}}{k_1! k_2! \cdots k_n!} + R_N(\mathbf{b}), \qquad (10.4)$$

where

$$R_N(\mathbf{b}) = \sum_{|\mathbf{k}| = N+1} \frac{\partial^{|\mathbf{k}|} f}{\partial \mathbf{x}^{\mathbf{k}}}(\mu) \frac{(\mathbf{b} - \mathbf{a})^{\mathbf{k}}}{k_1! k_2! \cdots k_n!}.$$

Proof: Define a function $\phi : \mathbb{R} \to \mathbb{R}$ by $\phi(t) = f(\mathbf{a} + t(\mathbf{b} - \mathbf{a})) = (f \circ \psi)(t)$, where $\psi(t) = \mathbf{a} + t(\mathbf{b} - \mathbf{a})$ for all $t \in [0, 1]$. Observe first that in matrix form

$$[\phi'(t)] = [f'(\psi(t))][\psi'(t)]$$

$$= \left[\frac{\partial f}{\partial x_1}(\psi(t)) \quad \cdots \quad \frac{\partial f}{\partial x_n}(\psi(t)) \right] \begin{bmatrix} b_1 - a_1 \\ \vdots \\ b_n - a_n \end{bmatrix}$$

$$= \left((b_1 - a_1)\frac{\partial}{\partial x_1} + \cdots + (b_n - a_n)\frac{\partial}{\partial x_1} \right) f \bigg|_{\psi(t)}.$$

Repeating this argument k times for each of the n summands in the last line above, we see that

$$\phi^{(k)}(t) = \left(\sum_{j=1}^{n} (b_j - a_j)\frac{\partial}{\partial x_j} \right)^k f \bigg|_{\psi(t)}$$

The rest of the proof of formula (10.3) is a matter of applying Taylor's Theorem (Theorem 4.6.1) for one variable to $\phi(t)$. In order to prove formula (10.4), we expand the powers of the differential operator in formula (10.3), using the expansion given by the so-called *multinomial* theorem:

$$(c_1 + \cdots + c_n)^K = \sum_{k_1 + \cdots + k_n = K} \binom{K}{k_1, \ldots, k_n} c_1^{k_1} \cdots c_n^{k_n},$$

where the coefficients

$$\binom{K}{k_1,\ldots,k_n} = \frac{K!}{k_1!k_2!\cdots k_n!}$$

are the *multinomial* coefficients of degree K. (See [2].) However, in order to carry out this expansion, we need to know Clairaut's theorem. This theorem, which is proven as Exercise 11.46, assures under the hypothesis of continuity of the partial derivatives that there is independence of the order of composition of the differential operators. That is, continuity of the partial derivatives assures that

$$\frac{\partial^2 f}{\partial x_i \partial x_j} = \frac{\partial^2 f}{\partial x_j \partial x_i}.$$

This permits the regrouping and collecting of the *like* partial derivatives in the expansions utilized above.   ■

We remark that sometimes it is useful to know that the sum of the *multinomial* coefficients of order N is

$$\sum_{|\mathbf{k}|=N} \binom{N}{k_1,\ldots,k_n} = n^N.$$

This follows from the fact that the sum represents the number of ways to select N things from a set of n things. Equivalently, it is the number of elements in the set of functions from a set of N elements to a set of n elements.

EXERCISES

10.44 Suppose $\mathbf{f} : \mathbb{E}^2 \to \mathbb{E}^2$ and $\mathbf{g} : \mathbb{E}^3 \to \mathbb{E}^2$ are both differentiable. If $\mathbf{g}(\mathbf{0}) = \mathbf{x}_0$,

$$\mathbf{g}'(\mathbf{0}) = \begin{pmatrix} 1 & -1 & 0 \\ -2 & 2 & 3 \end{pmatrix},$$

and

$$\mathbf{f}'(\mathbf{x}_0) = \begin{pmatrix} 1 & 2 \\ 2 & -1 \end{pmatrix},$$

find the matrix $[(\mathbf{f} \circ \mathbf{g})'(\mathbf{0})]$ using the standard bases.

10.45 Suppose that $T \in \mathcal{L}(\mathbb{E}^m, \mathbb{E}^p)$ and that $\mathbf{f} : \mathbb{E}^n \to \mathbb{E}^m$ is differentiable.
   a) Prove that $T \circ \mathbf{f}$ is differentiable at each $\mathbf{x} \in \mathbb{E}^n$ and that

$$(T \circ \mathbf{f})'(\mathbf{x}) \equiv T \circ \mathbf{f}'(\mathbf{x}).$$

   b) † If $\mathbf{a} \in \mathbb{E}^m$ is a constant vector, prove that $(\mathbf{a} \cdot \mathbf{f})'(\mathbf{x})$ exists and equals $\mathbf{a} \cdot \mathbf{f}'(\mathbf{x})$.

10.46 Suppose $\mathbf{f} \in \mathcal{C}^1(\mathbb{E}^3, \mathbb{E}^2)$ and the matrix

$$[\mathbf{f}'(\mathbf{0})] = \begin{pmatrix} 1 & 2 & 3 \\ 4 & 5 & 6 \end{pmatrix}.$$

If $(a, b) \in \mathbb{E}^2$, express the matrix $\left[((a, b) \cdot \mathbf{f})'(\mathbf{0})\right]$ in terms of a and b.

10.47 Suppose $\mathbf{f} : D \to \mathbb{E}^m$ is differentiable at $\mathbf{x} \in D$ and suppose there exists an *inverse function* $\mathbf{g} : \mathbf{f}(D) \to D$ that is differentiable at $\mathbf{f}(\mathbf{x})$. Find the relationship between $\mathbf{f}'$ and $\mathbf{g}'$.

10.48 Suppose $\mathbf{y} = \mathbf{g}(\mathbf{x})$ is differentiable in an open ball around $\mathbf{x} \in \mathbb{E}^n$. Suppose $\mathbf{z} = \mathbf{f}(\mathbf{y})$ is differentiable in an open ball around $\mathbf{g}(\mathbf{x})$, where $\mathbf{g} : \mathbb{E}^n \to \mathbb{E}^m$ and $\mathbf{f} : \mathbb{E}^m \to \mathbb{E}^p$. Apply the chain rule to the differentiable function $\mathbf{f} \circ \mathbf{g} : \mathbb{E}^n \to \mathbb{E}^p$ to compute $\frac{\partial z_i}{\partial x_j}$ for all $1 \le i \le p$ and all $1 \le j \le n$.

10.49 † Let $\mathbf{f} : \mathbb{E}^1 \to \mathbb{E}^2$ be defined by $\mathbf{f}(t) = (\cos t, \sin t)$. Show that there does *not* exist a point $\bar{t} \in [0, 2\pi]$ for which

$$\mathbf{f}(2\pi) - \mathbf{f}(0) = \mathbf{f}'(\bar{t})(2\pi - 0).$$

10.50 Suppose that $\mathbf{f}'(\mathbf{x})$ exists and that $\|\mathbf{f}'(\mathbf{x})\|$ is *bounded* on a *convex* set $D \subseteq \mathbb{E}^n$, where $\mathbf{f} : D \to \mathbb{E}^m$. Prove that $\mathbf{f}$ is *uniformly continuous* on D, as defined in Exercise 9.41.

10.51 Suppose $\mathbf{f} : \mathbb{E}^1 \to \mathbb{E}^2$ by $\mathbf{f}(x) = (\cos x, \sin x)$. Prove or give a counterexample: For *all* real numbers a and b we have

$$\|\mathbf{f}(b) - \mathbf{f}(a)\| \le |b - a|.$$

10.52 Suppose $\mathbf{f}'(\mathbf{x})$ exists for all $\mathbf{x}$ in a nonempty *open* set $D \subseteq \mathbb{E}^n$, where $\mathbf{f} : D \to \mathbb{E}^m$.

    a) Suppose D is *convex*. If $\mathbf{f}'(\mathbf{x}) \equiv 0 \in \mathcal{L}(\mathbb{E}^n, \mathbb{E}^m)$, prove that $\mathbf{f}$ is a constant function on D.

    b) Suppose next that D is an *open connected* set in $\mathbb{E}^n$, but not necessarily convex. If

$$\mathbf{f}'(\mathbf{x}) \equiv 0 \in \mathcal{L}(\mathbb{E}^n, \mathbb{E}^m)$$

on D, prove that $\mathbf{f}$ is a constant function on D. (Hint: Show that if $\mathbf{x} \in D$, then there exists an $r > 0$ such that $\mathbf{f}$ remains constant on $B_r(\mathbf{x})$. Show then that the set $\mathbf{f}^{-1}(\mathbf{c})$ is an open subset of $\mathbb{E}^n$ for each $\mathbf{c} \in \mathbb{E}^m$.)

10.53 † Let D be an open subset of $\mathbb{E}^n$ and $\mathbf{f} \in C^1(D, \mathbb{E}^m)$. Thus

$$\mathbf{f}' : D \to \mathcal{L}(\mathbb{E}^n, \mathbb{E}^m).$$

We can denote

$$\mathbf{f}'(\mathbf{x}) = \sum_{i,j} \frac{\partial f_i}{\partial x_j}(\mathbf{x}) T_{ij},$$

where T_{ij} is defined in Exercise 10.13. We say $\mathbf{f} \in C^2(D, \mathbb{E}^m)$ if and only if $\mathbf{f}'$ has a continuous derivative. Using $\{T_{ij}\}$ as a basis of $\mathcal{L}(\mathbb{E}^n, \mathbb{E}^m)$ to show that $\mathbf{f} \in C^2(D, \mathbb{E}^m)$ if and only if $\frac{\partial^2 f_i}{\partial x_j \partial x_k}$ exists and is continuous for all i, j, and k.

10.54   Let $D = \{\mathbf{x} \in \mathbb{E}^2 \mid \|\mathbf{x}\| > 1\} \cap \{\mathbf{x} \in \mathbb{E}^2 \mid x_2 \neq 0 \text{ if } x_1 \leq 0\}$.

    a) Show that D is open, but not convex.

    b) Apply Exercise 5.62 to show that there exists $g \in \mathcal{C}^\infty(\mathbb{R})$ such that $g(\theta) \equiv 1$ if $\theta \geq \frac{\pi}{2}$, $g(\theta) \equiv -1$ if $\theta \leq -\frac{\pi}{2}$, and $g'(\theta)$ is bounded on $\mathbb{R}$.

    c) Let $f : D \to \mathbb{R}$ be defined by

$$f(\mathbf{x}) = \begin{cases} g\left(\tan^{-1}\left(\frac{x_2}{x_1}\right)\right) & \text{if } \mathbf{x} \in D \text{ with } x_1 > 0, \\ 1 & \text{if } \mathbf{x} \in D \text{ with } x_1 \leq 0, x_2 > 0, \\ -1 & \text{if } \mathbf{x} \in D \text{ with } x_1 \leq 0, x_2 < 0. \end{cases}$$

    Show that $f \in \mathcal{C}^1(D)$ and that f' is bounded on D.

    d) Show that there does not exist $M < \infty$ such that

$$|f(\mathbf{x}) - f(\mathbf{x}')| \leq M\|\mathbf{x} - \mathbf{x}'\|$$

    identically on D.

10.55   Let $f : \mathbb{E}^2 \to \mathbb{R}$ be defined by $f(\mathbf{x}) = \sin(x_1 + x_2)$. Selecting $\mathbf{a} = \mathbf{0}$ in Taylor's Theorem, use either formula (10.3) or the expanded version to prove that $R_N(\mathbf{b}) \to 0$ as $N \to \infty$.

10.56   The following exercise works in any vector space equipped with a scalar product, but the reader may let $\mathbf{f} \in \mathbb{E}^m$ and let $\{\mathbf{e}_k \mid k = 1, \ldots, n\}$ be an *orthonormal* subset of $\mathbb{E}^m$, where $n \leq m$. Note that $\langle \mathbf{e}_j, \mathbf{e}_k \rangle = 0$ if $j \neq k$ but equals 1 if $j = k$. Prove that

$$\Phi(\mathbf{a}) = \left\| \mathbf{f} - \sum_{k=1}^n a_k \mathbf{e}_k \right\|_2^2$$

has an absolute minimum value on $\mathbb{E}^n$, which is achieved by selecting $a_k = \langle \mathbf{f}, \mathbf{e}_k \rangle$ for all $k = 1, \ldots, n$. Hint: Apply Theorems 10.2.5 and 10.3.3 to

$$\Phi(\mathbf{a}) = \left\langle \mathbf{f} - \sum a_k \mathbf{e}_k, f - \sum a_k \mathbf{e}_k \right\rangle.$$

We remark that this exercise can be applied usefully in function spaces such as $\mathcal{R}[a, b]$, giving rise to the standard formulas for Fourier coefficients and also coefficients of other orthogonal-function expansions.

10.4   INVERSE FUNCTIONS

Even for differentiable functions $f : \mathbb{R} \to \mathbb{R}$, there is no need for f to be invertible. And if the function f has an inverse, there is no need for that inverse to be differentiable. For example, $f(x) = x^2$ is continuously differentiable on $\mathbb{R}$, yet it is not one-to-one and hence has no inverse. Another good example is this one: Let $f(x) = x^3$. Now f is one-to-one on $\mathbb{R}$, but $f^{-1}(x) = \sqrt[3]{x}$, which has no derivative at the origin. On the other hand, suppose D is an open subset of $\mathbb{R}$ and suppose that

$f \in C^1(D, \mathbb{R})$ and $f'(a) \neq 0$ at some $a \in D$. By continuity of f' we know that f' remains nonzero on some interval $I = (a - \delta, a + \delta) \subseteq D$. By the Mean Value Theorem, the restriction $f|_I$ of f to I is one-to-one and thus has an inverse. We will generalize this theorem to $\mathbb{E}^n$ and we will prove a theorem that shows under suitable hypotheses that the inverse exists and is also continuously differentiable on a suitably small open ball. We begin with the following theorem which provides an interesting contrast to the Mean Value Theorem (Theorem 10.3.2).

Theorem 10.4.1 (Magnification Theorem) *Let $\mathbf{f} \in C^1(D, \mathbb{E}^n)$, where D is an open subset of $\mathbb{E}^n$. If $\mathbf{x} \in D$ is such that $\det \mathbf{f}'(\mathbf{x}) \neq 0$, then there exists an open ball $B_r(\mathbf{x})$ and a constant $\alpha > 0$ such that for all $\mathbf{y}$ and $\mathbf{z}$ in $B_r(\mathbf{x})$ we have*

$$\|\mathbf{f}(\mathbf{y}) - \mathbf{f}(\mathbf{z})\| \geq \alpha \|\mathbf{y} - \mathbf{z}\|.$$

Consequently, $\mathbf{f}$ is one-to-one and thus invertible on $B_r(\mathbf{x})$.

Proof: Begin by denoting $T = \mathbf{f}'(\mathbf{x}) \in \mathcal{L}(\mathbb{E}^n, \mathbb{E}^n) \subset C^1(\mathbb{E}^n, \mathbb{E}^n)$. Observe that T is invertible and that both T and T^{-1} have strictly positive norms. For all $\mathbf{y}$ and $\mathbf{z}$ in $\mathbb{E}^n$ we have

$$\|\mathbf{y} - \mathbf{z}\| = \left\|T^{-1}(T(\mathbf{y}) - T(\mathbf{z}))\right\| \leq \left\|T^{-1}\right\| \|T(\mathbf{y}) - T(\mathbf{z})\|.$$

Letting $\alpha = \frac{1}{2\|T^{-1}\|}$, we have

$$\|T(\mathbf{y}) - T(\mathbf{z})\| \geq 2\alpha \|\mathbf{y} - \mathbf{z}\|.$$

Suppose we could show that for sufficiently small $r > 0$ and for all $\mathbf{y}$ and $\mathbf{z}$ in $B_r(\mathbf{x})$ we have

$$\|(\mathbf{f}(\mathbf{y}) - \mathbf{f}(\mathbf{z})) - (T(\mathbf{y}) - T(\mathbf{z}))\| < \alpha \|\mathbf{y} - \mathbf{z}\|. \tag{10.5}$$

Then it would follow that

$$\|\mathbf{f}(\mathbf{y}) - \mathbf{f}(\mathbf{z})\| > \alpha \|\mathbf{y} - \mathbf{z}\|$$

since if that were not true, then it would follow that

$$\|T(\mathbf{y}) - T(\mathbf{z})\| < 2\alpha \|\mathbf{y} - \mathbf{z}\|,$$

which is false. Hence we will know that $\mathbf{f}$ is both one-to-one and invertible on $B_r(\mathbf{x})$ once we have proven Equation 10.5.

However, for all $\mathbf{y}$ and $\mathbf{z}$ in $B_r(\mathbf{x})$ we have

$$\|(\mathbf{f}(\mathbf{y}) - \mathbf{f}(\mathbf{z})) - (T(\mathbf{y}) - T(\mathbf{z}))\| = \|(\mathbf{f} - T)(\mathbf{y}) - (\mathbf{f} - T)(\mathbf{z})\|$$
$$\leq M\|\mathbf{y} - \mathbf{z}\|$$

where

$$M = \sup_{\mathbf{w} \in B_r(\mathbf{x})} \|(\mathbf{f} - T)'(\mathbf{w})\| = \sup_{\mathbf{w} \in B_r(\mathbf{x})} \|\mathbf{f}'(\mathbf{w}) - \mathbf{f}'(\mathbf{x})\|.$$

Here we are using the Mean Value Theorem 10.3.2 together with Exercises 10.30 and 10.32 for the basic properties of differentiation, as well as the convexity of a ball. Since $\mathbf{f}'$ is continuous, we can choose $r > 0$ so as to insure that $M < \alpha$. ∎

The next theorem will complete our preparation for the Inverse Function Theorem.

Theorem 10.4.2 (Open Mapping Theorem) *Let* $\mathbf{f} \in C^1(D, \mathbb{E}^n)$, *where D is an open subset of* $\mathbb{E}^n$. *Suppose that for all* $\mathbf{x} \in D$ *we have* $\det \mathbf{f}'(\mathbf{x}) \neq 0$. *Then* $\mathbf{f}(D)$ *is an open subset of* $\mathbb{E}^n$.

Proof: Let $\mathbf{x} \in D$ and $\mathbf{y} = \mathbf{f}(\mathbf{x})$. We seek an open ball $B_\rho(\mathbf{y}) \subseteq \mathbf{f}(D)$. Thanks to the Magnification Theorem, we can find $r > 0$ such that $\mathbf{f}$ is one-to-one even on a suitably chosen closed ball $\bar{B}_r(\mathbf{x}) \subseteq D$. Denote by $S^{n-1} = \partial B_r(\mathbf{x})$, the sphere that is the boundary of the n-dimensional ball $B_r(\mathbf{x})$. Thus $\mathbf{f}\left(S^{n-1}\right)$ is a compact subset of $\mathbb{E}^n$. Since the distance function $d(\mathbf{y}, \mathbf{u}) = \|\mathbf{y} - \mathbf{u}\|$ is continuous as a function of $\mathbf{u} \in \mathbf{f}\left(S^{n-1}\right)$, it follows that

$$R = \min \left\{ \|\mathbf{y} - \mathbf{u}\| \mid \mathbf{u} \in \mathbf{f}\left(S^{n-1}\right) \right\} > 0$$

since $\mathbf{y} \notin S^{n-1}$ and $\mathbf{f}$ is one-to-one. We claim that $\rho = \frac{R}{2}$ satisfies the requirements of the theorem.

Let $\mathbf{z} \in B_\rho(\mathbf{y})$. We need to prove $\mathbf{z} \in \mathbf{f}(D)$. We note that the distance of $\mathbf{z}$ from each point in $\mathbf{f}\left(S^{n-1}\right)$ is greater than ρ, whereas $\|\mathbf{z} - \mathbf{y}\| < \rho$. If we let

$$g(\mathbf{v}) = \|\mathbf{z} - \mathbf{f}(\mathbf{v})\|^2 = (\mathbf{z} - \mathbf{f}(\mathbf{v})) \cdot (\mathbf{z} - \mathbf{f}(\mathbf{v}))$$

for all $\mathbf{v} \in \bar{B}_r(\mathbf{x})$, then the minimum value of g must be less than ρ and must occur at an interior point of $\bar{B}_r(\mathbf{x})$. At that minimum point we must have

$$\mathbf{g}'(\mathbf{v}) = 0 \in \mathcal{L}(\mathbb{E}^n, \mathbb{R}).$$

By Theorem 10.2.5 we know that $\mathbf{g}'(\mathbf{v}) = 0$ implies

$$(\mathbf{z} - \mathbf{f}(\mathbf{v})) \cdot \mathbf{f}'(\mathbf{v})\mathbf{h} = 0$$

for all $\mathbf{h} \in \mathbb{E}^n$. Since $\mathbf{f}'(v)$ is a nonsingular linear transformation of $\mathbb{E}^n \to \mathbb{E}^n$ it is also onto $\mathbb{E}^n$. Thus $\mathbf{z} - \mathbf{f}(\mathbf{v}) = \mathbf{0}$, which implies that $\mathbf{z} \in \mathbf{f}(D)$. ∎

Observe that under the hypotheses of the Open Mapping Theorem, if $\mathbf{f}$ is invertible then it $\mathbf{f}^{-1}$ is continuous. We are ready to state and prove the Inverse Function Theorem.

Theorem 10.4.3 (Inverse Function Theorem) *Let* $\mathbf{f} \in C^1(D, \mathbb{E}^n)$, *where D is an open subset of* $\mathbb{E}^n$. *Suppose that for all* $\mathbf{x} \in D$ *we have* $\det \mathbf{f}'(\mathbf{x}) \neq 0$. *If $\mathbf{f}$ is one-to-one on D, then* $\mathbf{f}^{-1} \in C^1(\mathbf{f}(D), D)$ *and*

$$\left(\mathbf{f}^{-1}\right)'(\mathbf{f}(\mathbf{x})) = (\mathbf{f}'(\mathbf{x}))^{-1}. \tag{10.6}$$

Remark 10.4.1 We mention that the function $\det \mathbf{f}'(\mathbf{x})$ is often called the *Jacobian* of $\mathbf{f}$, so that the condition in the theorem is that the Jacobian not vanish on D. If the Jacobian does not vanish, the mapping $\mathbf{f}$ is called *nonsingular*. In Exercise 10.65 you will show that the Jacobian is independent of the choice of basis for $\mathbb{E}^n$.

Proof of the Theorem. By hypothesis $\mathbf{f}^{-1}$ exists and is continuous by the Open Mapping Theorem. (Alternatively, this could have been shown using Theorem 9.3.3 applied to a suitable *closed* ball.) We need to prove that $\mathbf{f}^{-1} \in C^1(\mathbf{f}(D), \mathbb{E}^n)$ and that its derivative is given by the formula in the theorem.

Denote $\mathbf{y} = \mathbf{f}(\mathbf{x})$ and $\mathbf{y} + \mathbf{k} = \mathbf{f}(\mathbf{x} + \mathbf{h})$. Write

$$\mathbf{f}^{-1}(\mathbf{y} + \mathbf{k}) - \mathbf{f}^{-1}(\mathbf{y}) = (\mathbf{f}'(\mathbf{x}))^{-1}\mathbf{k} + \epsilon(\mathbf{k}), \tag{10.7}$$

which defines $\epsilon(\mathbf{k})$. To show that $\mathbf{f}^{-1}$ differentiable at $\mathbf{f}(\mathbf{x})$ and that $(\mathbf{f}'(\mathbf{x}))^{-1}$ serves as the derivative of $\mathbf{f}^{-1}$ at $\mathbf{y}$ it is necessary and sufficient to show that

$$\frac{\|\epsilon(\mathbf{k})\|}{\|\mathbf{k}\|} \to 0$$

as $\mathbf{k} \to \mathbf{0}$. Since $\mathbf{f}$ is invertible, $\mathbf{h} \neq \mathbf{0}$ if and only if $\mathbf{k} \neq \mathbf{0}$. Applying the linear map $\mathbf{f}'(\mathbf{x})$ to both sides of Equation (10.7), we obtain

$$\mathbf{f}'(\mathbf{x})\mathbf{h} = \mathbf{f}(\mathbf{x} + \mathbf{h}) - \mathbf{f}(\mathbf{x}) + \mathbf{f}'(\mathbf{x})\epsilon(\mathbf{k}).$$

Since $\mathbf{f}$ is differentiable at $\mathbf{x}$, we know that

$$\frac{\|\mathbf{f}'(\mathbf{x})\epsilon(\mathbf{k})\|}{\|\mathbf{h}\|} \to 0$$

as $\mathbf{h} \to \mathbf{0}$, which is equivalent to $\mathbf{k} \to \mathbf{0}$ since both $\mathbf{f}$ and $\mathbf{f}^{-1}$ are continuous. It is not hard to show using a composition of a linear transformation with its own inverse that

$$\|\mathbf{f}'(\mathbf{x})\epsilon(\mathbf{k})\| \geq \frac{1}{\|(\mathbf{f}'(\mathbf{x}))^{-1}\|}\|\epsilon(\mathbf{k})\|,$$

which implies that

$$\frac{\|\epsilon(\mathbf{k})\|}{\|\mathbf{h}\|} \to 0.$$

By the Magnification Theorem we have the bound $\alpha\|\mathbf{h}\| \leq \|\mathbf{k}\|$ for some suitable constant $\alpha > 0$ and for $\mathbf{h}$ of sufficiently small norm. This inequality implies that

$$\frac{\|\epsilon(\mathbf{k})\|}{\|\mathbf{k}\|} \to 0$$

as $\mathbf{k} \to \mathbf{0}$. This proves that $(\mathbf{f})^{-1}$ is differentiable at $\mathbf{y}$.

The formula in Equation (10.6) follows immediately from application of the Chain rule to

$$\mathbf{f}^{-1} \circ \mathbf{f}(\mathbf{x}) \equiv \mathbf{x},$$

recognizing that both functions in the composition are differentiable. Thus we obtain

$$\left(\mathbf{f}^{-1}\right)'\left(\mathbf{f}(\mathbf{x})\right)\mathbf{f}'(\mathbf{x}) \equiv I \in \mathcal{L}(\mathbb{E}^n, \mathbb{E}^n),$$

the identity transformation. Hence

$$\left(\mathbf{f}^{-1}\right)'\left(\mathbf{f}(\mathbf{x})\right) = \left(\mathbf{f}'(\mathbf{x})\right)^{-1}.$$

It remains only to show that $\left(\mathbf{f}^{-1}\right)' \in \mathcal{C}(D, \mathcal{L}(\mathbb{E}^n))$. We can rewrite Equation (10.6) in the form

$$\left(\mathbf{f}^{-1}\right)'(\mathbf{x}) = \left(\mathbf{f}'\left(\mathbf{f}^{-1}(\mathbf{x})\right)\right)^{-1}.$$

We note that this expresses $\left(\mathbf{f}^{-1}\right)'$ as the composition of three maps. The first map is $\mathbf{f}^{-1}$, which was just shown to be continuous. The second map is $\mathbf{f}'$, which is assumed to be continuous. Finally, the map that sends an invertible linear transformation to its own inverse is a continuous mapping of $\mathcal{GL}(n, \mathbb{R})$ to itself, as is shown in Exercise 10.67. Since compositions of continuous maps are continuous, this completes the proof. ∎

EXERCISES

10.57   Let $f : \mathbb{E}^1 \to \mathbb{E}^1$ be defined by

$$f(x) = \begin{cases} x^3 \sin \frac{1}{x} & \text{if } x \neq 0, \\ 0 & \text{if } x = 0. \end{cases}$$

Show that $f \in \mathcal{C}^1\left(\mathbb{E}^1, \mathbb{E}^1\right)$ but that f is not invertible in any open interval $(-\delta, \delta)$ around the origin. Explain why the Magnification Theorem fails to insure invertibility of f in this exercise.

10.58   Let $f \in \mathcal{C}^1\left(\mathbb{E}^1, \mathbb{E}^1\right)$ be defined by $f(x) = \sin x$ for all $x \in \mathbb{E}^1$. Is $f\left(\mathbb{E}^1\right)$ open? If yes, prove it. If no, explain which hypothesis of the Open Mapping Theorem fails to be true in this example.

10.59   Suppose $D \subset \mathbb{E}^n$ and $S \subset \mathbb{E}^m$ are both open sets. Suppose $\mathbf{f} : D \to S$ is differentiable, one-to-one and onto S, and suppose $\mathbf{f}^{-1}$ is differentiable also. Prove that $n = m$. Conclude that if $m \neq n$, then $\mathbb{E}^n$ and $\mathbb{E}^m$ are *not diffeomorphic*, meaning that there is no differentiable one-to-one map of $\mathbb{E}^n$ onto $\mathbb{E}^m$ with differentiable inverse. (Hint: Apply the Chain Rule to the composition of $\mathbf{f}$ and $\mathbf{f}^{-1}$ in either order. If $T \in \mathcal{L}(\mathbb{E}^n, \mathbb{E}^m)$, recall a theorem from linear algebra concerning the *rank* and the *nullity* of T.)

10.60   For each of the following functions, find: (i) all points $\mathbf{x}$ at which the Jacobian of $\mathbf{f}$ does not vanish; (ii) A ball $B_r(\mathbf{x})$ on which $\mathbf{f}$ has a differentiable inverse, for those $\mathbf{x}$ identified in part (i).

    a) $\mathbf{f} : \mathbb{E}^2 \to \mathbb{E}^2$ by $\mathbf{f}(\mathbf{x}) = (x_1 \cos x_2, x_1 \sin x_2)$.

    b) $\mathbf{f} : \mathbb{E}^2 \to \mathbb{E}^2$ by $\mathbf{f}(\mathbf{x}) = (e^{x_1} \cos 2, e^{x_1} \sin x_2)$.

    c) $\mathbf{f} : \mathbb{E}^3 \rightarrow \mathbb{E}^3$ by $\mathbf{f}(\mathbf{x}) = (x_1 \cos x_2, x_1 \sin x_2, x_3)$.

    d) $\mathbf{f} : \mathbb{E}^3 \rightarrow \mathbb{E}^3$ by $\mathbf{f}(\mathbf{x}) = (x_1 \sin x_3 \cos x_2, x_1 \sin x_3 \sin x_2, x_1 \cos x_3)$.

10.61 Suppose $f \in C\left(\mathbb{E}^1, \mathbb{E}^1\right)$ and that f is *locally injective*, meaning that for each $x \in \mathbb{E}^1$ there is a corresponding $r > 0$ such that f restricted to $B_r(x)$ is injective (meaning one-to-one). Prove that f must be injective on $\mathbb{E}^1$. (Hint: Suppose false and deduce a contradiction.)

10.62 Give an example of a function $\mathbf{f} \in C^1\left(\mathbb{E}^2, \mathbb{E}^2\right)$ that is locally injective at each point $\mathbf{x} \in \mathbb{E}^2$ for which $x_1 \neq 0$ and that has $\det \mathbf{f}'(\mathbf{x}) \neq 0$ (that is, nonvanishing Jacobian) if $x_1 \neq 0$, yet for which $\mathbf{f}$ is not injective on $\left\{\mathbf{x} \in \mathbb{E}^2 \mid x_1 > 0\right\}$.

10.63 Suppose $f \in C^1\left(\mathbb{E}^1, \mathbb{E}^1\right)$ and that f' is nowhere zero on $\mathbb{E}^1$. Prove that f must be invertible on $\mathbb{E}^1$ and that $f^{-1} \in C^1\left(f\left(\mathbb{E}^1\right), \mathbb{E}^1\right)$.

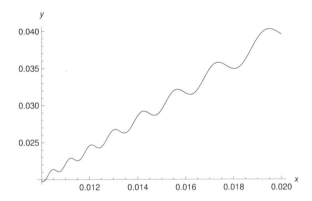

Figure 10.4 $2x + 4x^2 \sin \frac{1}{x}$.

10.64 Let $f : \mathbb{E}^1 \rightarrow \mathbb{E}^1$ be defined by

$$f(x) = \begin{cases} 2x + 4x^2 \sin \frac{1}{x} & \text{if } x \neq 0, \\ 0 & \text{if } x = 0. \end{cases}$$

Show that f is not injective (and therefore not invertible) in any open interval $(-r, r)$ with $r > 0$, although f' exists and is bounded in the open interval $(-1, 1)$. Which hypothesis of the Magnification Theorem fails, saving that theorem from being contradicted by the noninvertibility of f on $(-r, r)$ for all $r > 0$? Justify your conclusion. (Be sure to find the value of $f'(0)$ and to justify your result. It may help to apply Exercise 4.22. See Fig. 10.4.)

10.65 If $\mathbf{f} : \mathbb{E}^n \rightarrow \mathbb{E}^n$ is differentiable, prove that the Jacobian of $\mathbf{f}$ is independent of the choice of basis used in $\mathbb{E}^n$. (Hint: Use a change-of-basis matrix and a property of determinants that you learned in a course in linear algebra.)

10.66 Let $\mathbf{g} \in C^1(\mathbb{E}^n, \mathbb{E}^n)$ and $\mathbf{f} \in C^1(\mathbb{E}^n, \mathbb{E}^n)$. Denote the *Jacobian*

$$J_{\mathbf{g}}(\mathbf{x}) = \det \mathbf{g}'(\mathbf{x}).$$

Express the Jacobian $J_{\mathbf{f} \circ \mathbf{g}}$ in terms of the Jacobians of $J_{\mathbf{f}}$ and $J_{\mathbf{g}}$. Justify your conclusion.

10.67 † Show that the mapping $T \to T^{-1}$ is a continuous mapping from $\mathcal{GL}(n, \mathbb{R})$ to itself. (Hint: It is sufficient to prove that each matrix coefficient of the inverse matrix of T is a continuous function of the original matrix. You may use a formula from linear algebra for matrix inversion using determinants.)

10.68 Give an alternative coordinate-free proof of Exercise 10.67 by using the result of Exercise 10.11.

10.5 IMPLICIT FUNCTIONS

In elementary calculus courses we learn a process called *implicit differentiation*. We are taught to take an equation of the form $f(x, y) = 0$ and differentiate both sides using the chain rule, supposing that y is a function of x that satisfies the equation. That is, $y = g(x)$ and $f(x, g(x)) \equiv 0$. This yields the result

$$g'(x) = -\frac{\frac{\partial f}{\partial x}}{\frac{\partial f}{\partial y}},$$

where we assume the denominator is not zero. Although this is a very useful process, it can lead easily to nonsense. For example, consider the equation

$$x^2 + y^2 + 1 = 0.$$

We could perform implicit differentiation mechanically to obtain the apparent result

$$\frac{dy}{dx} = -\frac{x}{y}$$

whenever $y \neq 0$. If we do not think about what we are doing, we may imagine that there is a function $y = g(x)$ such that $x^2 + g(x)^2 + 1 = 0$ with the given derivative. Of course *there is no such function!* There is not even a single pair of real numbers x and y such that $x^2 + y^2 + 1 = 0$.

This shows us that we need a theorem that will enable us to know whether or not there really is a differentiable function $y = g(x)$ satisfying the given equation $f(x, y) = 0$ and with the derivative obtained according to the method of implicit differentiation. This theorem, the Implicit Function Theorem, is stated and proven below. It is very powerful. One need only consider what a formidable problem it is to solve even a polynomial equation of degree 5 or higher to appreciate the significance of being able to find the derivative of a function $y = g(x)$ even though it is very unlikely that we can solve the equation $f(x, y) = 0$ explicitly to express y in terms of x using only finitely many elementary operations.

We will adopt the following notation for the work of this section. Let $\mathbf{f}$ be in $\mathcal{C}^1(D, \mathbb{E}^m)$ where D is an open subset of $\mathbb{E}^{n+m}$, and denote $(\mathbf{x}, \mathbf{y}) \in \mathbb{E}^{n+m}$ where

$\mathbf{x} \in \mathbb{E}^n$ and $\mathbf{y} \in \mathbb{E}^m$. Denote the *Jacobian*

$$\frac{\partial(f_1, \ldots, f_m)}{\partial(y_1, \ldots, y_m)}(\mathbf{x}_0, \mathbf{y}_0) = \det \left(\frac{\partial f_i}{\partial y_j}(\mathbf{x}_0, \mathbf{y}_0) \right)_{m \times m}.$$

Theorem 10.5.1 (Implicit Function Theorem) *Suppose* $\mathbf{f} \in \mathcal{C}^1(D, \mathbb{E}^m)$, *where* D *is an open subset of* $\mathbb{E}^{n+m}$. *Suppose there is a point* $(\mathbf{x}_0, \mathbf{y}_0)$ *in* D *such that* $\mathbf{f}(\mathbf{x}_0, \mathbf{y}_0) = \mathbf{0}$. *Suppose that the Jacobian*

$$\frac{\partial(f_1, \ldots, f_m)}{\partial(y_1, \ldots, y_m)}(\mathbf{x}_0, \mathbf{y}_0) \neq 0.$$

Then there exist open sets U *and* V *in* $\mathbb{E}^n$ *and* $\mathbb{E}^m$, *respectively, such that* $\mathbf{x}_0 \in U$ *and* $\mathbf{y}_0 \in V$ *with the following properties:*

   i. For each $\mathbf{x} \in U$ *there exists a unique* $\mathbf{y} \in V$, *denoted by* $\mathbf{y} = \mathbf{g}(\mathbf{x})$, *such that* $\mathbf{f}(\mathbf{x}, \mathbf{g}(\mathbf{x})) \equiv \mathbf{0}$.

   ii. The function $\mathbf{g} \in \mathcal{C}^1(U, \mathbb{E}^m)$.

Proof: The proof of this theorem is not much longer than its statement. We will apply the Inverse Function Theorem.

Define $\tilde{\mathbf{f}}(\mathbf{x}, \mathbf{y}) = (\mathbf{x}, \mathbf{f}(\mathbf{x}, \mathbf{y})) \in \mathbb{E}^{n+m}$. Then the $(n+m) \times (n+m)$ *matrix*

$$\left[\tilde{\mathbf{f}}'(\mathbf{x}_0, \mathbf{y}_0) \right] = \left(\begin{array}{cc} I_{n \times n} & 0_{n \times m} \\ \left(\frac{\partial f_i}{\partial x_j}(\mathbf{x}_0, \mathbf{y}_0) \right)_{m \times n} & \left(\frac{\partial f_i}{\partial y_j}(\mathbf{x}_0, \mathbf{y}_0) \right)_{m \times m} \end{array} \right).$$

Therefore

$$\det \tilde{\mathbf{f}}'(\mathbf{x}_0, \mathbf{y}_0) = \frac{\partial(f_1, \ldots, f_m)}{\partial(y_1, \ldots, y_m)}(\mathbf{x}_0, \mathbf{y}_0) \neq 0$$

because of the *block-triangular* form of the matrix. Also, it is easy to see that $\tilde{\mathbf{f}} \in \mathcal{C}^1(D, \mathbb{E}^{n+m})$ and that the determinant of its derivative is a continuous function. By the Inverse Function Theorem together with the Magnification Theorem, there exists an open ball of the form

$$B_{r\sqrt{2}}(\mathbf{x}_0, \mathbf{y}_0) \supset B_r(\mathbf{x}_0) \times B_r(\mathbf{y}_0),$$

on which the Jacobian remains nonvanishing and on which $\tilde{\mathbf{f}}$ has a continuously differentiable inverse. Applying Exercise 10.83 we see that $\tilde{\mathbf{f}}(B_r(\mathbf{x}_0) \times B_r(\mathbf{y}_0))$ is open in $\mathbb{E}^{n+m}$. Thus there exists $\rho > 0$ such that

$$B_\rho(\mathbf{x}_0) \times B_\rho(\mathbf{0}) \subseteq \tilde{\mathbf{f}}(B_r(\mathbf{x}_0) \times B_r(\mathbf{y}_0)).$$

Thus for each $\mathbf{x} \in B_\rho(\mathbf{x}_0)$ there exists a unique $\mathbf{y} \in B_r(\mathbf{y}_0)$ such that

$$(\mathbf{x}, \mathbf{y}) = \tilde{\mathbf{f}}^{-1}(\mathbf{x}, \mathbf{0}).$$

If we denote $\mathbf{g}(\mathbf{x}) = \mathbf{y}$, we see that $\mathbf{f}(\mathbf{x}, \mathbf{g}(\mathbf{x})) \equiv \mathbf{0}$. In the notation of the theorem, we can take $U = B_\rho(\mathbf{x}_0)$ and $V = B_r(\mathbf{y}_0)$, the Cartesian product of these two sets being in $B_r(\mathbf{x}_0, \mathbf{y}_0)$. ∎

Remark 10.5.1 If one is presented with an $m \times (n + m)$ matrix, it is difficult to see by inspection whether or not there exist m column vectors from this matrix that together would form a $m \times m$ matrix with nonzero determinant. Thus it is hard to judge by inspection whether or not the conditions of the implicit function theorem are satisfied.

It is helpful to review the concept of rank from linear algebra. The $\mathrm{rank}(T)$ of a linear transformation $T : V \to W$ is $\dim T(V)$, the dimension of the image of T. T can be represented as a matrix in many ways, depending on the chosen bases for V and for W, the domain and range vector spaces. But for any matrix, it is not hard to prove that the rank of the corresponding linear transformation is the dimension of the span of the set of all the column vectors of the matrix, which is called the column rank of the matrix. There is also a concept of row rank, which is the dimension of the span of the set of all the rows of the matrix. It is a fundamental theorem in linear algebra that $\mathrm{rank}(T)$, the column rank of $[T]$, and the row rank of $[T]$ must all be the same number. If the row rank or the column rank of $[T]_{m \times (n+m)}$ is equal to m, then there must be m linearly independent columns and the student will then be able to look for those columns, whose determinant will yield the needed nonzero Jacobian as in the implicit function theorem. The row rank will be m if and only if the set of all m rows of the matrix is linearly independent.

We remark also that the open sets U and V can be challenging to identify in any given example, unless it is a relatively simple example such as one in which it is easy to solve for the implicitly determined function explicitly. The delicacy of the problem is reflected in the difficulty of establishing the existence of U and V in the proof of the implicit function theorem.

Let us consider in detail an example in which we apply the implicit function theorem to a specific real-valued function of three real variables.

■ **EXAMPLE 10.4**

Let $F : \mathbb{E}^3 \to \mathbb{R}$ by

$$F(\mathbf{x}) = x_1^{\frac{2}{3}} + x_2^{\frac{2}{3}} + x_3^{\frac{2}{3}} - 1.$$

For which points on the surface defined by $F(\mathbf{x}) = 0$ is it true that at least one variable is determined as a continuously differentiable function of the others, restricted to suitable open sets? Fig. 10.5 shows just the *upper half* of this surface.

The reader can see in the figure that the entire surface defined by $F(\mathbf{x}) = 0$ has twelve *sharp edges* and six *sharp points, or cusps*. We will begin our analysis of the problem of this example by seeing what information we can get from the implicit function theorem. To apply this theorem, we need to have

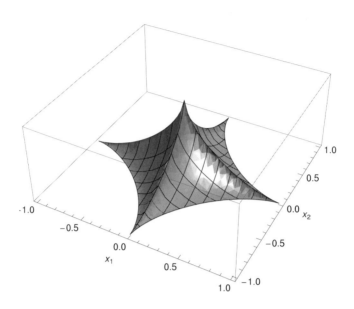

Figure 10.5 *Upper half of $x_1^{\frac{2}{3}} + x_2^{\frac{2}{3}} + x_3^{\frac{2}{3}} = 1$, $x_3 \geq 0$.*

$F \in \mathcal{C}^1$. But with respect to the standard basis, the matrix of the derivative of F is

$$[F'(\mathbf{x})] = \frac{2}{3}\left[x_1^{-\frac{1}{3}}, x_2^{-\frac{1}{3}}, x_3^{-\frac{1}{3}} \right],$$

provided that $x_1 x_2 x_3 \neq 0$. One sees that F' exists only in the open set $D = \{\mathbf{x} \mid x_1 x_2 x_3 \neq 0\}$, which is the union of the eight *open octants* of three-dimensional space which exclude the three coordinate planes. Moreover, $F \in \mathcal{C}^1(D, \mathbb{E}^1)$. Since $\frac{\partial F}{\partial x_i} \neq 0$ on D for any $i = 1, 2, 3$, each of the variables can be expressed as a $\mathcal{C}^1$ function of the other two, in suitably restricted open sets. For example, suppose $F(\mathbf{x}_0) = 0$, where $\mathbf{x}_0 = (a, b, c)$. If we solve for x_3 in terms of the first two, we can restrict (x_1, x_2) to the open set U, which is that part of the open quadrant of the $x_1 x_2$-plane to which (a, b) belongs and is strictly inside the curve

$$x_1^{\frac{2}{3}} + x_2^{\frac{2}{3}} = 1.$$

In order to insure the existence of a *unique* continuously differentiable solution for x_3 over the domain U chosen as above, we restrict x_3 to the open set $V = (0, 1)$ or else $V = (-1, 0)$, choosing the interval to which c belongs. The reader sees that uniqueness of the solution is a delicate matter here, since if (x_1, x_2, x_3) is on the *full* graph, then so is $(x_1, x_2, -x_3)$, which wrecks uniqueness unless x_3 is suitably restricted.

Finally, we mention that we have not shown yet that unique $\mathcal{C}^1$ solutions *fail* to exist if at least one of the three coordinates *is* zero. That information does not come from the implicit function theorem, but depends on elementary calculations, which we leave to Exercise 10.76.

We can translate the Implicit Function Theorem into a statement about the partial derivatives of the component functions g_i of the implicitly determined function $\mathbf{g}(\mathbf{x})$. Since $\mathbf{f}(\mathbf{x}, \mathbf{g}(\mathbf{x})) = \mathbf{0} \in \mathbb{E}^m$ for all $x \in U \subseteq \mathbb{E}^n$, it is natural to define $\tilde{\mathbf{g}}(\mathbf{x}) = (\mathbf{x}, \mathbf{g}(\mathbf{x})) \in \mathbb{E}^{n+m}$. Observe that

$$\mathbb{E}^n \xrightarrow{\tilde{\mathbf{g}}} \mathbb{E}^{n+m} \xrightarrow{\mathbf{f}} \mathbb{E}^m.$$

The derivative $(\mathbf{f} \circ \tilde{\mathbf{g}})'(\mathbf{x}) = \mathbf{f}'(\tilde{\mathbf{g}}(\mathbf{x}))\tilde{\mathbf{g}}'(\mathbf{x}) = 0 \in \mathcal{L}(\mathbb{E}^n, \mathbb{E}^m)$ for all $\mathbf{x} \in U$. Expressed in terms of matrices this composition of linear transformations becomes the following matrix equation.

$$\begin{pmatrix} \frac{\partial f_1}{\partial x_1} & \cdots & \frac{\partial f_1}{\partial y_m} \\ \vdots & & \vdots \\ \frac{\partial f_m}{\partial x_1} & \cdots & \frac{\partial f_m}{\partial y_m} \end{pmatrix}_{m \times (n+m)} \begin{pmatrix} 1 & \cdots & 0 \\ \vdots & & \vdots \\ 0 & \cdots & 1 \\ \frac{\partial g_1}{\partial x_1} & \cdots & \frac{\partial g_1}{\partial x_n} \\ \vdots & & \vdots \\ \frac{\partial g_m}{\partial x_1} & \cdots & \frac{\partial g_m}{\partial x_n} \end{pmatrix}_{(n+m) \times n} = 0_{m \times n}.$$

$$(10.8)$$

Suppose we fix a value of j between 1 and n, and we wish to calculate $\frac{\partial g_i}{\partial x_j}$, for each value of $i = 1, \ldots, m$. For this we consider the jth column of the product matrix, obtaining a system of m equations in m unknowns, $\frac{\partial g_1}{\partial x_j}, \ldots, \frac{\partial g_m}{\partial x_j}$, as follows.

$$\frac{\partial f_1}{\partial x_j} + \frac{\partial f_1}{\partial y_1}\frac{\partial g_1}{\partial x_j} + \cdots + \frac{\partial f_1}{\partial y_m}\frac{\partial g_m}{\partial x_j} = 0,$$

$$\vdots$$

$$\frac{\partial f_m}{\partial x_j} + \frac{\partial f_m}{\partial y_1}\frac{\partial g_1}{\partial x_j} + \cdots + \frac{\partial f_m}{\partial y_m}\frac{\partial g_m}{\partial x_j} = 0.$$

One can calculate the unique solution to this system of equations using Cramer's Rule from linear algebra, since the Jacobian

$$\frac{\partial(f_1, \ldots, f_m)}{\partial(y_1, \ldots, y_m)} \neq 0.$$

We mention however that for large m it is generally simpler to compute the solutions to the system of equations using the method of row reduction, taught in all introductory courses in linear algebra.

If we denote

$$\left(\frac{\partial f_i}{\partial x_j}(\mathbf{x}, \mathbf{g}(\mathbf{x})) \right)_{m \times n} = \left(\frac{d\mathbf{f}}{d\mathbf{x}} \right)$$

and

$$\left(\frac{\partial f_i}{\partial y_j}(\mathbf{x}, \mathbf{g}(\mathbf{x})) \right)_{m \times m} = \left(\frac{d\mathbf{f}}{d\mathbf{y}} \right),$$

then we can rewrite Equation (10.8) as follows.

$$\left(\frac{d\mathbf{f}}{d\mathbf{x}} \right)_{m \times n} + \left(\frac{d\mathbf{f}}{d\mathbf{y}} \right)_{m \times m} \left(\frac{d\mathbf{g}}{d\mathbf{x}} \right)_{m \times n} = 0_{m \times n}. \qquad (10.9)$$

This yields the full solution for the partial derivatives $\frac{\partial g_i}{\partial x_j}$ in matrix form as

$$\left(\frac{d\mathbf{g}}{d\mathbf{x}} \right)_{m \times n} = -\left(\frac{d\mathbf{f}}{d\mathbf{y}} \right)_{m \times m}^{-1} \left(\frac{d\mathbf{f}}{d\mathbf{x}} \right)_{m \times n}. \qquad (10.10)$$

Although Equation (10.10) has theoretical significance, we remark again that when one solves for the partial derivatives of the implicitly defined functions it is often more convenient to solve the systems of equations expressed either in the form of Equation (10.8) or Equation (10.9) by the elementary method of row reduction, without inverting the matrix for $\frac{d\mathbf{f}}{d\mathbf{y}}$.

■ **EXAMPLE 10.5**

We will describe a curve in $\mathbb{E}^3$ as the intersection of two surfaces, defined by the equations $f_1(x_1, x_2, x_3) = 0$ and $f_2(x_1, x_2, x_3) = 0$. Suppose f_1 and f_2 are in $C^1(\mathbb{E}^3, \mathbb{R})$.

Such descriptions of curves as intersections of two surfaces in $\mathbb{E}^3$ are common in introductory courses in the calculus of several variables. We will see how such a description can arise from the implicit function theorem, by denoting $\mathbf{f} = (f_1, f_2) \in C^1(\mathbb{E}^3, \mathbb{E}^2)$. If at a particular point in the intersection of surfaces we have

$$\frac{\partial(f_1, f_2)}{\partial(x_2, x_3)} \neq 0,$$

then the implicit function theorem guarantees the existence locally of solutions $x_2 = x_2(x_1)$ and $x_3 = x_3(x_1)$ to the vector equation $\mathbf{f}(\mathbf{x}) = \mathbf{0}$. We *could* apply Equation (10.9) to find $\frac{dx_2}{dx_1}$ and $\frac{dx_3}{dx_1}$ along the curve of intersection, which is parameterized in this way by the variable x_1. However, as an alternative to remembering Equation (10.9) the student can apply the Chain Rule directly to find the derivatives once the implicit function theorem has been used to assure the existence of differentiable solutions. Since $\mathbf{f}(x_1, x_2(x_1), x_3(x_1)) \equiv \mathbf{0}$, we denote $\tilde{\mathbf{g}}(x_1) = (x_1, x_2(x_1), x_3(x_1))$, so that $\mathbf{f} \circ \tilde{\mathbf{g}}(x_1) \equiv \mathbf{0}$. Next we apply the Chain Rule to differentiate with respect to x_1 on both sides of the latter equation. This yields a matrix equation

$$[f'(\tilde{\mathbf{g}}(x_1))]_{2 \times 3} [\tilde{\mathbf{g}}'(x_1)]_{3 \times 1} = [0]_{2 \times 1}.$$

Thus we have

$$\begin{pmatrix} \frac{\partial f_1}{\partial x_1} & \frac{\partial f_1}{\partial x_2} & \frac{\partial f_1}{\partial x_3} \\ \frac{\partial f_2}{\partial x_1} & \frac{\partial f_2}{\partial x_2} & \frac{\partial f_2}{\partial x_3} \end{pmatrix} \begin{pmatrix} 1 \\ x_2'(x_1) \\ x_3'(x_1) \end{pmatrix} = \begin{pmatrix} 0 \\ 0 \end{pmatrix}.$$

This matrix equation yields two real equations which can be solved for $x_2'(x_1)$ and $x_2'(x_1)$, because

$$\frac{\partial(f_1, f_2)}{\partial(x_2, x_3)} \neq 0.$$

In general, when we are given a function $\mathbf{f} \in \mathcal{C}^1(\mathbb{E}^{n+m})$, we may not know whether m variables $\mathbf{y}$ can be selected for which there exist differentiable solutions of the equation $f(\mathbf{x}, \mathbf{y}) = \mathbf{0}$ in terms of the remaining n variables $\mathbf{x}$ in a differentiable manner, at least locally in a neighborhood of some point in the solution set. If the rank of the linear transformation $\mathbf{f}'(\mathbf{p})$ is m, then there exist m columns of the matrix of the derivative that collectively form an $m \times m$ nonsingular matrix, which means that the determinant of that square matrix is not zero. Those m selected columns correspond to the components of the vector $\mathbf{y}$ for which there exists a locally differentiable and unique solution in terms of the remaining variables, which we could then label collectively as the vector $\mathbf{x}$.

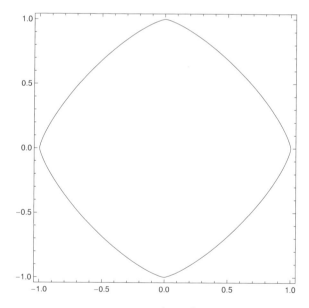

Figure 10.6 $x_1^{\frac{4}{3}} + x_2^{\frac{4}{3}} = 1$.

EXERCISES

10.69   Define $\mathbf{f} \in \mathcal{C}^1\left(\mathbb{E}^3, \mathbb{E}^2\right)$ by

$$\mathbf{f}(x_1, x_2, x_3) = (x_1 x_2 \cos x_3, \ x_2 \sin x_3).$$

Find the Jacobian $\frac{\partial(f_1, f_2)}{\partial(x_2, x_3)}$.

10.70 Define $\mathbf{f} \in C^1\left(\mathbb{E}^3, \mathbb{E}^2\right)$ by

$$\mathbf{f}(x_1, x_2, x_3) = \left(x_1 + x_2 x_3, \; x_2^2 - 2x_3\right).$$

Find the Jacobian $\frac{\partial(f_1, f_2)}{\partial(x_1, x_3)}$.

10.71 Let $f \in C^1\left(\mathbb{E}^4, \mathbb{E}^1\right)$ be such that the matrix

$$[f'(\mathbf{x})]_{1 \times 4} = [x_2, \; x_1, \; x_4, \; x_3].$$

Find all $\mathbf{x}_0 = (x_1, x_2, x_3, x_4)$ at which the implicit function theorem guarantees that there is a *local* C^1 solution of the equation $f(\mathbf{x}) = f(\mathbf{x}_0)$ for x_3 in terms of the other three variables.

10.72 Many laws of nature, such as the *ideal gas law*, can be written in the form $F(x_1, x_2, x_3) = 0$. Suppose that $F \in C^1\left(\mathbb{E}^3, \mathbb{E}^1\right)$ and that for all $i = 1, 2, 3$ we have $\frac{\partial F}{\partial x_i} \neq 0$. Prove that

$$\frac{\partial x_1}{\partial x_2} \frac{\partial x_2}{\partial x_3} \frac{\partial x_3}{\partial x_1} = -1.$$

State the hypotheses needed on the points (x_1, x_2, x_3) to insure that this conclusion is valid.

10.73 Suppose $F(x_1, \ldots, x_n) = 0$, where $F \in C^1\left(\mathbb{E}^n, \mathbb{E}^1\right)$. Suppose that for all $i = 1, \ldots, n$ we have $\frac{\partial F}{\partial x_i} \neq 0$. Find the numerical value of the product

$$\prod_{i=1,\ldots,n} \frac{\partial x_i}{\partial x_{i+1}},$$

where we make the notational agreement that x_{n+1} means x_1. State the hypotheses needed on the points $(x_1, x_2, \ldots, x_n)$ to insure that this conclusion is valid.

10.74 Let $F : \mathbb{E}^2 \to \mathbb{R}$ by

$$F(\mathbf{x}) = x_1^{\frac{4}{3}} + x_2^{\frac{4}{3}}.$$

For which points on the curve defined by $F(\mathbf{x}) = 1$ is it true that at least one variable is determined as a continuously differentiable function of the other, restricted to suitable open sets U and V? Find U and V and justify your conclusion. (See Fig. 10.6.)

10.75 Let $F : \mathbb{E}^3 \to \mathbb{R}$ by

$$F(\mathbf{x}) = x_1^{\frac{4}{3}} + x_2^{\frac{4}{3}} + x_3^{\frac{4}{3}}.$$

For which points on the surface defined by $F(\mathbf{x}) = 1$ is it true that at least one variable is determined as a continuously differentiable function of the others, restricted to

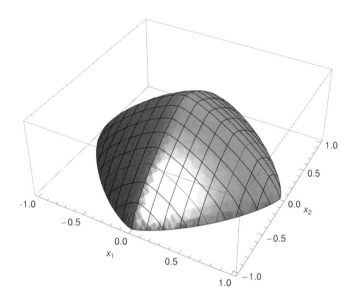

Figure 10.7 Upper half of $x_1^{\frac{4}{3}} + x_2^{\frac{4}{3}} + x_3^{\frac{4}{3}} = 1$, $x_3 \geq 0$.

suitable open sets U and V, which you should find? Justify your conclusion. (See Fig. 10.7.)

10.76 Let
$$F(\mathbf{x}) = x_1^{\frac{2}{3}} + x_2^{\frac{2}{3}} + x_3^{\frac{2}{3}} - 1$$
as in Example 10.4. Suppose that $c = 0$ for the point $\mathbf{x}_0 = (a, b, c)$. Show that the equation $F(\mathbf{x}) = \mathbf{0}$ cannot be solved uniquely for x_3 as a $\mathcal{C}^1$ function in terms of x_1, x_2, even when the variables are restricted to suitable open neighborhoods of $c \in \mathbb{R}$ and of $(a, b) \in \mathbb{E}^2$. At that same point, are the other two variables, x_1 and x_2, locally unique $\mathcal{C}^1$ functions of the other two? Justify your conclusions.

10.77 Suppose that $\mathbf{f} = (f_1, f_2) \in \mathcal{C}^1 \left(\mathbb{E}^3, \mathbb{E}^2 \right)$ and $\mathbf{x}_0 = (a, b, c) \in \mathbb{E}^3$ is in the solution set of the equation $\mathbf{f}(\mathbf{x}) = \mathbf{0}$. Suppose this equation can be solved for $x_2 = x_2(x_1)$ and $x_3 = x_3(x_1)$, both $\mathcal{C}^1$ functions in a neighborhood of $\mathbf{x}_0$ with $x_2(a) = b$, and $x_3(a) = c$. If the matrix
$$[\mathbf{f}'(\mathbf{x}_0)] = \begin{pmatrix} 1 & 1 & 2 \\ -1 & 0 & 3 \end{pmatrix},$$
then find $x_2'(a)$ and $x_3'(a)$. (Hint: Apply the Chain Rule to differentiate
$$\mathbf{f}(x_1, x_2(x_1), x_3(x_1)) = \mathbf{0}$$

with respect to x_1 on both sides.)

10.78 Consider the curve C of intersection in $\mathbb{E}^3$ of the two surfaces with the equations $f_1(\mathbf{x}) = 0$ and $f_2(\mathbf{x}) = 0$, where

$$f_1(x_1, x_2, x_3) = x_1 x_2 - 1$$

and

$$f_2(x_1, x_2, x_3) = x_2^2 + x_3^2 - 1.$$

(See Fig. 10.8.) Show that the point $\left(2, \frac{1}{2}, \frac{\sqrt{3}}{2}\right)$ lies on the curve C and that in a neighborhood of this point the curve C can be parameterized as $x_1 = x_1(x_3)$ and $x_2 = x_2(x_3)$. Find $x_1'\left(\frac{\sqrt{3}}{2}\right)$ and $x_2'\left(\frac{\sqrt{3}}{2}\right)$.

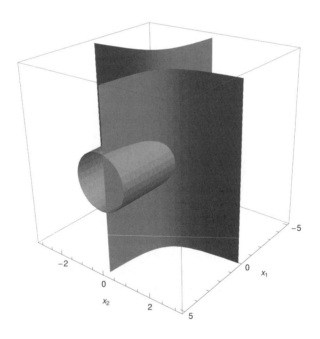

Figure 10.8 Intersection of cylinder with hyperbolic cylinder.

10.79 Consider the curve of intersection in $\mathbb{E}^3$ of the two surfaces with the equations $f_1(\mathbf{x}) = 0$ and $f_2(\mathbf{x}) = 0$, where

$$f_1(x_1, x_2, x_3) = x_2 + x_3 - 1$$

and

$$f_2(x_1, x_2, x_3) = x_1 - x_2^2 - x_3^2.$$

a) Explain why $\mathbf{f} = (f_1, f_2) \in C^1\left(\mathbb{E}^3, \mathbb{E}^2\right)$.

b) Find all points (x_2, x_3) for which $\frac{\partial(f_1, f_2)}{\partial(x_2, x_3)} \neq 0$.

c) At each point identified in item (b), apply Equation (10.10) to find $\frac{dx_2}{dx_1}$ and $\frac{dx_3}{dx_1}$ along the curve of intersection described as $(x_2, x_3) = \mathbf{g}(x_1)$.

10.80   Describe the 3-sphere S^3 in $\mathbb{E}^4$ as the solution set of $f(\mathbf{x}) = 0$, where

$$f(\mathbf{x}) = \|\mathbf{x}\|^2 - 1.$$

If $\mathbf{x} = (x_1, x_2, x_3, x_4) \in S^3$, *show by applying the Implicit Function Theorem that for at least one value of i the coordinate x_i can be expressed as a continuously differentiable function of the three other coordinates, with x_i and the triplet of other coordinates restricted to suitable open sets V and U which you should identify explicitly. Then find the solution for x_i explicitly, without using the Implicit Function Theorem.*

10.81   Let $\mathcal{SL}(2, \mathbb{R})$ denote the set of all two-by-two real matrices

$$X = \left(\begin{array}{cc} x_1 & x_2 \\ x_3 & x_4 \end{array} \right)$$

for which $\det X = 1$. If $X \in \mathcal{SL}(2, \mathbb{R})$, *show by applying the Implicit Function Theorem that at least one of the four coordinates x_1, x_2, x_3, x_4 can be expressed as a continuously differentiable function of the three other coordinates, with the variables restricted to suitable open sets which you should identify explicitly. Then solve explicitly for that identified coordinate, without using the Implicit Function Theorem.*

10.82   Define the so-called *orthogonal group* of the plane by

$$\mathcal{O}(2) = \left\{ X \in \mathcal{GL}\left(\mathbb{E}^2\right) \mid XX^t = I \right\},$$

where X^t denotes the transpose of X and I is the identity transformation. Denote the matrix

$$[X] = \left(\begin{array}{cc} x_1 & x_2 \\ x_3 & x_4 \end{array} \right).$$

a) Prove that $\mathcal{O}(2)$ is a group by showing that it is closed under the operations of multiplication and taking inverses.

b) Prove that $X \in \mathcal{O}(2) \Leftrightarrow \mathbf{f}(\mathbf{x}) = \mathbf{0} \in \mathbb{E}^3$, where

$$\mathbf{f}(\mathbf{x}) = \left(x_1 x_3 + x_2 x_4, x_1^2 + x_2^2 - 1, x_3^2 + x_4^2 - 1 \right)$$

for each $\mathbf{x} \in \mathbb{E}^4$.

c) Prove that the *rank* of the matrix $[\mathbf{f}'(\mathbf{x})]$ is 3 if $X \in \mathcal{O}(2)$. (Hint: How many *linearly independent rows* does this matrix have?)

d) Use the Implicit Function Theorem to identify three of the four coordinates of $\mathbf{x}$ that can be expressed as C^1 functions of the remaining coordinate, in a neighborhood of the identity element $I \in O(2)$.

e) Use Equation (10.10) to find the derivatives at the identity element of the three chosen dependent variables with respect to the chosen independent one.

f) This part is not about the implicit function theorem but it is interesting. Use your knowledge of determinants to show as follows that $\mathcal{O}(2)$ is *not connected*.

i. Show that if $A \in \mathcal{O}(2)$, then $\det A = \pm 1$ and show that both cases actually occur.

ii. Denote by $\mathcal{SO}(2)$ the set of elements of the orthogonal group with determinant equal to 1. Show that $\mathcal{SO}(2)$ is a (sub)group of $\mathcal{O}(2)$. (It is called the *special orthogonal group* of the plane.)

iii. Prove that the set of elements of $\mathcal{O}(2)$ that have determinant equal to -1 is not closed under multiplication.

10.83   Suppose U is an open set in $\mathbb{E}^n$ and V is an open set in $\mathbb{E}^m$. Prove that the Cartesian product $U \times V = \{(\mathbf{x}, \mathbf{y}) \mid \mathbf{x} \in U, \mathbf{y} \in V\}$ is an open set in $\mathbb{E}^{n+m}$.

10.6 TANGENT SPACES AND LAGRANGE MULTIPLIERS[30]

If a differentiable function $\mathbf{G} = (G_1, \ldots, G_k) : \mathbb{E}^{n+k} \to \mathbb{E}^k$, then the surface S defined by

$$S = \{\mathbf{x} \mid \mathbf{G}(\mathbf{x}) = \mathbf{v}\}$$

is called the *level surface* for $\mathbf{G}(\mathbf{x}) = \mathbf{v}$.

Note that each of the functions $G_i : \mathbb{E}^{n+k} \to \mathbb{R}$. If we denote by S_i the level surface for the equation $G_i(\mathbf{x}) = v_i$, then

$$S = \bigcap_{i=1}^{k} S_i.$$

Suppose that $\mathbf{x}^0 = (x_1^0, \ldots, x_{n+k}^0) \in S$ and that

$$\mathbf{G}'\left(\mathbf{x}^0\right) \in \mathcal{L}(\mathbb{E}^{n+k}, \mathbb{E}^k)$$

has rank k. Let $\delta_{i,j} = 1$ if $i = j$ and 0 if $i \neq j$. With respect to the standard basis

$$\left\{\mathbf{e}_j = (\delta_{1,j}, \ldots, \delta_{n+k,j}) \mid j = 1, 2, \ldots, n + k\right\}$$

for $\mathbb{E}^{n+k}$ and the analogous smaller basis for $\mathbb{E}^k$, we note that the *matrix*

$$\left[\mathbf{G}'\left(\mathbf{x}^0\right)\right]_{k \times (n+k)}$$

[30]This section can be omitted without disturbing the continuity of the book.

has its whole set of k row vectors *linearly independent*, and these row vectors are the *gradient vectors*

$$\nabla G_1\left(\mathbf{x}^0\right), \nabla G_2\left(\mathbf{x}^0\right), \ldots, \nabla G_k\left(\mathbf{x}^0\right).$$

Let $\phi : \mathbb{R} \to S$ be a differentiable function for which $\phi(0) = \mathbf{x}^0$. Then we call the vector $\mathbf{v} = \phi'(0)$ a *tangent vector* to S at $\mathbf{x}^0$. Of course there are infinitely many tangent vectors at any one point on a differentiable surface because there are so many different differentiable curves in the surface passing through that point.

Definition 10.6.1 *The* tangent space $T_{\mathbf{x}^0}(S)$ at $\mathbf{x}^0 \in S$ *is the set of* all *tangent vectors to S at* $\mathbf{x}^0$. *The* translate

$$\mathbf{x}^0 + T_{\mathbf{x}^0} = \left\{\mathbf{x}^0 + \mathbf{v} \,\middle|\, \mathbf{v} \in T_{\mathbf{x}^0}(S)\right\}$$

is called the tangent plane *to the surface S, with point of tangency at* $\mathbf{x}^0$.

In the next theorem, we will prove that the tangent space is always a vector subspace of $\mathbb{E}^{n+k}$. A translate $\mathbf{a} + V = \{\mathbf{a} + \mathbf{x} \mid \mathbf{x} \in V\}$ of a vector subspace V of $\mathbb{E}^n$ is called an *affine* subspace of $\mathbb{E}^n$. An affine subspace is a vector subspace if and only if $\mathbf{a} \in V$. (See Exercise 10.84.)

Theorem 10.6.1 *Let* $\mathbf{G} : \mathbb{E}^{n+k} \to \mathbb{E}^k$ *be a differentiable function. Let*

$$\mathbf{x}^0 \in S = \left\{\mathbf{x} \,\middle|\, \mathbf{G}(\mathbf{x}) = \mathbf{v}\right\}.$$

Suppose $\mathbf{G}'\left(\mathbf{x}^0\right)$ *has rank k. Then the tangent space* $T_{\mathbf{x}^0}(S)$ *is the* vector subspace

$$T_{\mathbf{x}^0}(S) = \left(\operatorname{span}_{\mathbb{R}}\left\{\nabla G_1\left(\mathbf{x}^0\right), \ldots, \nabla G_k\left(\mathbf{x}^0\right)\right\}\right)^{\perp}$$

of dimension n. In words, $T_{\mathbf{x}^0}(S)$ *is the orthogonal complement of the span of the k gradient vectors* $\nabla G_1\left(\mathbf{x}^0\right), \ldots, \nabla G_k\left(\mathbf{x}^0\right)$.

Proof: Suppose first that $\mathbf{v} = \phi'(0) \in T_{\mathbf{x}^0}(S)$. Here ϕ maps into each level surface $G_i(\mathbf{x}) = v_i$. We will show that $\mathbf{v} \perp \nabla G_i\left(\mathbf{x}^0\right)$ for each $i = 1, \ldots, k$. In fact, $G_i(\phi(t)) \equiv v_i$, a real constant. We differentiate using the Chain Rule to find that

$$G_i'(\phi(0))\phi'(0) = 0.$$

In terms of the standard matrix representation of the left side of the latter equation, we have $\nabla G_i\left(\mathbf{x}^0\right) \cdot \phi'(0) = 0$, so that $\mathbf{v} \perp \nabla G_i\left(\mathbf{x}^0\right)$. This shows that $T_{\mathbf{x}^0}(S)$ is a subset of $\nabla G_i\left(\mathbf{x}^0\right)^{\perp}$ for each i. This implies that

$$T_{\mathbf{x}^0}(S) \subseteq \left(\operatorname{span}_{\mathbb{R}}\left\{\nabla G_1\left(\mathbf{x}^0\right), \ldots, \nabla G_k\left(\mathbf{x}^0\right)\right\}\right)^{\perp}.$$

The hypothesis that rank $\left(\mathbf{G}'\left(\mathbf{x}^0\right)\right) = k$ implies that $\dim T_{\mathbf{x}^0}(S) \leq n$. If we can show that the tangent space is at least n-dimensional, then it will have to be the entire orthogonal complement of the span of the gradient vectors as claimed. Thus it will suffice to produce a linearly independent set of n vectors in the tangent space.

Because the rank of a matrix is also the number of linearly independent column vectors, it follows that the matrix $\left[\mathbf{G}'\left(\mathbf{x}^0\right)\right]$ has k independent columns. We can rearrange the order of the n elements of the standard basis of $\mathbb{E}^{n+k}$ to arrange that the first k columns are linearly independent. By the Implicit Function Theorem, there exists an open set $U \subset \mathbb{E}^k$ containing $\left(x_1^0, \ldots, x_k^0\right)$ and an open set $V \subset \mathbb{E}^n$ containing $\left(x_{k+1}^0, \ldots, x_{k+n}^0\right)$ such that there are unique differentiable functions

$$
\begin{aligned}
x_1 &= \psi_1(x_{k+1}, \ldots, x_{k+n}) \\
&\vdots \\
x_k &= \psi_k(x_{k+1}, \ldots, x_{k+n})
\end{aligned}
$$

solving the equation

$$\mathbf{G}\big(\psi_1(x_{k+1}, \ldots, x_{k+n}), \ldots, \psi_k(x_{k+1}, \ldots, x_{k+n}), x_{k+1}, \ldots, x_{k+n}\big) = \mathbf{v}.$$

Next we define n differentiable curves on S by the equations

$$
\begin{aligned}
\phi_1(t) &= \Big(\psi_1\big(x_{k+1}^0 + t, x_{k+2}^0, \ldots, x_{k+n}^0\big), \ldots, \psi_k\big(x_{k+1}^0 + t, x_{k+2}^0, \ldots, x_{k+n}^0\big); \\
&\qquad x_{k+1}^0 + t, x_{k+2}^0, \ldots, x_{k+n}^0\Big) \\
&\vdots \\
\phi_n(t) &= \Big(\psi_1\big(x_{k+1}^0, \ldots, x_{n+k-1}^0, x_{k+n}^0 + t\big), \ldots, \psi_k\big(x_{k+1}^0, \ldots, x_{k+n-1}^0, \\
&\qquad x_{k+n}^0 + t\big); \; x_{k+1}^0, \ldots x_{k+n-1}^0, x_{k+n}^0 + t\Big)
\end{aligned}
$$

In comparing the vectors $\phi_i'(0)$ for $i = 1, \ldots, n$, observe that for each of these vectors the *final* n entries, which follow the semicolon, are all 0 except for a single entry that is 1. The location of the 1 is different for each of these vectors. Thus the n vectors are independent and the theorem is proved. ∎

Corollary 10.6.1 *Let* $\mathbf{G} : \mathbb{E}^{k+n} \to \mathbb{E}^k$ *be a differentiable function and let*

$$\mathbf{x}^0 \in S = \{\mathbf{x} \mid \mathbf{G}(\mathbf{x}) = \mathbf{v}\}.$$

Suppose $\mathbf{x}^0$ *is a local extreme point of a differentiable function* $f : S \to \mathbb{R}$ *and that* $\mathbf{G}'\left(\mathbf{x}^0\right)$ *has rank* k. *Then there exist numbers* $\lambda_1, \ldots, \lambda_k$ *such that*

$$\nabla f\left(\mathbf{x}^0\right) = \lambda_1 \nabla G_1\left(\mathbf{x}^0\right) + \cdots + \lambda_k \nabla G_k\left(\mathbf{x}^0\right). \tag{10.11}$$

The numbers $\lambda_1, \ldots, \lambda_k$ *are called* Lagrange multipliers.

Proof: If $\phi : \mathbb{R} \to S$ is a differentiable curve on S with $\phi(0) = \mathbf{x}^0$, let

$$\psi(t) = f(\phi(t)).$$

Since this function has an extreme point at 0, we have

$$\psi'(0) = \nabla f(\phi(0)) \cdot \phi'(0) = 0.$$

It follows from Theorem 10.6.1 that $\nabla f\left(\mathbf{x}^0\right)$ is orthogonal to the tangent space $T_{\mathbf{x}^0}(S)$. Since the co-dimension of $T_{\mathbf{x}^0}(S)$ is k, it follows that $\nabla f\left(\mathbf{x}^0\right)$ lies in the span of the k vectors $\nabla G_1\left(\mathbf{x}^0\right), \ldots, \nabla G_k\left(\mathbf{x}^0\right)$. This proves the corollary.   ■

The method of Lagrange multipliers permits an optimization problem to be replaced by a problem of solving a system of equations. From the $k + n$ components of the vectors in Equation (10.11), we obtain a system of $k + n$ equations in the $n + 2k$ unknowns $x_1, \ldots, x_{k+n}, \lambda_1, \ldots, \lambda_k$. We get k additional equations from the k components of the equation $\mathbf{G}(\mathbf{x}) = \mathbf{v}$. Thus we obtain a system of $n + 2k$ equations in $n + 2k$ unknowns. Although we have replaced a calculus problem with an algebraic problem, the algebraic problem can be challenging. Nevertheless, the method of Lagrange multipliers is a powerful tool for optimization problems.

■ **EXAMPLE 10.6**

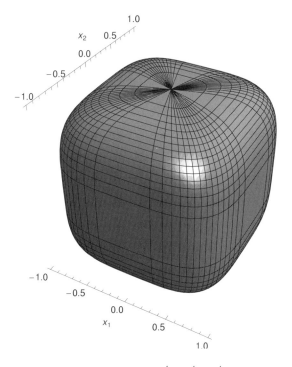

Figure 10.9 $x_1^4 + x_2^4 + x_3^4 = 1$.

We will begin with a three-dimensional example. Consider the surface S defined by the equation

$$x_1^4 + x_2^4 + x_3^4 = 1$$

in $\mathbb{E}^3$, shown in Fig. 10.9. We will find both the maximum and the minimum values of the function

$$f(\mathbf{x}) = x_1^2 + x_2^2 + x_3^2 = 1$$

on S. (In effect, we are determining the closest and furthest distances from the origin on S.) In this example, we denote $\mathbf{x} = (x, x_2, x_3)$. Observe that if we define

$$G(\mathbf{x}) = x_1^4 + x_2^4 + x_3^4$$

then $S = G^{-1}(\{1\})$. Hence S is closed because G is continuous. S is also bounded. (Why?) Hence the function f must achieve both a maximum and a minimum value somewhere on S. Since S is smooth at all points and since ∇G is nonvanishing on S, the extreme points must occur at those points for which $\nabla f(\mathbf{x}) = \lambda \nabla G(\mathbf{x})$. This yields the following system of equations.

$$
\begin{aligned}
x_1 \left(1 - 2\lambda x_1^2\right) &= 0, \\
x_2 \left(1 - 2\lambda x_2^2\right) &= 0, \\
x_3 \left(1 - 2\lambda x_3^2\right) &= 0, \\
x_1^4 + x_2^4 + x_3^4 &= 1.
\end{aligned}
$$

The reader should check the following statements by making the necessary calculations.

- If none of the three variables is zero, then

$$x_1^2 = x_2^2 = x_3^2 = \frac{1}{2\lambda}$$

  showing that $\lambda = \pm\frac{\sqrt{3}}{2}$. This implies that $f(x_1, x_2, x_3) = \sqrt{3}$.

- If exactly one of the three variables is zero, then at a point satisfying the system of equations we must have $f(x_1, x_2, x_3) = \sqrt{2}$.

- If exactly two of the variables are zero, then at a point satisfying the system we must have $f(x_1, x_2, x_3) = 1$.

It follows that the maximum value of f on S is $\sqrt{3}$. But the reader should be able to explain why at least one of the variables must be nonzero. Thus the minimum value is 1. There is also an easy way to explain even from the outset why $f(x_1, x_2, x_3) \geq 1$ everywhere on S.

EXERCISES

10.84   Prove that the tangent plane $\mathbf{x}^0 + T_{\mathbf{x}^0}(S)$ is a vector subspace of $\mathbb{E}^n$ if and only if $\mathbf{x}^0 \in T_{\mathbf{x}^0}(S)$.

10.85   Describe both the tangent space and the tangent plane to the sphere

$$S^{n-1} = \left\{ \mathbf{x} \in \mathbb{E}^n \mid \|\mathbf{x}\| = 1 \right\}$$

at the point $\mathbf{x}^0 = \left(\frac{1}{\sqrt{n}}, \frac{1}{\sqrt{n}}, \ldots, \frac{1}{\sqrt{n}} \right)$.

10.86   The *sphere* $S^3 \subset \mathbb{E}^4$ is defined by

$$S^3 = \left\{ \mathbf{x} \mid \sum_{i=1}^{4} x_i^2 = 1 \right\}.$$

Define $f : S^3 \rightarrow \mathbb{R}$ by

$$f(\mathbf{x}) = \sum_{i=1}^{4} a_i x_i$$

where a_i is a constant for each $i \in \{1, 2, 3, 4\}$. Show that the maximum and minimum values of f on S^3 are

$$\pm \sqrt{\sum_{i=1}^{4} a_i^2}.$$

10.87   The group $S\mathcal{L}(2, \mathbb{R})$ of matrices was defined in Exercise 10.81. Let f, mapping $S\mathcal{L}(2, \mathbb{R})$ to $\mathbb{R}$, be defined by

$$f(\mathbf{x}) = x_1^2 + x_2^2 + x_3^2 + x_4^2,$$

where we identify the matrix

$$X = \begin{pmatrix} x_1 & x_2 \\ x_3 & x_4 \end{pmatrix} \in S\mathcal{L}(2, \mathbb{R})$$

with the vector $\mathbf{x} = (x_1, x_2, x_3, x_4)$ constrained to the surface S in $\mathbb{E}^4$ defined by the equation

$$x_1 x_4 - x_2 x_3 = 1.$$

  a) Prove that f achieves a minimum value on S but that it has no maximum.

  b) Use the method of Lagrange multipliers to find the minimum value of f on S.

10.88   Let

$$f(\mathbf{x}) = \sum_{i=1}^{4} x_i^2$$

for all $\mathbf{x} \in \mathbb{E}^4$. Let S_1 be the surface in determined by

$$x_1 x_4 - x_2 x_3 = 1$$

and let S_2 be the surface defined by

$$\sum_{i=1}^{4} x_i = 2.$$

Let $S = S_1 \cap S_2$.

    a) Prove that $f(\mathbf{x})$ has a minimum value on S but no maximum.

    b) Find the minimum value of $f(\mathbf{x})$ on S.

10.7   TEST YOURSELF

EXERCISES

10.89   Let A and B be in $\mathcal{L}\left(\mathbb{E}^2\right)$ with matrices in the standard basis given by

$$[A] = \begin{pmatrix} 1 & 0 \\ 0 & 0 \end{pmatrix} \text{ and } [B] = \begin{pmatrix} 0 & 0 \\ 0 & 1 \end{pmatrix}.$$

Find $\|A\|$, $\|B\|$, and $\|A + B\|$.

10.90   True or False: It is possible to give an example of $X \in \mathcal{L}\left(\mathbb{E}^2\right)$ for which $\|X + X\| < \|X\| + \|X\|$.

10.91   Define $S \subset \mathcal{L}\left(\mathbb{E}^2\right)$ by letting $S = \left\{A \in \mathcal{L}\left(\mathbb{E}^2\right) \mid a_{11}^2 > 1\right\}$, where the matrix $[A] = [a_{ij}]_{2 \times 2}$. True or False: S is a *connected* subset of $\mathcal{L}\left(\mathbb{E}^2\right)$.

10.92   Give an example of $T \in \mathcal{L}\left(\mathbb{E}^2\right)$ such that in which $\left\|T^2\right\| < \|T\|^2$.

10.93   Let $\mathbf{f} \in \mathcal{C}\left(D, \mathbb{E}^2\right)$, where D is an open subset of $\mathbb{E}^2$. Suppose also that $\mathbf{f}' \in \mathcal{C}\left(D, \mathcal{L}\left(\mathbb{E}^2, \mathbb{E}^2\right)\right)$. True or False:

$$\mathbf{f}'\left(c\mathbf{x_1} + \mathbf{x_2}\right) = c\mathbf{f}'\left(\mathbf{x_1}\right) + \mathbf{f}'\left(\mathbf{x_2}\right)$$

for all $\mathbf{x}_1$ and $\mathbf{x}_2$ in D and $c \in \mathbb{R}$.

10.94   Let $\mathbf{f} : \mathbb{E}^2 \to \mathbb{E}^2$ be defined by $\mathbf{f}(\mathbf{x}) = \left(e^{x_1} \cos x_2, e^{x_1} \sin x_2\right)$. Calculate $D_{\mathbf{v}}\mathbf{f}\left(1, \frac{\pi}{6}\right)$, where $\mathbf{v} = (2, 1)$.

10.95   Suppose $\mathbf{f} : \mathbb{E}^2 \to \mathbb{E}^2$ and $\mathbf{g} : \mathbb{E}^3 \to \mathbb{E}^2$ are both differentiable. Let $\mathbf{g}(\mathbf{0}) = \mathbf{x}_0$,

$$[\mathbf{g}'(\mathbf{0})] = \begin{pmatrix} 1 & -1 & 0 \\ -2 & 3 & -2 \end{pmatrix}$$

and

$$[\mathbf{f}'(\mathbf{x}_0)] = \begin{pmatrix} 1 & 2 \\ 3 & -1 \end{pmatrix}.$$

in the standard bases. Find the matrix $[(\mathbf{f} \circ \mathbf{g})'(\mathbf{0})]$ using the standard bases.

10.96   True or give a Counterexample: If $f \in \mathcal{C}\left(\mathbb{E}^1, \mathbb{E}^1\right)$ is *locally injective*, meaning that for each $x \in \mathbb{E}^1$ there is a corresponding $r > 0$ such that f restricted to $B_r(x)$ is injective (meaning one-to-one), then f must be injective on $\mathbb{E}^1$.

10.97   Find all points $\mathbf{x}$ at which the Jacobian of $\mathbf{f}$ does not vanish, and for each such $\mathbf{x}$ a ball $B_r(\mathbf{x})$ on which $\mathbf{f}$ has a differentiable inverse, if $\mathbf{f} : \mathbb{E}^3 \to \mathbb{E}^3$ by $\mathbf{f}(\mathbf{x}) = (x_1 \cos x_2, x_1 \sin x_2, x_3)$.

10.98   Let $f \in \mathcal{C}^1\left(\mathbb{E}^4, \mathbb{E}^1\right)$ be such that the matrix

$$[f'(\mathbf{x})]_{1 \times 4} = [x_2,\ x_1,\ x_4,\ x_3].$$

Find all $\mathbf{x}_0 = (x_1, x_2, x_3, x_4)$ at which the implicit function theorem guarantees that there is a *local* $\mathcal{C}^1$ solution of the equation $f(\mathbf{x}) = f(\mathbf{x}_0)$ for x_3 in terms of the other three variables.

10.99   Give an example of a *locally* injective map $\mathbf{f} : \mathbb{E}^1 \to \mathbb{E}^2$ which is *not* injective.

10.100   Suppose that $\mathbf{f} = (f_1, f_2) \in \mathcal{C}^1\left(\mathbb{E}^3, \mathbb{E}^2\right)$ and $\mathbf{x}_0 = (a, b, c) \in \mathbb{E}^3$ is in the solution set of the equation $\mathbf{f}(\mathbf{x}) = \mathbf{0}$. Suppose this equation can be solved for $x_2 = x_2(x_1)$ and $x_3 = x_3(x_1)$, both $\mathcal{C}^1$ functions in a neighborhood of $\mathbf{x}_0$ with $x_2(a) = b$, and $x_3(a) = c$. If the matrix

$$[\mathbf{f}'(\mathbf{x}_0)] = \begin{pmatrix} 1 & 1 & 3 \\ -1 & 0 & 2 \end{pmatrix},$$

then find $x_2'(a)$ and $x_3'(a)$. (Hint: Apply the Chain Rule to differentiate

$$\mathbf{f}(x_1, x_2(x_1), x_3(x_1)) = \mathbf{0}$$

with respect to x_1 on both sides.)

CHAPTER 11

RIEMANN INTEGRATION IN EUCLIDEAN SPACE

11.1 DEFINITION OF THE INTEGRAL

For functions $f : [a, b] \rightarrow \mathbb{R}$ the Riemann integral was motivated by the desire to represent the *signed* area trapped between the graph of f and the x-axis. We saw in Chapter 3 that every Riemann integrable function on a closed, finite interval $[a, b]$ must be bounded. One can motivate integrals of bounded real-valued functions mapping a bounded domain of definition $D_f \subset \mathbb{E}^n$ to $\mathbb{R}$ in terms of signed volumes for $n = 2$, or in terms of mass (as the integral of a density function) for $n = 3$ if $f(\mathbf{x}) \geq 0$ for all $x \in D_f$. For $n > 3$ the applications are less visual though still important. The fundamental concept of the integral remains the same as it was for $n = 1$. We would like to partition the domain of f into very small pieces. In each piece we select an arbitrary evaluation point at which to evaluate f. Then we would like the finite sums of the lengths (respectively area, volume, etc) of the pieces weighted by the values of f at the chosen evaluation points to converge as the mesh of the partition approaches zero. Such a limit, if it exists, is called a Riemann integral. We saw for Riemann integrals in $\mathbb{R}^1$ that the existence of a Riemann integral could be characterized very conveniently and easily using the Darboux Criterion (Section 3.2). This will be our starting point for Riemann integration in $\mathbb{E}^n$: we will begin

Advanced Calculus: An Introduction to Linear Analysis. By Leonard F. Richardson
Copyright © 2008 John Wiley & Sons, Inc.

by defining upper and lower integrals for functions defined in $\mathbb{E}^n$ much as we did for $\mathbb{R}^1$.

The reader will recall from a first course in the calculus of several variables that functions of interest are seldom defined on a rectangular domain. Most often, useful examples of functions of several variables are defined on domains that are bounded by several intersecting curves (in the plane), or by several intersecting surfaces (in three-dimensional space). We will begin, however, by considering functions defined on a rectangular block. At the end of this section we will show how to extend Riemann integration on $\mathbb{E}^n$ to functions defined on more general bounded domains.

Definition 11.1.1 *Let* $\mathbf{a}$ *and* $\mathbf{b} \in \mathbb{E}^n$. *We define a* closed rectangular block

$$B = [\mathbf{a}, \mathbf{b}] = [a_1, b_1] \times \ldots \times [a_n, b_n]$$

to be a Cartesian product of closed finite intervals on the n *axes. Note that if for any* i *we have* $a_i > b_i$ *then* $[a_i, b_i] = \phi$, *the empty set, which is both closed and open. We denote the* interior *of a closed rectangular block as the* open rectangular block

$$B^o = (\mathbf{a}, \mathbf{b}) = (a_1, b_1) \times \ldots \times (a_n, b_n).$$

(See Exercise 11.1.) *In either case we define the* measure *of the block* B *by*

$$\mu(B) = \prod_{1 \le i \le n} (b_i - a_i),$$

if $a_i < b_i$ *for all* i. *If* $a_i \ge b_i$ *for at least one* i, *then we define* $\mu(B) = 0$.

Note that B degenerates into the empty set if $a_i > b_i$ and has at most $n - 1$ strictly positive dimensions if $a_i = b_i$ for some i. Note also that $\mu(B)$ is called length in $\mathbb{E}^1$, area in $\mathbb{E}^2$, and volume in $\mathbb{E}^3$. Next, we define what is meant by a *partition* of a block B and by the *mesh* of a partition.

Definition 11.1.2 *A partition* $\mathcal{P}$ *of a block* $B = [\mathbf{a}, \mathbf{b}]$ *is a finite set*

$$\mathcal{P} = \big\{ B_i = [\mathbf{a}_i, \mathbf{b}_i] \,\big|\, i = 1, \ldots, N \big\}$$

such that $B = \bigcup_{i=1,\ldots,N} B_i$ *and* $B_i^o \cap B_j^o = \emptyset$ *for all pairs* $\{i, j\}$ *such that* $i \ne j$. *We define the* mesh *of* $\mathcal{P}$ *by*

$$\|\mathcal{P}\| = \max_{[\mathbf{a},\mathbf{b}] \in \mathcal{P}} \sqrt{\sum_{1 \le i \le N} (b_i - a_i)^2}.$$

A partition $\mathcal{P}'$ *is called a* refinement *of a partition* $\mathcal{P}$, *provided that for each block* $B_i' \in \mathcal{P}'$, B_i' *is a subset of some block* B_j *of the partition* $\mathcal{P}$.

We continue our generalization to $\mathbb{E}^n$ of the contents of Section 3.2 by defining the *upper* and *lower sums* for a bounded function f on a closed rectangular block B.

Definition 11.1.3 *Let* $f : B \to \mathbb{R}$, *where* B *is a closed rectangular block in* $\mathbb{E}^n$. *Let* $\mathcal{P} = \{B_i \mid i = 1, \dots, N\}$ *be any partition of* B. *For each* $B_i \in \mathcal{P}$ *with* $\mu(B_i) > 0$, *we denote* $M_i = \sup\{f(\mathbf{x}) \mid \mathbf{x} \in B_i\}$ *and* $m_i = \inf\{f(\mathbf{x}) \mid \mathbf{x} \in B_i\}$. *We define the* upper sum *by*

$$U(f, \mathcal{P}) = \sum_{1 \le i \le N} M_i \mu(B_i)$$

and the lower sum *by*

$$L(f, \mathcal{P}) = \sum_{1 \le i \le N} m_i \mu(B_i)$$

where both sums are over those i *between* 1 *and* N *with* $\mu(B_i) > 0$.

Theorem 11.1.1 *Let* $\mathcal{P}$ *and* $\mathcal{P}'$ *be any two partitions of a closed rectangular block* B *and let* f *be any bounded function on* B. *Then* $L(f, \mathcal{P}') \le U(f, \mathcal{P})$.

Proof: Suppose first that a block $B_i' \subseteq B_j$ happens to occur. The suprema and infima of f on the two blocks are related as follows:

$$M_i' \le M_j \text{ and } m_i' \ge m_j.$$

Next we observe that for all $B_i' \in \mathcal{P}'$ and $B_j \in \mathcal{P}$ the intersection $B_i' \cap B_j$ is again a rectangular block, though of course it could be a block of measure zero, whether empty or not. Since $\mathcal{P}'$ is a partition of B, $\{B_i' \cap B_j \mid B_i' \in \mathcal{P}'\}$ is itself a partition of B_j. Let $M_{i,j}'$ be the supremum of f on $B_i' \cap B_j$. Then

$$\sum_i M_{i,j}' \mu \left(B_i' \cap B_j \right) \le M_j \mu(B_j).$$

Now define the *mutual refinement* $\mathcal{P}''$ of $\mathcal{P}$ and $\mathcal{P}'$ to be

$$\mathcal{P}'' = \left\{ B_i' \cap B_j \mid B_i' \in \mathcal{P}', B_j \in \mathcal{P} \right\}.$$

Then we see that

$$L(f, P') \le L(f, P'') \le U(f, P'') \le U(f, P),$$

which is what we needed to prove. ∎

Just as in Section 3.2, we see now that the family of all upper sums (for the various possible partitions) is bounded below by each lower sum, and vice versa. Hence we may make the following definition.

Definition 11.1.4 *Let* f *be any bounded function on a closed rectangular block* B. *Define the* upper integral

$$\overline{\int}_B f = \inf\{U(f, \mathcal{P}) \mid \mathcal{P} \text{ is a partition of } B\}$$

and define the lower integral

$$\underline{\int_B} f = \sup\{L(f, \mathcal{P}) \mid \mathcal{P} \text{ is a partition of } B\}.$$

Theorem 11.1.2 *For each bounded function f on a block B, we have*

$$\underline{\int_B} f \leq \overline{\int_B} f.$$

The proof is left to Exercise 11.2.

Definition 11.1.5 *We call a bounded function f integrable on a closed block B and write $f \in \mathcal{R}(B)$ if and only if $\underline{\int_B} f = \overline{\int_B} f$. If $f \in \mathcal{R}(B)$, we define*

$$\int_B f(\mathbf{x}) \, d\mathbf{x} = \underline{\int_B} f = \overline{\int_B} f.$$

Theorem 11.1.3 (Darboux Integrability Criterion)
   Let $f : B \to \mathbb{R}$ be a bounded function on a rectangular closed block B. Then $f \in \mathcal{R}(B)$ if and only for each $\epsilon > 0$ there exists a partition $\mathcal{P}$ for which $U(f, \mathcal{P}) - L(f, \mathcal{P}) < \epsilon$.

The proof is left to Exercise 11.4.

Theorem 11.1.4 *The set $\mathcal{R}[\mathbf{a}, \mathbf{b}]$ is a vector space, and the map $T : \mathcal{R}[\mathbf{a}, \mathbf{b}] \to \mathbb{R}$ defined by $T(f) = \int_B f(\mathbf{x}) \, d\mathbf{x}$ is linear.*

The proof is left to Exercise 11.5.

Theorem 11.1.5 *If $f \in \mathcal{C}[\mathbf{a}, \mathbf{b}]$, then $f \in \mathcal{R}[\mathbf{a}, \mathbf{b}]$.*

Proof: Since f is continuous on the block $B = [\mathbf{a}, \mathbf{b}]$, which is both closed and bounded, we know that f is *uniformly continuous* on B. Thus if $\epsilon > 0$, there exists a $\delta > 0$ such that $\|\mathbf{x} - \mathbf{x}'\| < \delta$ implies that

$$|f(\mathbf{x}) - f(\mathbf{x}')| < \frac{\epsilon}{2\mu(B)},$$

where we assume $\mu(B) > 0$. Now if we partition B with a finite set of blocks $\mathcal{P} = \{B_i \mid i = 1, \ldots, p\}$ with $\|\mathcal{P}\| < \delta$, it follows that on each block B_i we have

$$|M_i - m_i| \leq \frac{\epsilon}{2\mu(B)}.$$

Hence

$$U(f, \mathcal{P}) - L(f, \mathcal{P}) \leq \frac{\epsilon}{2} < \epsilon.$$

It follows from the Darboux Criterion that $f \in \mathcal{R}(B)$.   ∎

We turn our attention now to the definition of the Riemann integral for functions defined on *any* bounded subset $D_f \subset \mathbb{E}^n$. Then the domain D_f is contained in some sufficiently large rectangular block $[\mathbf{a}, \mathbf{b}]$. We would *like to make the following definition*.

Definition 11.1.6 *Suppose that the domain D_f of a function f is contained in some sufficiently large rectangular block $[\mathbf{a}, \mathbf{b}]$. We will extend the definition of f so that $f(\mathbf{x}) = 0$ at each point $\mathbf{x} \in [\mathbf{a}, \mathbf{b}] \setminus D_f$. We define*

$$\int_{D_f} f(\mathbf{x})\, d\mathbf{x} = \int_{[\mathbf{a},\mathbf{b}]} f(\mathbf{x})\, d\mathbf{x},$$

provided that this integral exists.

However, for this definition to make sense, we must show that the definition is *independent of the choice of block* $[\mathbf{a}, \mathbf{b}] \supseteq D_f$. For this purpose, it suffices to prove the following theorem.

Theorem 11.1.6 *Let f be any bounded real-valued function on a bounded domain $D_f \subset \mathbb{E}^n$. Suppose $D_f \subseteq [\mathbf{a}, \mathbf{b}]$ and $D_f \subseteq [\mathbf{a}', \mathbf{b}']$. Extend the definition of f to be zero identically on the complement of D_f in each block. Then*

$$\overline{\int}_{[\mathbf{a},\mathbf{b}]} f = \overline{\int}_{[\mathbf{a}',\mathbf{b}']} f \text{ and } \underline{\int}_{[\mathbf{a},\mathbf{b}]} f = \underline{\int}_{[\mathbf{a}',\mathbf{b}']} f.$$

Proof: Let $[\mathbf{A}, \mathbf{B}]$ be any closed rectangular block that contains both $[\mathbf{a}, \mathbf{b}]$ and $[\mathbf{a}', \mathbf{b}']$. It will suffice to prove that

$$\overline{\int}_{[\mathbf{a},\mathbf{b}]} f = \overline{\int}_{[\mathbf{A},\mathbf{B}]} f \text{ and } \underline{\int}_{[\mathbf{a},\mathbf{b}]} f = \underline{\int}_{[\mathbf{A},\mathbf{B}]} f.$$

The two arguments are so similar that we will present a proof of only the first. Let $\epsilon > 0$. There exists a partition $\mathcal{P}$ of $[\mathbf{A}, \mathbf{B}]$ for which

$$U(f, \mathcal{P}) - \overline{\int}_{[\mathbf{A},\mathbf{B}]} f < \frac{\epsilon}{3}.$$

Without disturbing this inequality, we can refine $\mathcal{P}$ in such a way that no block $B_k \in \mathcal{P}$ intersects both the interior of $[\mathbf{a}, \mathbf{b}]$ and the interior of its complement. Thus the collection $\mathcal{P}'$ consisting of those blocks of $\mathcal{P}$ that lie in the smaller block $[\mathbf{a}, \mathbf{b}]$ is a partition of $[\mathbf{a}, \mathbf{b}]$ that has the property that $U(f, \mathcal{P}') - \overline{\int}_{[\mathbf{a},\mathbf{b}]} f < \frac{\epsilon}{3}$.

In order to establish that

$$\left| \overline{\int}_{[\mathbf{a},\mathbf{b}]} f - \overline{\int}_{[\mathbf{A},\mathbf{B}]} f \right| < \epsilon,$$

it will suffice to show that for $\mathcal{P}$ suitably selected we have

$$|U(f,\mathcal{P}) - U(f,\mathcal{P}')| < \frac{\epsilon}{3}.$$

Since f vanishes on the complement of $[\mathbf{a}, \mathbf{b}]$, the difference under consideration consists of a sum over only those blocks of $\mathcal{P}$ that are outside $[\mathbf{a}, \mathbf{b}]$ but have one face lying on a face of $[\mathbf{a}, \mathbf{b}]$. Thus

$$|U(f,\mathcal{P}) - U(f,\mathcal{P}')| \le \|f\|_{\sup} A \|\mathcal{P}\|,$$

where A denotes the $(n-1)$-dimensional surface area of the block $[\mathbf{a}, \mathbf{b}]$. By refining $\mathcal{P}$ we can insure that $\|\mathcal{P}\|$ is small enough to accomplish this goal. ∎

EXERCISES

11.1 † Prove that $B^o = (\mathbf{a}, \mathbf{b})$ is the interior of the rectangular block $B = [\mathbf{a}, \mathbf{b}]$, where *interior* is defined in Exercise 8.26.

11.2 Prove that

$$\underline{\int_B} f \le \overline{\int_B} f$$

for all bounded functions $f : B \to \mathbb{R}$ defined on a closed rectangular block B.

11.3 † Let f and g be bounded functions on a rectangular block B, and let $c \in \mathbb{R}$.
  a) Prove that $\overline{\int_B}(f + g) \le \overline{\int_B} f + \overline{\int_B} g$.
  b) Prove that $\underline{\int_B}(f + g) \ge \underline{\int_B} f + \underline{\int_B} g$.
  c) Compare $\overline{\int_B} cf$ and $\underline{\int_B} cf$ with $\overline{\int_B} f$ and $\underline{\int_B} f$. Take note of the effect of c being either positive or negative.

11.4 † Prove Theorem 11.1.3. (Hint: Apply Definition 11.1.5 directly. Compare with Theorem 3.2.4.)

11.5 † The following parts will prove Theorem 11.1.4, which can be compared with Theorem 3.1.2. Apply the parts of Exercise 11.3 to show each of the following parts.
  a) $\mathcal{R}[\mathbf{a}, \mathbf{b}]$ is closed under addition.
  b) $\mathcal{R}[\mathbf{a}, \mathbf{b}]$ is closed under scalar multiplication.
  c) The map $T : \mathcal{R}[\mathbf{a}, \mathbf{b}] \to \mathbb{R}$ defined by $T(f) = \int_B f(\mathbf{x})\, d\mathbf{x}$ is linear.

11.6 Let

$$f(\mathbf{x}) = \begin{cases} 1 & \text{if } x_1 = 1, \\ 0 & \text{if } x_1 \in [0, 2] \setminus \{1\} \end{cases}$$

be a function defined on a block $B \subset \mathbb{E}^2$. Find a partition $\mathcal{P}$ of $B = [0, 2] \times [0, 1]$ for which

$$U(f,\mathcal{P}) - L(f,\mathcal{P}) < \frac{1}{8}.$$

11.7   † Let f be any real-valued function on a rectangular block $B \subseteq \mathbb{E}^n$. Define

$$f^+(\mathbf{x}) = \begin{cases} f(\mathbf{x}) & \text{if } f(\mathbf{x}) \geq 0, \\ 0 & \text{if } f(\mathbf{x}) < 0 \end{cases}$$

and let

$$f^-(\mathbf{x}) = \begin{cases} -f(\mathbf{x}) & \text{if } f(\mathbf{x}) < 0, \\ 0 & \text{if } f(\mathbf{x}) \geq 0. \end{cases}$$

for all $\mathbf{x} \in B$. Prove that

$$f(\mathbf{x}) = f^+(\mathbf{x}) - f^-(\mathbf{x}) \text{ and } |f(\mathbf{x})| = f^+(\mathbf{x}) + f^-(\mathbf{x})$$

for all $\mathbf{x} \in D$. (Hint: Just check the cases based on the sign of $f(\mathbf{x})$.)

11.8   † Suppose $f \in \mathcal{R}[\mathbf{a}, \mathbf{b}]$. Prove: f^+ and f^- are in $\mathcal{R}[\mathbf{a}, \mathbf{b}]$. Hint: Show that

$$U(f^+, \mathcal{P}) - L(f^+, \mathcal{P}) \leq U(f, \mathcal{P}) - L(f, \mathcal{P}).$$

11.9   If $f \in \mathcal{R}[\mathbf{a}, \mathbf{b}]$, prove $|f| \in \mathcal{R}[\mathbf{a}, \mathbf{b}]$. (Hint: Use Exercises 11.7 and 11.8 above.)

11.10   Give an example of f such that $|f| \in \mathcal{R}[\mathbf{a}, \mathbf{b}]$ yet $f \notin \mathcal{R}[\mathbf{a}, \mathbf{b}]$.

11.11   † If $f \in \mathcal{R}[\mathbf{a}, \mathbf{b}]$, prove that

$$\left| \int_{[\mathbf{a}, \mathbf{b}]} f(\mathbf{x}) \, dx \right| \leq \int_{[\mathbf{a}, \mathbf{b}]} |f(\mathbf{x})| \, dx.$$

(Hint: Write the left side by expressing $f = f^+ - f^-$ and use the fact that f^+ and f^- are both nonnegative functions.)

11.12   Let the linear map $T : \mathcal{R}[\mathbf{a}, \mathbf{b}] \to \mathbb{R}$ be defined by $T(f) = \int_{[\mathbf{a}, \mathbf{b}]} f(\mathbf{x}) \, dx$. Prove that T is a *bounded* linear functional.

11.13   Suppose $f \in \mathcal{C}(\mathbf{a}, \mathbf{b})$ and also that f is bounded on $[\mathbf{a}, \mathbf{b}]$. Prove that f is Riemann integrable on $[\mathbf{a}, \mathbf{b}]$.

11.14   Let $f : [(0, 0), (1, 1)] \to \mathbb{R}$ be defined by

$$f(\mathbf{x}) = \begin{cases} \frac{1}{2} \sin \frac{\pi}{x_1 x_2} & \text{if } 0 < x_i \leq 1, i = 1, 2, \\ 0 & \text{if } x_1 x_2 = 0. \end{cases}$$

Prove that f is Riemann integrable on the closed rectangular box

$$[(0, 0), (1, 1)] = [0, 1]^{\times 2}.$$

(See Fig. 11.1.) Show also that f is not continuous at any point in the uncountable set for which $x_1 x_2 = 0$.

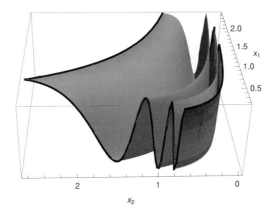

Figure 11.1 $f(\mathbf{x}) = \frac{1}{2} \sin \frac{\pi}{x_1 x_2}$.

11.15   Let $\mathbf{a} = (0,0)$ and $\mathbf{b} = (1,1) \in \mathbb{E}^2$. Let $f : [\mathbf{a}, \mathbf{b}] \to \mathbb{R}$ by

$$f(\mathbf{x}) = \begin{cases} 1 & \text{if } x_1 \in \mathbb{Q} \text{ and } x_2 \in \mathbb{Q}, \\ 0 & \text{otherwise.} \end{cases}$$

Prove: $f \notin \mathcal{R}[\mathbf{a}, \mathbf{b}]$.

11.2 LEBESGUE NULL SETS AND JORDAN NULL SETS

Even in the study of Riemann integrability on the real line, it was difficult to see what exactly is the connection between the Riemann integrability of f and continuity of f. We know that every continuous function is Riemann integrable, but so is every step function and every monotone function, and even every function of bounded variation. In Exercise 3.14 we saw a Riemann integrable function with a countably infinite set of discontinuities. And in Exercise 7.10 we saw a function $f \in \mathcal{R}[0,1]$ that had a discontinuity at every rational point of $[0,1]$. Thus we saw that a function can be Riemann integrable even with its points of discontinuity constituting a set that is both countably infinite and dense. Finally, in Exercise 11.14 we saw a function $f \in \mathcal{R}[(0,0),(1,1)]$ that was discontinuous at each point in the unit square for which either $x_1 = 0$ or $x_2 = 0$. This set of points of discontinuity is uncountably infinite. It is therefore with good reason that we say the relationship between continuity and Riemann integrability appears to this point in our study to be mysterious. The precise formulation of a necessary and sufficient criterion for Riemann integrability of a bounded function on a bounded domain in terms of continuity was established early in the twentieth century by the French mathematician Henri Lebesgue. Lebesgue's name is associated most often with a more refined concept of integration called the Lebesgue Integral, and related to this is the concept of the Lebesgue measure of a set. We will not delve into these graduate-level topics in this course. But both the concept of a Lebesgue null set and the related concept of a Jordan null set are accessible to us at this point and they are essential to understanding Riemann integrability in terms of continuity.

Definition 11.2.1 *A set $S \subseteq \mathbb{E}^n$ is called a* Lebesgue null set *if and only if for each $\epsilon > 0$ there exists a* countable *family of open blocks B_i^o such that*

$$S \subseteq \bigcup_{i \in \mathbb{N}} B_i^o \ \ and \ \ \sum_{i=1}^{\infty} \mu(B_i) < \epsilon.$$

In Example 1.15 we saw that the set of rational numbers is a Lebesgue null set, because we showed there how to cover $\mathbb{Q}$ with a countable sequence of open intervals with the sum of the lengths being less than and preassigned $\epsilon > 0$. This argument can be generalized to any countable set, as the reader will see in the Exercises.

Closely related to the concept of a Lebesgue null set is that of a Jordan null set, which we define next.

Definition 11.2.2 *A set $S \subseteq \mathbb{E}^n$ is called a* Jordan null set *if and only if for each $\epsilon > 0$ there exists a* finite *family of open blocks B_i^o such that*

$$S \subseteq \bigcup_{i=1}^{p} B_i^o \ \ and \ \ \sum_{i=1}^{p} \mu(B_i) < \epsilon.$$

In the Exercises the reader will show that every finite set is a Jordan null set, but that the set of rational numbers is not a Jordan null set. The following theorem establishes a useful relationship between Lebesgue and Jordan null sets.

Theorem 11.2.1 *If a compact set $K \subset \mathbb{E}^n$ is a Lebesgue null set, then K is a Jordan null set.*

Proof: Let $\epsilon > 0$. By hypothesis, there exists a countable family of open blocks B_n^o such that $K \subseteq \bigcup_{n \in \mathbb{N}} B_n^o$ and such that $\sum_{n \in \mathbb{N}} \mu(B_n) < \epsilon$. Since K is compact, there exists a finite subcollection $\{B_{j_1}, \ldots, B_{j_p}\}$ the union of which still covers K. But then $\sum_{k=1}^{p} \mu(B_{j_k}) < \epsilon$.   ∎

We will apply the concepts of Lebesgue and Jordan null sets especially to certain sets of points closely related the concept of continuity of a function. Specifically, we will need the concept of the *oscillation* of a function at a point. Intuitively, oscillation at a jump discontinuity should be the height of the jump, and oscillation equal to zero should mean continuity.

Definition 11.2.3 *Let* $f : D \to \mathbb{R}$, *where* $D \subseteq \mathbb{E}^n$. *We define the* oscillation *of* f *at a point* $\mathbf{x} \in D$ *to be*

$$o(f, \mathbf{x}) = \lim_{r \to 0+} \left(\sup_{\mathbf{y} \in D \cap B_r(\mathbf{x})} f(\mathbf{y}) - \inf_{\mathbf{y} \in D \cap B_r(\mathbf{x})} f(\mathbf{y}) \right).$$

It is left to the reader to show that this limit exists in Exercise 11.25.

Theorem 11.2.2 *Let* $f : D \to \mathbb{R}$, *where* $D \subseteq \mathbb{E}^n$. *Then* f *is continuous at* $\mathbf{x} \in D$ *if and only if* $o(f, \mathbf{x}) = 0$.

Proof: First, suppose f is continuous at $\mathbf{x}$. Then if $\epsilon > 0$ there exists $\delta > 0$ such that $0 < r < \delta$ implies

$$\sup_{\mathbf{y} \in D \cap B_r(\mathbf{x})} f(\mathbf{y}) - \inf_{\mathbf{y} \in D \cap B_r(\mathbf{x})} f(\mathbf{y}) \leq \epsilon.$$

Since this is true for all $\epsilon > 0$, it follows that $o(f, \mathbf{x}) = 0$.

Next, suppose that $o(f, \mathbf{x}) = 0$. We must show that f is continuous at $\mathbf{x}$. Let $\epsilon > 0$. Then there exists $\delta > 0$ such that $0 < r < \delta$ implies

$$\sup_{\mathbf{y} \in D \cap B_r(\mathbf{x})} f(\mathbf{y}) - \inf_{\mathbf{y} \in D \cap B_r(\mathbf{x})} f(\mathbf{y}) < \epsilon.$$

It follows that for all $\mathbf{y} \in D \cap B_r(\mathbf{x})$ we have $|f(\mathbf{y}) - f(\mathbf{x})| < \epsilon$. Thus f is continuous at $\mathbf{x}$.   ∎

The following theorem will be useful in establishing Lebesgue's Criterion for Riemann integrability.

Theorem 11.2.3 *Let* $f : D \to \mathbb{R}$, *where* $D \subseteq \mathbb{E}^k$ *and let* $\delta > 0$. *Then the set* $E = \{\mathbf{x} \in D \mid o(f, \mathbf{x}) \geq \delta\}$ *is a closed set.*

Proof: Suppose $\mathbf{x}_n \in E$ for all $n \in \mathbb{N}$ and suppose $\mathbf{x}_n \to \mathbf{x}$ as $n \to \infty$. We need to prove that $\mathbf{x} \in E$. For each $\epsilon > 0$ there exists $\mathbf{x}_n \in B_\epsilon(\mathbf{x})$, and there exists $r > 0$ such that $B_r(\mathbf{x}_n) \subseteq B_\epsilon(\mathbf{x})$. Also,

$$\sup_{\mathbf{y} \in D \cap B_r(\mathbf{x}_n)} f(\mathbf{y}) - \inf_{\mathbf{y} \in D \cap B_r(\mathbf{x}_n)} f(\mathbf{y}) \geq \delta.$$

However, this implies that

$$\sup_{\mathbf{y} \in D \cap B_\epsilon(\mathbf{x})} f(\mathbf{y}) - \inf_{\mathbf{y} \in D \cap B_\epsilon(\mathbf{x})} f(\mathbf{y}) \geq \delta,$$

which completes the proof since this is true for all $\epsilon > 0$.   ∎

The following theorem is useful when studying the Jacobian Theorem for changes of variables in Riemann integrals on $\mathbb{E}^n$.

Theorem 11.2.4 *Let $\mathcal{O} \subset \mathbb{E}^n$ be an open set, B a closed block in $\mathcal{O}$ and let ϕ be in $\mathcal{C}^1(\mathcal{O}, \mathbb{E}^n)$.*
(i) If $S \subset B^\circ$ is a Jordan null set, then $\phi(S)$ is also a Jordan null set.
(ii) If $S \subset B^\circ$ is a Lebesgue null set, then $\phi(S)$ is also a Lebesgue null set.

Proof: Denote $M = \|\phi'\|_{\sup} < \infty$ and let $\epsilon > 0$. We begin by proving item *(i)*. By hypothesis, there exists a finite sequence of closed blocks $B_1, \ldots, B_N$ such that

$$S \subset \bigcup_{k=1}^{N} B_k^\circ \text{ and } \sum_{k=1}^{N} \mu\left(B_k\right) < \frac{\epsilon}{2M^n n^{\frac{n}{2}}}.$$

Note that the intersection of two rectangular blocks (with edges parallel to the axes) must again be a rectangular block. Thus without loss of generality we can assume that each block $B_k \subseteq B$.

It is easy to see that any closed rectangular block R is contained in a union of finitely many cubes for which the sum of the volumes of the cubes is less than twice $\mu(R)$. This is true even though the edges of R need not be commensurable, because we can slightly enlarge the edges of R to make all the (extended) edges commensurable. Thus, without loss of generality, we can assume that each block B_k used above is actually a cube. If we denote by D_k the diagonal measurement of B_k and s_k its edge-length, then $D_k = s_k \sqrt{n}$ and

$$\mu(B_k) = \frac{D_k^n}{n^{\frac{n}{2}}}.$$

On the other hand, by Theorem 10.3.2, each individual coordinate function of the image $\phi(B_k)$ is limited to at most an interval of length MD_k. Hence $\phi(B_k)$ is contained in a cube B_k' such that $\mu(B_k') \le M^n n^{\frac{n}{2}} \mu(B_k)$. By slightly enlarging the cubes B_k' we can assure that $\phi(B_k) \subset (B_k')^\circ$ and that

$$\mu\left(B_k'\right) \le 2M^n n^{\frac{n}{2}} \mu\left(B_k\right).$$

Thus

$$\phi(S) \subseteq \bigcup_{k=1}^{N} \phi\left(B_k^\circ\right) \subset \bigcup_{k=1}^{N} \left(B_k'\right)^\circ \text{ and } \sum_{k=1}^{N} \mu\left(B_k'\right) < \epsilon.$$

Item (ii) is very similar and is left as an exercise for the reader.   ∎

EXERCISES

11.16   † Suppose $E_k \subset \mathbb{E}^n$ is a Lebesgue null set, for each $k \in \mathbb{N}$. Prove that $\bigcup_{k \in \mathbb{N}} E_k$ is also a Lebesgue null set. (Hint: See Exercise 1.88.)

11.17   Let $S \subset \mathbb{E}^n$ be any countable set. Show that S is a Lebesgue null set. (Hint: See Example 1.15.)

11.18   Prove that the union of finitely many Jordan null sets is a Jordan null set. Give an example to show that the union of countably many Jordan null sets need not be a Jordan null set.

11.19   Show that if $F \subset \mathbb{E}^n$ is a finite set, then F is a Jordan null set.

11.20   Prove that the set of rational numbers in the interval $[0, 1]$ is not a Jordan null set, but that it is a Lebesgue null set.

11.21   Give an example of a countable subset of $\mathbb{E}^2$ that is not a Jordan null set.

11.22   † Prove that every Jordan null set is a Lebesgue null set.

11.23   Prove that every *convergent* sequence $\mathbf{x}_n$ is a Jordan null set.

11.24   Prove that an interval I of strictly positive length is never a Lebesgue null set.

11.25   Prove that the limit in Definition 11.2.3 exists. (Hint: Show that the function of r in the definition is monotone increasing on $(0, \infty)$. Compare with Exercise 2.28.)

11.26   Prove that $o(f, \mathbf{x}) \geq 0$ for all $\mathbf{x}$ and for all $f : D \to \mathbb{R}, D \subseteq \mathbb{E}^n$.

11.27   † Prove that the set of points of discontinuity of $f : D \to \mathbb{R}, D \subseteq \mathbb{E}^n$, is the

$$\bigcup_{n \in \mathbb{N}} \left\{ \mathbf{x} \in D \,\middle|\, o(f, \mathbf{x}) \geq \frac{1}{n} \right\}.$$

11.28   Prove that every subset of a Lebesgue null set is a Lebesgue null set.

11.29   † Suppose that the set E of points of discontinuity of $f : D \to \mathbb{R}, D \subseteq \mathbb{E}^n$, is a Lebesgue null set, and suppose that D is a *bounded* subset of $\mathbb{E}^n$. Let

$$E_k = \left\{ \mathbf{x} \in D \,\middle|\, o(f, \mathbf{x}) \geq \frac{1}{k} \right\}.$$

Prove that E_k is a Jordan null set. (Hint: Use Theorem 11.2.3.)

11.3   LEBESGUE'S CRITERION FOR RIEMANN INTEGRABILITY

The main objective of this section is to prove the following theorem.

Theorem 11.3.1 (Lebesgue's Criterion for Riemann Integrability) *Let $f : B \to \mathbb{R}$ be any bounded function defined on a closed, bounded rectangular block $B \subset \mathbb{E}^n$. Let $S = \{\mathbf{x} \in B \mid f$ is discontinuous at $\mathbf{x}\}$. Then $f \in \mathcal{R}(B)$ if and only if S is a Lebesgue null set.*

Proof: First suppose that $f \in \mathcal{R}(B)$. We will prove that S is a Lebesgue null set. By Exercise 11.27 we know that $S = \bigcup_{k \in \mathbb{N}} S_k$, where

$$S_k = \left\{ \mathbf{x} \in B \,\middle|\, o(f, \mathbf{x}) \geq \frac{1}{k} \right\}.$$

By Exercise 11.16, we see that it suffices to prove for each $k \in \mathbb{N}$ that S_k is a Lebesgue null set. By Exercise 11.22, it suffices to prove that S_k is a Jordan null set.

Let $\epsilon > 0$. There exists a partition $P = \{B_1, \ldots, B_p\}$ of B for which

$$U(f, P) - L(f, P) < \frac{\epsilon}{k}.$$

Denote

$$P' \overset{(\text{def})}{=} \bigcup_{i=1}^{p} B_i^o$$

and observe that $B \setminus P'$ is a Jordan null set, being a union of degenerate blocks, and so is $S_k \setminus P'$. Thus it suffices to prove that

$$S_k' \overset{(\text{def})}{=} S_k \cap P'$$

is a Jordan null set. If $\mathbf{x} \in S_k'$, then $\mathbf{x}$ lies in the interior of exactly one block B_i of the partition $\mathcal{P}$, and on that block we have $M_i - m_i \geq o(f, \mathbf{x})$. Hence

$$U(f, \mathcal{P}) - L(f, \mathcal{P}) \geq \frac{1}{k} \sum_{B_i \cap S_k' \neq \emptyset} \mu(B_i).$$

It follows that

$$\sum_{B_i \cap S_k' \neq \emptyset} \mu(B_i) < \epsilon,$$

and thus S_k' is a Jordan null set as claimed.

For the opposite direction of implication in the theorem, we suppose that S is a Lebesgue null set, and we must prove that $f \in \mathcal{R}(B)$. By hypothesis, each set S_k is a Lebesgue null set. By Exercise 11.29, each S_k is a Jordan null set. Let $\epsilon > 0$. We need to show there exists a partition $\mathcal{P}$ of B such that

$$U(f, \mathcal{P}) - L(f, \mathcal{P}) < \epsilon.$$

Select $k \in \mathbb{N}$ such that

$$k > \frac{4\mu(B)}{\epsilon}.$$

Without loss of generality, suppose $\|f\|_{\sup} > 0$. Since S_k is a Jordan null set, there exists a finite set of blocks $B_1, \ldots, B_l$ such that $B_i^o \cap B_j^o = \emptyset$ if $i \neq j$

$$S_k \subseteq \bigcup_{i=1}^{l} B_i^o \quad \text{and} \quad \sum_{i=1}^{l} \mu(B_i) < \frac{\epsilon}{4\|f\|_{\sup}}.$$

Hence

$$\sum_{i=1}^{l} (M_i - m_i)\mu(B_i) < \frac{\epsilon}{2}.$$

Let

$$K = B \setminus \bigcup_{i=1}^{l} B_i^o$$

be a compact set. For each $\mathbf{x} \in K$ we have

$$o(f, \mathbf{x}) < \frac{1}{k}.$$

Thus there exists an open block $B_{\mathbf{x}}^o$ containing $\mathbf{x}$ such that on this block we have

$$M_{\mathbf{x}} - m_{\mathbf{x}} < \frac{2}{k}.$$

There exists a finite subcollection $B_{l+1}, \ldots, B_N$ that covers K, by the Heine–Borel Theorem. Without loss of generality, we can now intersect each B_j with B itself to insure that each $B_j \subseteq B$, and we can refine this set to make a partition, with $B_{j_1}^o \cap B_{j_2}^o = \emptyset$ for all $j_1 \neq j_2$ and to insure that each $B_j \subseteq K$. And

$$\sum_{j=l+1}^{N} (M_j - m_j)\mu(B_j) < \frac{2}{k}\mu(B) < \frac{\epsilon}{2}.$$

Hence $U(f, \mathcal{P}) - L(f, \mathcal{P}) < \epsilon$, so $f \in \mathcal{R}(B)$.   ∎

EXERCISES

11.30   For each of the following functions on $\mathbb{E}^1$, determine whether or not the function is Riemann integrable on $[0, 1]$, and prove your conclusion.

  a) Let $f(x) = 1_{\mathbb{Q} \cap [0,1]}$, the indicator of the rational numbers in the unit interval.

  b) Let
  $$f(x) = \begin{cases} \sin \frac{\pi}{x} & \text{if } x \in (0, 1], \\ 0 & \text{if } x = 0. \end{cases}$$

  c) Let
  $$f(x) = \begin{cases} (-1)^n & \text{if } x \in \left(\frac{1}{n+1}, \frac{1}{n} \right], n \in \mathbb{N}, \\ 0 & \text{if } x = 0. \end{cases}$$

  (See Fig. 11.2, in which $[1/x]$ denotes the *integer part* of $1/x$.)

11.31   For each of the following functions on $\mathbb{E}^2$, determine whether or not the function is Riemann integrable on $[0, 1]^{\times 2}$, the unit square, and prove your conclusion.

  a) Let $f(\mathbf{x}) = 1_{\mathbb{Q}^{\times 2} \cap [0,1]^{\times 2}}(\mathbf{x})$, the indicator of the rational number pairs in the unit square.

  b) Let
  $$f(\mathbf{x}) = \begin{cases} \sin \frac{\pi}{x_1 x_2} & \text{if } x_i \in (0, 1], i = 1, 2, \\ 0 & \text{if } x_1 x_2 = 0. \end{cases}$$

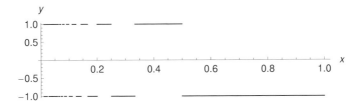

Figure 11.2 $(-1)^{\lfloor 1/x \rfloor}$. Note the ambiguity of the plot near $x = 0$.

(See Fig. 11.1.)

11.32   Prove that the product of two Riemann integrable functions on $B \subset E^n$ is Riemann integrable.

11.33   † Let B be a closed rectangular block and $S \subset B$ a Jordan null set.
  a) If h is a bounded function that is zero everywhere on B except on S, Prove: $h \in \mathcal{R}(B)$ and

$$\int_B h(\mathbf{x}) \, dx = 0.$$

  b) Suppose f and $g \in \mathcal{R}(B)$ and suppose $f(\mathbf{x}) = g(\mathbf{x})$ for all $\mathbf{x} \in B \setminus S$. Prove

$$\int_B f(\mathbf{x}) \, dx = \int_B g(\mathbf{x}) \, dx.$$

  c) Let $P = \{B_j \mid j = 1, \ldots, n\}$ be a partition of B. Prove that

$$\int_B f(\mathbf{x}) \, dx = \sum_{j=1}^{n} \int_{B_j} f(\mathbf{x}) \, dx.$$

  $\left(\text{Hint: Let } f_j(\mathbf{x}) = f(\mathbf{x}) 1_{B_j}(\mathbf{x}) \text{ and consider } \sum_{j=1}^{n} f_j.\right)$

11.34   † Suppose $f(\mathbf{x}) \geq 0$ for all $\mathbf{x} \in B \subset E^n$. Suppose $f \in \mathcal{R}(B)$, where B is a bounded rectangular block. Suppose $\int_B f(\mathbf{x}) \, dx = 0$. Prove that

$$S = \{\mathbf{x} \in B \mid f(\mathbf{x}) > 0\}$$

is a Lebesgue null set. (Hint: Show that

$$S_n = \left\{\mathbf{x} \in B \,\middle|\, f(\mathbf{x}) \geq \frac{1}{n}\right\}$$

is a Jordan null set.)

11.35   Let f and $g \in \mathcal{R}(B)$ where $B = [0, 1]^{\times n} \subset E^n$. Extend the domain of f, making $f(\mathbf{x})$ identically zero off B. Prove that the *convolution*

$$f \star g(\mathbf{x}) \stackrel{\text{(def)}}{=} \int_B f(\mathbf{x} - \mathbf{y}) g(\mathbf{y}) \, dy$$

exists for all $\mathbf{x} \in \mathbb{E}^n$. (Hint: Use Exercise 11.32.)

11.36   A set $E \subseteq B$, a bounded closed block in $\mathbb{E}^n$, is called *Jordan measurable* provided that $1_E \in \mathcal{R}(B)$, where the *indicator function* 1_E is defined on page 83. If E is Jordan measurable, we define its *Jordan content* to be

$$\nu(E) = \int_B 1_E(\mathbf{x}) \, d\mathbf{x}.$$

Prove: A bounded set $E \subset \mathbb{E}^n$ is Jordan measurable if and only if its boundary, ∂E, is a Lebesgue null set.

11.37   Let $B \subset \mathbb{E}^n$ be a closed, bounded rectangular block, and let $E \subset B$ be a Jordan null set. Prove that E is Jordan measurable, as defined in Exercise 11.36, and that the Jordan content $\nu(E) = 0$.

11.38   Let $B \subset \mathbb{E}^n$ be a closed, bounded rectangular block, and let $E \subset B$ be a Jordan measurable set for which the Jordan content $\nu(E) = 0$. Prove that E is a Jordan null set.

11.4   FUBINI'S THEOREM

In an elementary course in the calculus of functions of several real variables, students learn to calculate double and triple integrals by the method of iteration. A theorem attributed to Fubini may have been mentioned without proof, stating that if $B = [a_1, b_1] \times [a_2, b_2]$, then

$$\iint_B f(x, y) d(x, y) = \int_{a_1}^{b_1} \left(\int_{a_2}^{b_2} f(x, y) dy \right) dx$$

$$= \int_{a_2}^{b_2} \left(\int_{a_1}^{b_1} f(x, y) \, dx \right) dy.$$

Although this statement is correct for *continuous* functions on B, many Riemann integrable functions are far from continuous. The following example illustrates one of the difficulties that can be encountered. Specifically, it is not clear that for any given fixed value of x the one-variable integral $\int_{a_2}^{b_2} f(x, y) dy$ will exist. If it exists, there is still the question of whether or not the outer integral with respect to x exists.

■ **EXAMPLE 11.1**

Let $f : [0, 1]^{\times 2} \to \mathbb{R}$ be defined by

$$f(x_1, x_2) = \begin{cases} 0 & \text{if } x_1 \neq 0.5, \\ 1 & \text{if } x_2 \in \mathbb{Q} \text{ and } x_1 = 0.5, \\ 0 & \text{if } x_2 \notin \mathbb{Q} \text{ and } x_1 = 0.5. \end{cases}$$

We see that

$$\int_0^1 f(0.5, x_2) \, dx_2$$

does not exist. We see also that $\int_0^1 \left(\int_0^1 f(x_1, x_2) \, dx_1 \right) dx_2$ exists and equals 0. Moreover, Lebesgue's Criterion tells us that $\iint_{[0,1] \times 2} f(\mathbf{x}) \, d\mathbf{x}$ does exist and is easily seen to be equal to 0. Phenomena of this type motivate the use of upper and lower integrals in the following statement of Fubini's Theorem.

Theorem 11.4.1 (Fubini's Theorem) *Let* $[\mathbf{a}, \mathbf{b}]$ *be a closed rectangular block in* $\mathbb{E}^m$ *and let* $[\mathbf{c}, \mathbf{d}]$ *be a closed rectangular block in* $\mathbb{E}^n$, *so that* $B = [\mathbf{a}, \mathbf{b}] \times [\mathbf{c}, \mathbf{d}]$ *is a closed rectangular block in* $\mathbb{E}^{m+n}$. *Let* $f : B \to \mathbb{R}$ *be a bounded function. Then, denoting* $\mathbf{x} \in \mathbb{E}^m$ *and* $\mathbf{y} \in \mathbb{E}^n$, *we have*

i.

$$\underline{\int_B} f(\mathbf{x}, \mathbf{y}) \, d(\mathbf{x}, \mathbf{y}) \overset{(i)}{\leq} \int_{\mathbf{a}}^{\mathbf{b}} \overline{\left(\int_{\mathbf{c}}^{\mathbf{d}} f(\mathbf{x}, \mathbf{y}) \, d\mathbf{y} \right)} \, d\mathbf{x}$$

$$\leq \overline{\int_{\mathbf{a}}^{\mathbf{b}}} \left(\int_{\mathbf{c}}^{\mathbf{d}} f(\mathbf{x}, \mathbf{y}) \, d\mathbf{y} \right) \, d\mathbf{x}$$

$$\overset{(ii)}{\leq} \overline{\int_B} f(\mathbf{x}, \mathbf{y}) d(\mathbf{x}, \mathbf{y}),$$

ii.

$$\underline{\int_B} f(\mathbf{x}, \mathbf{y}) \, d(\mathbf{x}, \mathbf{y}) \leq \int_{\mathbf{a}}^{\mathbf{b}} \underline{\left(\int_{\mathbf{c}}^{\mathbf{d}} f(\mathbf{x}, \mathbf{y}) \, d\mathbf{y} \right)} \, d\mathbf{x}$$

$$\leq \overline{\int_{\mathbf{a}}^{\mathbf{b}}} \left(\underline{\int_{\mathbf{c}}^{\mathbf{d}}} f(\mathbf{x}, \mathbf{y}) \, d\mathbf{y} \right) \, d\mathbf{x}$$

$$\leq \overline{\int_B} f(\mathbf{x}, \mathbf{y}) d(\mathbf{x}, \mathbf{y}).$$

iii. If $f \in \mathcal{R}(B)$, *then we have*

$$\int_B f(\mathbf{x}, \mathbf{y}) \, d(\mathbf{x}, \mathbf{y}) = \int_{\mathbf{a}}^{\mathbf{b}} \overline{\left(\int_{\mathbf{c}}^{\mathbf{d}} f(\mathbf{x}, \mathbf{y}) \, d\mathbf{y} \right)} \, d\mathbf{x}$$

$$= \int_{\mathbf{a}}^{\mathbf{b}} \underline{\left(\int_{\mathbf{c}}^{\mathbf{d}} f(\mathbf{x}, \mathbf{y}) \, d\mathbf{y} \right)} \, d\mathbf{x}.$$

Remarks. In this theorem, we could just as well have reversed the order of iteration from $d\mathbf{y} d\mathbf{x}$ to $d\mathbf{x} d\mathbf{y}$. The student will be most familiar with this theorem in the special case in which the blocks $[\mathbf{a}, \mathbf{b}]$ and $[\mathbf{c}, \mathbf{d}]$ are replaced by intervals $[a, b]$ and $[c, d]$, so that $m = n = 1$. The theorem as stated above can be used in any Euclidean space $\mathbb{E}^p$ to successively reduce an integral over a p-dimensional block to p iterated integrals over intervals. This decomposition is simplest to write if all the lower-dimensional integrals in the iteration exist, so that it is not necessary to resort to lower or upper integrals, which always exist for a bounded integrand on a closed, bounded block. The latter condition is certainly met if $f \in \mathcal{C}(B)$.

Proof: In order to analyze $\underline{\int}_B f(\mathbf{x}, \mathbf{y}) \, d(\mathbf{x}, \mathbf{y})$ and $\overline{\int}_B f(\mathbf{x}, \mathbf{y}) \, d(\mathbf{x}, \mathbf{y})$ we need to consider arbitrary partitions $\mathcal{P}$ of B. We will show first that actually it suffices to take the supremum and infimum of the lower and upper sums, respectively, over those partitions that arise as Cartesian products of arbitrary partitions $\mathcal{P}_1$ of $[\mathbf{a}, \mathbf{b}]$ and $\mathcal{P}_2$ of $[\mathbf{c}, \mathbf{d}]$ as follows. We denote

$$\mathcal{P}_1 \times \mathcal{P}_2 = \{R_i \times S_j \mid R_i \in \mathcal{P}_1, S_j \in \mathcal{P}_2\},$$

which is a partition of B, though not an arbitrary partition of B. However, define *orthogonal projections*

$$\pi_1 : \mathbb{E}^{m+n} \to \mathbb{E}^m$$

by $\pi_1(\mathbf{x}) = (x_1, \ldots, x_m)$ and

$$\pi_2 : \mathbb{E}^{m+n} \to \mathbb{E}^n$$

by $\pi_2(\mathbf{x}) = (x_{m+1}, \ldots, x_{m+n})$. If $\mathcal{P}$ denotes an arbitrary partition of B, define $\pi_1(\mathcal{P})$ to be the partition obtained by using all the nonempty intersections of all the blocks of the form $\pi_1(B')$, where $B' \in \mathcal{P}\}$. Define $\pi_2(\mathcal{P})$ to be the partition obtained by using all the nonempty intersections of the blocks of the form $\pi_2(B')$ where $B' \in \mathcal{P}\}$. Thus $\pi_1(\mathcal{P})$ is a partition of $[\mathbf{a}, \mathbf{b}]$ and $\pi_2(\mathcal{P})$ is a partition of $[\mathbf{c}, \mathbf{d}]$. Furthermore, $\pi_1(\mathcal{P}) \times \pi_2(\mathcal{P})$ is a *refinement* of $\mathcal{P}$, and refinements raise lower sums and lower upper sums. This justifies the claim that to prove (1) for example, it suffices to begin with a partition $\mathcal{P} = \mathcal{P}_1 \times \mathcal{P}_2$, where $\mathcal{P}_1$ is an arbitrary partition of $[\mathbf{a}, \mathbf{b}]$ and $\mathcal{P}_2$ is an arbitrary partition of $[\mathbf{c}, \mathbf{d}]$.

Part (ii) is very similar to part (i), and part (iii) follows from the first two parts. To prove part (i), we proceed as follows. First we note that there is need to prove only inequalities (i) and (ii).

Observe that

$$\overline{\int}_{[\mathbf{c},\mathbf{d}]} f(\mathbf{x}, \mathbf{y}) \, d\mathbf{y} = \sum_j \overline{\int}_{S_j} f(\mathbf{x}, \mathbf{y}) \, d\mathbf{y}$$

by Exercise 11.33.c. Next we apply Exercise 11.3 to see that

$$\overline{\int}_{[\mathbf{a},\mathbf{b}]} \left(\overline{\int}_{[\mathbf{c},\mathbf{d}]} f(\mathbf{x}, \mathbf{y}) \, d\mathbf{y} \right) d\mathbf{x} \leq \sum_j \overline{\int}_{[\mathbf{a},\mathbf{b}]} \left(\overline{\int}_{S_j} f(\mathbf{x}, \mathbf{y}) \, d\mathbf{y} \right) d\mathbf{x}$$

$$= \sum_{i,j} \overline{\int}_{R_i} \left(\overline{\int}_{S_j} f(\mathbf{x}, \mathbf{y}) \, d\mathbf{y} \right) d\mathbf{x}$$

$$\leq U(f, \mathcal{P})$$

Thus

$$\overline{\int_{[\mathbf{a},\mathbf{b}]}} \left(\overline{\int_{[\mathbf{c},\mathbf{d}]}} f(\mathbf{x},\mathbf{y})\, d\mathbf{y} \right) d\mathbf{x} \le U(f,\mathcal{P})$$

for all partitions $\mathcal{P}$ of B. Hence

$$\overline{\int_{[\mathbf{a},\mathbf{b}]}} \left(\overline{\int_{[\mathbf{c},\mathbf{d}]}} f(\mathbf{x},\mathbf{y})\, d\mathbf{y} \right) d\mathbf{x} \le \overline{\int_B} f(\mathbf{x},\mathbf{y})\, d(\mathbf{x},\mathbf{y})$$

which proves inequality (ii). The proof of inequality (i) is similar and is left to the Exercises. ∎

EXERCISES

11.39   Let

$$f(x_1,x_2) = \begin{cases} 0 & \text{if } x_1 \notin \mathbb{Q}, \\ 1 & \text{if } x_1 \in \mathbb{Q}. \end{cases}$$

Prove:

    a) $\int_0^1 \left(\int_0^1 f(x_1,x_2)\, dx_2 \right) dx_1$ does not exist.

    b) $\int_0^1 \left(\int_0^1 f(x_1,x_2)\, dx_1 \right) dx_2$ does not exist.

    c) $\int_{[0,1]^{\times 2}} f(\mathbf{x})\, d\mathbf{x}$ does not exist.

11.40   Let $f : [0,1]^{\times 2} \to \mathbb{R}$ be defined by

$$f(x_1,x_2) = \begin{cases} 1 & \text{if } x_1 \in \mathbb{Q}, \\ 2x_2 & \text{if } x_1 \notin \mathbb{Q}. \end{cases}$$

Prove:

    a) The function f is not continuous at any point of $[0,1]^{\times 2}$ except for those at which $x_2 = \frac{1}{2}$, so that $f \notin \mathcal{R}\left([0,1]^{\times 2}\right)$.

    b) $\int_0^1 \left(\int_0^1 f(x_1,x_2)\, dx_2 \right) dx_1 = 1$.

    c) Both

$$\int_0^1 \left(\underline{\int_0^1} f(x_1,x_2)\, dx_1 \right) dx_2$$

and

$$\int_0^1 \left(\overline{\int_0^1} f(x_1,x_2)\, dx_1 \right) dx_2$$

exist. Find their values.

11.41   Let $\mathbb{P}$ denote the set of all prime numbers. Let

$$S = \bigcup_{p \in \mathbb{P}} \left\{ \left(\frac{m}{p}, \frac{n}{p} \right) \,\middle|\, m \text{ and } n \in \{1,2,\ldots,p-1\} \right\}.$$

a) Prove: S is dense in the unit square $B = [0,1]^{\times 2}$. (Hint: The set $\mathbb{P}$ has no upper bound.)

b) Let $S^c = B \setminus S$ and let $f = 1_{S^c}$, the *indicator function* of S^c. Prove that $f \notin \mathcal{R}(B)$.

c) Prove that

$$\int_0^1 \left(\int_0^1 f(x_1, x_2)\, dx_2 \right) dx_1$$

and

$$\int_0^1 \left(\int_0^1 f(x_1, x_2)\, dx_1 \right) dx_2$$

both exist and equal 1. (Hint: You may assume the uniqueness of factorizations into primes.)

11.42   Suppose

$$\int_0^1 \left(\underline{\int_0^1 f(x_1, x_2)\, dx_2} \right) dx_1$$

and

$$\int_0^1 \left(\overline{\int_0^1 f(x_1, x_2)\, dx_2} \right) dx_1$$

both exist and are equal. Prove that the set of numbers x_1 for which

$$\int_0^1 f(x_1, x_2)\, dx_2$$

does not exist is a Lebesgue null set in the real line. (Hint: Use Exercise 11.34.)

11.43   Prove Inequality (i) of Theorem 11.4.1.

11.44   Prove Parts (2) and (3) of Theorem 11.4.1.

11.45   *(Differentiation across the integral sign)* Let $B = [a,b] \times [c,d] \subset \mathbb{E}^2$ and denote both (x,t) and $(x,y) \in \mathbb{E}^2$. Suppose that

- $f : B \to \mathbb{R}$.

- $f(\cdot, t) \in \mathcal{R}[a,b]$ for each fixed $t \in [c,d]$.

- $\frac{\partial f}{\partial t} \in \mathcal{R}(B)$.

- $\frac{\partial f}{\partial t}(x, \cdot) \in \mathcal{R}[c,d]$ for each fixed $x \in [a,b]$.

- $\int_a^b \frac{\partial f}{\partial t}(x,t)\, dx \in \mathcal{C}[c,d]$ as a function of t.

a) Prove that

$$\frac{\partial}{\partial y} \int_a^b f(x,y)\, dx$$

exists and equals

$$\int_a^b \frac{\partial f}{\partial y}(x, y)\, dx.$$

(Hint: Use Fubini's Theorem to prove that the difference

$$\int_a^b f(x, y)\, dx - \int_c^y \left(\int_a^b \frac{\partial f}{\partial t}(x, t)\, dx \right) dt$$

is independent of y, and apply the Fundamental Theorem of Calculus.)

b) The *Fourier transform* of a function $\phi \in \mathcal{R}[0, 1]$ may be defined to be the *complex-valued* function $\widehat{\phi} : \mathbb{R} \to \mathbb{C}$ defined by

$$\widehat{\phi}(\alpha) = \int_0^1 \phi(x) \cos(2\pi\alpha x)\, dx - i \int_0^1 \phi(x) \sin(2\pi\alpha x)\, dx.$$

Prove that the *derivative*

$$\frac{d\widehat{\phi}}{d\alpha}(\alpha) = -2\pi i \widehat{\psi}(\alpha),$$

where $\psi(x) = x f(x)$.

11.46   (Clairaut's Theorem) Suppose $f \in C^2(D, \mathbb{R})$, where D is an open subset of $\mathbb{E}^2$. Prove that

$$\frac{\partial^2 f}{\partial x_1 \partial x_2} = \frac{\partial^2 f}{\partial x_2 \partial x_1}.$$

(Hint: Suppose false. Show by continuity that there is a closed rectangular block $R \subset D$ on which one of the two mixed partial derivatives remains identically strictly larger than the other. Conclude that the double integral of the first over R is strictly greater than the double integral of the second. Now show the contradiction that both integrals must be equal by integrating each second order partial derivative over R using Fubini's theorem. Clairaut's theorem can be readily generalized to higher-order mixed partial derivatives.)

11.5   JACOBIAN THEOREM FOR CHANGE OF VARIABLES

In one-variable calculus, the student has learned how to change variables in a single integral. Suppose we are given a *surjection* (i.e., an onto mapping) $g : [c, d] \to [a, b]$ that is a C^1 function with nonvanishing derivative. Since $g'(x)$ is not zero for any value of $x \in [c, d]$, either $g'(x) > 0$ for all x or else $g'(x) < 0$ for all x. In either case, g is one-to-one. If $g'(x) > 0$ for all x and if $f \in \mathcal{R}[a, b]$, then $f \circ g \in \mathcal{R}[c, d]$ (why?) and

$$\int_c^d f \circ g(x) g'(x)\, dx = \int_a^b f(y)\, dy.$$

or, $\int_a^b f(y)\, dy = \int_{g^{-1}([a,b])} f \circ g(x)\, g'(x)\, dx$

On the other hand, if $g'(x) < 0$ for all x, then $f \circ g \in \mathcal{R}[c, d]$ and

$$-\int_c^d f \circ g(x) g'(x)\, dx = \int_a^b f(y)\, dy.$$

The reader should note the minus sign. In either case, we can write that

$$\int_{g^{-1}[a,b]} f \circ g(x) |g'(x)|\, dx = \int_{[a,b]} f(y)\, dy.$$

In this section we will prove a generalization of this *change-of-variables* theorem to Euclidean spaces of all finite dimensions $n \geq 1$.

Theorem 11.5.1 *Let* $\mathbf{g} \in \mathcal{C}^1([\mathbf{c}, \mathbf{d}], \mathbb{E}^n)$ *be a one-to-one nonsingular function defined on a closed rectangular block* $[\mathbf{c}, \mathbf{d}] \subset \mathbb{E}^n$. *Suppose* $[\mathbf{a}, \mathbf{b}] \subset \mathbf{g}(\mathbf{c}, \mathbf{d})$, *which is necessarily an open set. Let* $f \in \mathcal{R}[\mathbf{a}, \mathbf{b}]$, *and extend* f *to be identically zero on* $\mathbf{g}[\mathbf{c}, \mathbf{d}] \setminus [\mathbf{a}, \mathbf{b}]$. *Then* $f \circ \mathbf{g} \in \mathcal{R}[\mathbf{c}, \mathbf{d}]$ *and*

$$\int_{\mathbf{g}^{-1}[\mathbf{a},\mathbf{b}]} f \circ \mathbf{g}(\mathbf{x}) \left| \frac{\partial \mathbf{g}}{\partial \mathbf{x}} \right| d\mathbf{x} = \int_{[\mathbf{a},\mathbf{b}]} f(\mathbf{x})\, d\mathbf{x}$$

Remark 11.5.1 Here we define

$$\int_{\mathbf{g}^{-1}[\mathbf{a},\mathbf{b}]} f \circ \mathbf{g}(\mathbf{x}) \left| \frac{\partial \mathbf{g}}{\partial \mathbf{x}} \right| d\mathbf{x} = \int_{[\mathbf{c},\mathbf{d}]} f \circ \mathbf{g}(\mathbf{x}) \left| \frac{\partial \mathbf{g}}{\partial \mathbf{x}} \right| d\mathbf{x}$$

since $f \circ \mathbf{g}$ vanishes off $\mathbf{g}^{-1}[\mathbf{a}, \mathbf{b}]$, the boundary of which is a Jordan null set.

Proof: We observe that since $\mathbf{g}$ is a nonsingular $\mathcal{C}^1$-function that is invertible globally, the inverse is also a $\mathcal{C}^1$-function by Theorem 10.4.3. Note that $f \in \mathcal{R}[\mathbf{a}, \mathbf{b}]$, and the boundary of $[\mathbf{a}, \mathbf{b}]$ is a Lebesgue null set, as is the set of points of discontinuity of f itself. Hence the set of points of discontinuity of $f \circ \mathbf{g}$ in $[\mathbf{c}, \mathbf{d}]$ is a Lebesgue null set by Theorem 11.2.4. This shows that $f \circ \mathbf{g} \in \mathcal{R}[\mathbf{c}, \mathbf{d}]$.

Let $\epsilon > 0$. By Theorem 11.1.3 we know there exists a partition $\mathcal{P}$ for which

$$|U(f, \mathcal{P}) - L(f, \mathcal{P})| < \epsilon.$$

Corresponding to Definition 11.1.3 we can define *upper and lower step functions* $U(\mathbf{x})$ and $L(\mathbf{x})$ as follows. On each block $B_i \in \mathcal{P}$ we let $U(\mathbf{x}) = M_i$ and $L(\mathbf{x}) = m_i$ on B_i°. On the null set that is the boundary of B_i, we let

$$U(\mathbf{x}) = \max\{M_j \mid \mathbf{x} \in \partial B_j\}$$

and

$$L(\mathbf{x}) = \min\{m_j \mid \mathbf{x} \in \partial B_j\}.$$

Thus $L(\mathbf{x}) \leq f(\mathbf{x}) \leq U(\mathbf{x})$ for all $\mathbf{x} \in \mathbf{g}[\mathbf{c}, \mathbf{d}]$ and

$$\int_{[\mathbf{a},\mathbf{b}]} (U - L)(\mathbf{x}) d\mathbf{x} = \int_{\mathbf{g}[\mathbf{c},\mathbf{d}]} (U - L)(\mathbf{x}) d\mathbf{x} < \epsilon.$$

Now we will *suppose that we have already proven the change of variables theorem for indicator functions of closed rectangular blocks.* (We will prove this special case of the change of variables theorem as the last stage of the proof of Theorem 11.5.1.) Since the step functions U and L are linear combinations of finitely many indicator functions of blocks (except on a Jordan null set, which does not affect the Riemann integral), this will imply the validity of the change of variables theorem for U and L as well.

It follows that

$$\int_{[\mathbf{a},\mathbf{b}]} L(\mathbf{x})\,d\mathbf{x} = \int_{\mathbf{g}^{-1}[\mathbf{a},\mathbf{b}]} L \circ \mathbf{g}(\mathbf{x}) \left| \frac{\partial \mathbf{g}}{\partial \mathbf{x}} \right| d\mathbf{x}$$

$$\leq \int_{\mathbf{g}^{-1}[\mathbf{a},\mathbf{b}]} f \circ \mathbf{g}(\mathbf{x}) \left| \frac{\partial \mathbf{g}}{\partial \mathbf{x}} \right| d\mathbf{x}$$

$$\leq \int_{\mathbf{g}^{-1}[\mathbf{a},\mathbf{b}]} U \circ \mathbf{g}(\mathbf{x}) \left| \frac{\partial \mathbf{g}}{\partial \mathbf{x}} \right| d\mathbf{x}$$

$$= \int_{[\mathbf{a},\mathbf{b}]} U(\mathbf{x})\,d\mathbf{x}.$$

Hence

$$\left| \int_{\mathbf{g}^{-1}[\mathbf{a},\mathbf{b}]} f \circ \mathbf{g}(\mathbf{x}) \left| \frac{\partial \mathbf{g}}{\partial \mathbf{x}} \right| d\mathbf{x} - \int_{[\mathbf{a},\mathbf{b}]} f(\mathbf{x})\,d\mathbf{x} \right| < \epsilon.$$

Since $\epsilon > 0$ is arbitrary, the desired equality is established.

The proof the theorem will be complete once we have proven it for the special case of the indicator function of a rectangular block. Without loss of generality, we may assume that $f = 1_B$, an indicator function of a rectangular block $B = [\mathbf{A}, \mathbf{B}]$. Since the change of variables theorem is known to be true for $\mathbb{E}^n$ if $n = 1$, we proceed by induction. Thus we assume the theorem is true for $\mathbb{E}^k$ for all $k < n$ and we must establish the theorem for $\mathbb{E}^n$.

It is clear that the validity of the theorem is independent of the order in which the coordinate axes are listed in $\mathbb{E}^n$. Since at each point at least one of the partial derivatives $\frac{\partial g_1}{\partial x_j} \neq 0$, we can change the order in which we list the coordinate axes in the domain space of $\mathbf{g}$ to insure that $\frac{\partial g_1}{\partial x_1} \neq 0$. Since $\mathbf{g} \in \mathcal{C}^1$, we can insure by making B sufficiently small that $\frac{\partial g_1}{\partial x_1}$ has *constant sign* on $\mathbf{g}^{-1}(B)$. Moreover, the Jacobian $\frac{\partial \mathbf{g}}{\partial \mathbf{x}}$ must have constant sign on $\mathbf{g}^{-1}(B)$ since it cannot vanish and $\mathbf{g}^{-1}(B)$ is connected. Thus we can assume without loss of generality (by changing the order of the basis in the domain space) that $\frac{\partial \mathbf{g}}{\partial \mathbf{x}} > 0$ on $\mathbf{g}^{-1}(B)$.

In addition to the mapping $\mathbf{g} : \mathbf{g}^{-1}(B) \to B = [\mathbf{A}, \mathbf{B}]$, we define a mapping $\mathbf{h}$ from $\mathbf{g}^{-1}(B)$ onto an image that we call H by letting

$$\mathbf{h}(\mathbf{x}) = (g_1(\mathbf{x}), x_2, x_3, \dots, x_n).$$

Since $\mathbf{h}'(\mathbf{x}) = \frac{\partial g_1}{\partial x_1}$ has constant sign on its domain, and since only the first coordinate of $\mathbf{x}$ is altered by $\mathbf{h}$, we see that $\mathbf{h}$ is invertible on H and that the inverse is a $\mathcal{C}^1$

function. Now we consider the mapping

$$\mathbf{g} \circ \mathbf{h}^{-1} : H \to B$$

and we note that

$$\mathbf{g} \circ \mathbf{h}^{-1}(\mathbf{x}) = (x_1, g_2(\mathbf{x}), \ldots, g_n(\mathbf{x}))$$

so that the first coordinate is unaffected by this composite mapping. We denote by B_{x_1} the cross section of B determined by fixing the value of x_1, and we define the cross section H_{x_1} of H in the same way. Note that this cross section of H must be indexed by the subscript x_1 since it, unlike B_{x_1}, is not independent of x_1. Observe also that

$$\left(\mathbf{g} \circ \mathbf{h}^{-1}\right)' = 1 \cdot \frac{\partial(g_2, \ldots, g_n)}{\partial(x_2, \ldots, x_n)} = \left(\mathbf{g} \circ \mathbf{h}^{-1}\big|_{H_{x_1}}\right)'.$$

We are ready to complete the proof of the change of variables theorem as follows. We proceed by computing the volume of the box B, making use of the hypothesis that the theorem works already in $n - 1$ variables and in 1 variable. We denote by H_{x_1} the set of elements of H with the first coordinate fixed at the value x_1. Also, Let π project $\mathbb{E}^n$ orthogonally onto the hyperplane determined by setting $x_1 = 0$, and denote $H' = \pi(H) = \pi(\mathbf{g}^{-1}(B))$, since $\mathbf{h}$ alters only the first coordinate.

$$
\begin{aligned}
\int_B 1\, d\mathbf{x} &= \int_{A_1}^{B_1} \left(\int_{B_{x_1}} 1\, d(x_2, \ldots, x_n) \right) dx_1 \\[1ex]
&\overset{(i)}{=} \int_{A_1}^{B_1} \left(\int_{H_{x_1}} \left|(\mathbf{g} \circ \mathbf{h}^{-1})'\right| d(x_2, \ldots, x_n) \right) dx_1 \\[1ex]
&= \int_{A_1}^{B_1} \left(\int_{H_{x_1}} \left|(\mathbf{g}'\,(\mathbf{h}^{-1}(\mathbf{x}))\right| \left|(\mathbf{h}^{-1})'\,(\mathbf{x})\right| d(x_2, \ldots, x_n) \right) dx_1 \\[1ex]
&\overset{(ii)}{=} \int_{H'} \left(\overline{\int_{H_{(x_2, \ldots, x_n)}} \left|(\mathbf{g}'\,(\mathbf{h}^{-1}\,(\mathbf{x}))\right| \left|(\mathbf{h}^{-1})'\,(\mathbf{x})\right| dx_1} \right) d(x_2, \ldots, x_n) \\[1ex]
&\overset{(iii)}{=} \int_{\pi(\mathbf{g}^{-1}(B))} \left(\overline{\int_{\pi(\mathbf{g}^{-1}(B))_{(x_2, \ldots, x_n)}} \left|(\mathbf{g}'\,(\mathbf{h}^{-1}\,(\mathbf{x}))\right| dx_1} \right) d(x_2, \ldots, x_n) \\[1ex]
&= \int_{\mathbf{g}^{-1}(B)} |\mathbf{g}'(\mathbf{x})|\, d\mathbf{x}
\end{aligned}
$$

In the equality labeled (i), we have used the $(n - 1)$-dimensional version of the change of variables formula, noting that $\mathbf{g} \circ \mathbf{h}^{-1} : H_{x_1} \to B_{x_1}$. In the equality labeled (ii), the change in order of iteration is justified by Fubini's Theorem together with the existence of the Riemann integral on the product space. This integrability is a consequence of the boundary of the domain being a Jordan null set, by Theorem 11.2.4. The inner integral exists except perhaps on a Lebesgue null set in the outer variable, on which either the upper or lower form of the inner integral can be used with no effect on the value of the iterated integral. In the equality labeled (iii), we have used the one-dimensional version of the change of variables formula for *upper integrals* (Exercise 11.55), bearing in mind that $\mathbf{h}^{-1}$ alters only the first coordinate of $\mathbf{x}$. Also, we use here the definition that the Riemann integral of f over a bounded set is the Riemann integral of a trivial extension of f to any closed rectangular block containing that set. ∎

■ **EXAMPLE 11.2**

Consider *polar coordinates*, for which $\mathbf{g} : (r, \theta) \rightarrow (x_1, x_2)$ where

$$x_1 = r \cos \theta \quad \text{and} \quad x_2 = r \sin \theta.$$

Thus $\mathbf{g}$ maps the rectangle $[0, a] \times [0, 2\pi]$ onto a *punctured* circular disk of radius a. The area of this disk is

$$\int_0^a \int_0^{2\pi} 1 \, |\mathbf{g}'(r, \theta)| \, d\theta dr = \pi a^2$$

since $\mathbf{g}'(r, \theta) = r$, as the reader should verify easily.

EXERCISES

11.47 If $f \in \mathcal{R}[a, b]$ and if we have a surjection $g \in C^1([c, d], [a, b])$ such that g' is nonvanishing on $[a, b]$, prove that $f \circ g \in \mathcal{R}[c, d]$.

11.48 In *spherical coordinates*, $\mathbf{g}(\rho, \theta, \phi) = (x_1, x_2, x_3)$, where

$$x_1 = \rho \sin \phi \cos \theta, \ x_2 = \rho \sin \phi \sin \theta, \text{ and } x_3 = \rho \cos \phi.$$

Calculate $\mathbf{g}'(\rho, \theta, \phi)$ and use it to find the volume of a sphere of radius a.

11.49 If $B_1(\mathbf{0})$ denotes the unit ball in $\mathbb{E}^3$, calculate $\int_{B_1(\mathbf{0})} \frac{1}{x_1^2 + x_2^2 + x_3^2} \, d\mathbf{x}$.

11.50 Find the Jacobian of the transformation that corresponds to cylindrical coordinates in elementary calculus.

11.51 Let $S \subset \mathbb{E}^n$ be any *Jordan measurable* set and let $T \in \mathcal{L}(\mathbb{E}^n)$. Prove that $T(S)$ is Jordan measurable, and that

$$\mu(T(S)) = \det(T)\mu(S).$$

11.52 Let $T : \mathbb{E}^3 \rightarrow \mathbb{E}^3$ be the *affine* transformation defined by

$$T(x_1, x_2, x_3) = (x_1 + a, x_2 + b, x_3 + c + ax_2)$$

where a, b, and c are real constants. Find the volume of $T(B_1(\mathbf{0}))$, where $B_1(\mathbf{0})$ is the unit ball around the origin.

11.53 Prove that the change of variables theorem remains valid even if the transformation $\mathbf{g}$ has the property that $\mathbf{g}'$ vanishes on a Jordan null set. (Hint: What can you say about $\mathbf{g}(S)$, if S is a Jordan null set?)

11.54   Let $\bar{B}_1(0)$ denote the closed ball of radius 1 centered at $0 \in \mathbb{E}^n$. Let

$$I_n = \int_{\bar{B}_1(0)} \frac{1}{\|\mathbf{x}\|^2}\, d\mathbf{x},$$

which we define by

$$I_n = \lim_{r \to 0+} \int_{\{\mathbf{x} \in \mathbb{E}^n \,|\, r \le \|\mathbf{x}\| \le 1\}} \frac{1}{\|\mathbf{x}\|^2}\, d\mathbf{x}.$$

For parts (a) and (b) below, use polar or spherical coordinates.

a) Prove that $I_n = \infty$ if $n = 1$ or if $n = 2$.

b) Prove that $I_3 < \infty$.

c) Introduce *spherical coordinates* in $\mathbb{E}^n$ as follows. Let $\rho = \|\mathbf{x}\|$. Let

$$\phi_n = \cos^{-1} \frac{x_n}{\rho} \quad \text{and} \quad \rho_{n-1} = \rho \sin \phi_n$$

for $0 \le \phi_n \le \pi$. Let

$$\phi_{n-1} = \cos^{-1} \frac{x_{n-1}}{\rho_{n-1}} \quad \text{and} \quad \rho_{n-2} = \rho_{n-1} \sin \phi_{n-1}$$

for $0 \le \phi_{n-1} \le \pi, \ldots,$ and let

$$\phi_3 = \cos^{-1} \frac{x_3}{\rho_3}, \quad \text{and} \quad \rho_2 = \rho_3 \sin \phi_3$$

for $0 \le \phi_3 \le \pi$. Let $x_1 = \rho_2 \cos \theta, x_2 = \rho_2 \sin \theta$, with $0 \le \theta < 2\pi$. Prove that

$$\frac{\partial(x_1, \ldots, x_n)}{\partial(\rho, \theta, \phi_3, \ldots, \phi_n)}$$

is ρ^{n-1} times a polynomial in the sines and cosines of $\theta, \phi_3, \ldots, \phi_n$.

d) Prove that if $n > 3$, then $I_n < \infty$.

11.55   Let $g \in \mathcal{C}^1((a, b), \mathbb{E}^1)$ be nonsingular (meaning g' is nowhere zero) and let $S \subseteq [c', d'] \subset g(a, b)$ be any bounded subset of $g(a, b)$. Let f be any bounded real valued function on S. Prove:

$$\overline{\int_S} f(x)\, dx = \overline{\int_{g^{-1}(S)}} f(x)|g'(x)|\, dx.$$

(Hint: Interpret f as being identically zero off S. Consider any partition $\mathcal{P}$ of $[c', d'] = g[c, d]$, and compare upper sums over both domains.)

11.56   Fix any $\mathbf{a} \in \mathbb{E}^n$ and define $T \in \mathcal{L}(\mathbb{E}^n)$ by $T(\mathbf{x}) \equiv \mathbf{x} + \mathbf{a}$. If B is any closed finite box in $\mathbb{E}^n$ and if $f \in \mathcal{R}(B)$, then $f \in \mathcal{R}(T^{-1}B)$ and

$$\int_B f(\mathbf{x})\, d\mathbf{x} = \int_{T^{-1}B} f(T\mathbf{x})\, d\mathbf{x}.$$

That is, the Riemann integral on $\mathbb{E}^n$ is *invariant* under all *translations*.

11.57   Denote by $\mathcal{O}(n)$ the set of all $n \times n$ matrices A for which $AA^t = I$, the identity matrix, where A^t denotes the *transpose* of the matrix A. Prove:

a) $\mathcal{O}(n)$ is a group under the operation of matrix multiplication. This is called the *orthogonal group* on $\mathbb{E}^n$.

b) $|\det A| = 1$ for all $A \in \mathcal{O}(n)$.

c) If B is any closed finite box in $\mathbb{E}^n$ and if $f \in \mathcal{R}(B)$, then $f \in \mathcal{R}(A^{-1}B)$ and

$$\int_B f(\mathbf{x})\, d\mathbf{x} = \int_{A^{-1}B} f(A\mathbf{x})\, d\mathbf{x}.$$

That is, the Riemann integral on $\mathbb{E}^n$ is *invariant* under orthogonal transformations.

11.6   TEST YOURSELF

EXERCISES

11.58   Give an example of a function $f : B \to \mathbb{R}$ for which

$$\underline{\int_B} f = 0 \ \text{ and } \ \overline{\int_B} f = 1,$$

where $B = [0, 1]^{\times 2}$.

11.59   Give an example of functions f and g mapping $B \to \mathbb{R}$ such that

$$\overline{\int_B} (f + g) < \overline{\int_B} f + \overline{\int_B} g,$$

where $B = [0, 1]^{\times 2}$.

11.60   Give an example of a function $f : B \to \mathbb{R}$ for which f is not Riemann integrable but f^- is Riemann integrable, where $B = [0, 1]^{\times 2}$.

11.61   Let the linear map $T : \mathcal{R}[\mathbf{a}, \mathbf{b}] \to \mathbb{R}$ be defined by

$$T(f) = \int_{[\mathbf{a}, \mathbf{b}]} f(\mathbf{x})\, d\mathbf{x}.$$

Find $\|T\|$.

11.62   Let $f = 1_{\mathbb{Q} \cap [0, 1]}$, the indicator function of the set of all rational numbers in the unit interval of $\mathbb{E}^1$. Find the set of points of discontinuity of f.

11.63   Let $S = \{x \in \mathbb{E}^2 \mid x_1 \notin \mathbb{Q}, x_2 \notin \mathbb{Q}\}$. Is S Jordan measurable?

11.64   True or False: The set $\{(-1)^n + \frac{1}{n} \mid n \in \mathbb{N}\}$ is a Jordan null set in $\mathbb{E}^1$.

11.65   Let
$$f(x) = \begin{cases} \sin \frac{\pi}{x}, & x \neq 0, \\ 0, & x = 0. \end{cases}$$

Find the oscillation $o(f, 0)$.

11.66   Find the Jordan measure of the set of points of discontinuity of the function shown in Fig. 11.1.

11.67   Suppose $F(x_1, x_2) = f(x_1)1_{\mathbb{Q}}(x_2)$. Find all functions $f(x_1)$ for which $F \in \mathcal{R}[0, 1]^{\times 2}$.

11.68   Let $\mathbf{f}(\mathbf{x}) = (-x, -y, xy - z)$ for all $\mathbf{x} \in \mathbb{E}^3$.
  a) Find the matrix $[\mathbf{f}'(\mathbf{x})]$.
  b) Find the volume of the image of a sphere of radius 2 under the function $\mathbf{f}$ in this exercise.

APPENDIX A

SET THEORY

A.1 TERMINOLOGY AND SYMBOLS

When the author was young, it was easy for a professor to know whether or not students were familiar with set-theoretic symbols and language from high school. The answer was *no*. Thus professors took care to explain what these terms mean. Today it is less clear. High-schools teach some set theory, but maybe not very much. And since students seldom get to use this language in their high-school problem solving, they may not have developed facility with it. The same is true with introductory college courses. There is likely to be some use of set theory, but it is not clear what background is common to all students. The formal study of set theory refers particularly to the study of infinite sets (also called transfinite sets), and the reader is referred to [11] for a deep and serious study of that subject. Here we begin by defining the commonly used symbols that appear in this text and giving a few illustrative examples and theorems.

*

Advanced Calculus: An Introduction to Linear Analysis. By Leonard F. Richardson
Copyright © 2008 John Wiley & Sons, Inc.

For reasons that we will sketch briefly in Example A.7, logic requires us to limit the *frame of reference* in any set-theoretic discussion to some *universal set*, which is chosen for convenience or suitability to a given purpose. In the study of sets in the abstract, this universal set is commonly designated by the letter X. It is common to designate a *subset* A of X, denoted by $A \subset X$ by by writing that A is the set of all elements of X that possess some property.

■ **EXAMPLE A.1**

Let $A = \{x \in \mathbb{R} \mid -1 \leq x \leq 1\}$. This is read A *is the set of all those real numbers x that have the property that x lies between -1 and 1 inclusive.* It is common to write the set A in interval notation for the real numbers as $A = [-1, 1]$.

Some authors use the symbol $A \subset X$ to mean *proper subset*, which means that A is contained in X but A is not equal to X. The author of this book will usually write $A \subsetneq B$ for this however. If it is important to stress the possibility of a subset of X being equal to X itself, the author will often write $A \subseteq X$.

Two very important operations between sets are *union* and *intersection*.

Definition A.1.1 *The union of A and B, denoted by $A \cup B$ is the set of all $x \in X$ such that either $x \in A$ or $x \in B$. This can be written as*

$$A \cup B = \{x \in X \mid x \in A \text{ or } x \in B\}.$$

The intersection of A and B is

$$A \cap B = \{x \in X \mid x \in A \text{ and } x \in B\}.$$

Note that in mathematics we use the word *or* in the inclusive sense, not in the exclusive sense. Thus $x \in A \cup B$, provided that $x \in A$, $x \in B$, or both.

■ **EXAMPLE A.2**

Continuing Example A.1, in which $A = [-1, 1] \subset \mathbb{R}$ we can define $B = \{x \in \mathbb{R} \mid 0 \leq x < 2\} = [0, 2)$. Then we have $A \cup B = [-1, 2)$ in interval notation, and $A \cap B = [0, 1]$.

Definition A.1.2 *If $A \cap B = \emptyset$, the empty set, we call A and B disjoint sets.*

Another important operation is the *difference* between two sets.

Definition A.1.3 *The difference between two sets A and B, denoted by $A \setminus B$, is the set of all those elements of A that are not in B.*

■ **EXAMPLE A.3**

Continuing Example A.2, we have $A \setminus B = [-1, 0)$.

Definition A.1.4 *Within the context of a universal set X, we denote $X \setminus A = A^c$, which is called the* complement *of A.*

Thus the *complement* of a set has meaning only within the context of a previously specified universal set.

■ **EXAMPLE A.4**

Continuing Example A.2, the set $A^c = (-\infty, -1) \cup (1, \infty)$, provided we are working with reference to the universal set $\mathbb{R}$.

A different but sometimes useful concept of subtraction of two sets is the following.

Definition A.1.5 *The* symmetric difference *of two sets A and B is denoted and defined as follows: $A \triangle B = (A \setminus B) \cup (B \setminus A)$.*

■ **EXAMPLE A.5**

In terms of Example A.2, $A \triangle B = [-1, 0) \cup (1, 2)$.

Theorem A.1.1 $A \triangle B = (A \cup B) \setminus (A \cap B)$.

Proof: This proof illustrates a frequently useful approach to proving that two sets, $A \triangle B$ and $(A \cup B) \setminus (A \cap B)$, are the same. The plan is to show that each of the latter two sets is a subset of the other. We will begin by proving that

$$A \triangle B \subseteq (A \cup B) \setminus (A \cap B).$$

By Definition A.1.5, if $x \in A \triangle B$, either $x \in A$ but $x \notin B$, or vice versa. (Observe that in this case the union that defines $A \triangle B$ does not permit x to be in both sets.) If $x \in A$ but $x \notin B$, then $x \in A \cup B$ but $x \notin X \cap B$. Hence $x \in (A \cup B) \setminus (A \cap B)$. On the other hand, if $x \in B$ but $x \notin A$, we can still conclude that $x \in (A \cup B) \setminus (A \cap B)$. Therefore, $A \triangle B \subseteq (A \cup B) \setminus (A \cap B)$ as claimed.

We leave it to the reader to verify by similar reasoning that

$$A \triangle B \supseteq (A \cup B) \setminus (A \cap B).$$

■

In this book, we need to apply the concepts of union and intersection to *infinite* families of sets—that is, to infinite sets *of* sets. We will clarify this with an example.

■ **EXAMPLE A.6**

Suppose to each $x \in \mathbb{R}$ we associate an interval $A_x = (x-1, x+1)$. If $S \subset \mathbb{R}$ we can define the *union over* $x \in S$ of the sets A_x to be the set of all those real numbers that lie in at least one of the sets A_x with $x \in S$. Thus in the example

$$\bigcup_{x \in [0,3]} A_x = \bigcup\{A_x \mid x \in [0,3]\} = (-1, 4).$$

Similarly, the intersection of a set of sets is the set of all those elements that belong simultaneously to every one of the sets. Continuing the example of this paragraph, we have

$$\bigcap_{x \in (0,1.5)} A_x = (0.5, 1).$$

Theorem A.1.2 *(DeMorgan's Laws) Suppose for each $s \in S$, a set of indices, there is associated a set $A_s \subseteq X$. Then*

i.

$$\left(\bigcup_{s \in S} A_s \right)^c = \bigcap_{s \in S} A_s^c$$

ii.

$$\left(\bigcap_{s \in S} A_s \right)^c = \bigcup_{s \in S} A_s^c$$

Proof: To prove the first equality, we will prove that the set that is written as the left side is contained in the set that is the right side. Then we will prove the reverse inclusion, thereby establishing equality. So suppose that $x \in \left(\bigcup_{s \in S} A_s \right)^c$. This tells us that for all $s \in S$ we have $x \notin A_s$. This tells us in turn that $x \in A_s^c$ for each $s \in S$. Consequently, $x \in \bigcap_{s \in S} A_s^c$, which establishes that

$$\left(\bigcup_{s \in S} A_s \right)^c \subseteq \bigcap_{s \in S} A_s^c.$$

To prove the reverse inclusion, suppose that $x \in \bigcap_{s \in S} A_s^c$. This means that $x \in A_s^c$ for each $s \in S$. Hence $x \notin A_s$ for any $s \in S$. Thus $x \in \left(\bigcup_{s \in S} A_s \right)^c$. This tells us that

$$\left(\bigcup_{s \in S} A_s \right)^c \supseteq \bigcap_{s \in S} A_s^c,$$

which completes the proof of equality of the left and right sides of the first law. We leave the second law to the reader to prove as Exercise A.1.   ■

EXERCISES

A.1   Prove that

$$\left(\bigcap_{s \in S} A_s\right)^{\text{c}} = \bigcup_{s \in S} A_s^{\text{c}}.$$

A.2   If a set E_a is a closed subset of $\mathbb{R}$ for each $a \in A$, an index set, prove that $\bigcap_{a \in A} E_a$ must be a closed set.

A.2   PARADOXES

We turn next to a brief introduction to some of the logical problems that can arise in set theory and in the logical use of language.

■ **EXAMPLE A.7**

We explain in this example why it is necessary to fix a frame of reference, called a universal set, at the beginning of any logical discussion involving set theory. At first glance it might seem reasonable to define a set $S = \{A \mid A \text{ is a set}\}$. Thus we *appear* to have defined an extraordinarily big set S that has as its elements every set that exists. In other words, $A \in S \Leftrightarrow A$ is a set. (This is a concept that is reminiscent of the optimistic concept of a war to end all wars.) We will explain now why there can be *no* such set as S. We will show that the alleged definition of S introduces a *paradox*, known as Cantor's paradox, with the result that the claimed definition is not a legitimate definition but is only a misleading sequence of words. In order to do this we must use a theorem that we will not prove here, but which the reader can find in the book [11] by E. Kamke. We begin with a (legitimate!) definition.

Definition A.2.1   *The* power set *of any set X, denoted by $P(X)$, is defined to be the set of* all *subsets of X.*

Thus $A \in P(X)$ if and only if $A \subseteq X$. If X happens to be a finite set, with n elements, it can be proved that the number of elements in $P(X)$ must be 2^n. (The reader can find this fact in [11] or in [2].) However, there is a much more difficult theorem in set theory that states the following.

Theorem A.2.1   *For every set X, $P(X)$ has more elements than X itself.*

For a proof, the ambitious reader can consult Kamke's book [11]. (The proof is a powerful generalization of the method the reader will see on a simpler level in Cantor's Theorem 1.8.1.) This theorem is true even when X is an infinite set, and it means that it is not possible to pair each element of $P(X)$ one-to-one with the elements of X itself or with the elements of any subset of X itself. Now consider the *alleged* set S defined in the present Example. Since each element of $P(S)$ is itself a set in its own right, the elements of $P(S)$ are all elements of S itself! This means that $P(S) \subseteq S$ and the elements of $P(S)$ can certainly be paired with themselves on a one-to-one basis. This contradicts Theorem A.2.1, which concludes the proof that there cannot exist such a set as S.

As a consequence of logical problems such as are illustrated in Example A.7, mathematicians begin set-theoretic reasoning with some set that exists, and new sets are introduced using only legitimate operations of set theory, such as forming the power set of a given set. Not every sequence of words is a legitimate definition.

Any subject in mathematics begins with definitions and axioms, and within that framework theorems are proven. A theorem must be a sentence that can be called a *proposition* in the sense of logic. A proposition is a sentence that must be either true or false within the context of the given axiomatic system. (That there is no in-between status for a proposition is sometimes called the *law of the excluded middle*.) That not every sentence is a proposition was explained in an amusing way by Bertrand Russell, a very distinguished twentieth-century mathematical logician and philosopher.

■ EXAMPLE A.8

Here we present *Russell's Paradox*. A (male) barber puts a sign in his shop window advertising as follows: *I will shave all those men and only those men who do not shave themselves.* Although the claim is ambitious and may provoke skepticism, it is probably not apparent at first sight that the barber's sign bears only a meaningless sequence of words, and not a valid proposition. The problem is this: Must the barber shave himself? If he shaves himself, his own advertisement prohibits him from doing so. But if he does not shave himself, his advertisement requires him to do so.

Here is a more serious version of Russell's Paradox for the purposes of mathematics. Let $S = \{A \mid A \notin A\}$. In words, S is the set of all those sets that are not elements of themselves. The reader may be wondering about an example of a set that *is* an element of itself. Try this one: $B = \{A \mid A \text{ is } not \text{ an Edsel}\}$. It seems clear that $B \in B$ since the set B is not an Edsel–at least not literally. There is a logical paradox regarding the proposed set S. The question is this: Is $S \in S$? If the answer is *yes*, then S cannot be in S. But if the answer is *no*, then S must be in S. A rigorous treatment of symbolic logic was undertaken by Russell and many others during the twentieth century in order to develop formal methods of avoiding such paradoxes by excluding the possibility of constructing such a definition as the one we attempted to give in this example for the set S.

The student may have observed that mathematics professors try to be very careful in the use of language and that we tend to be quite critical of the student's mathematical writing. Centuries of experience have proven this to be a wise practice.

PROBLEM SOLUTIONS

SOLUTIONS FOR CHAPTER 1

1.8 We can use any value of δ satisfying the double inequality $0 < \delta < \frac{\epsilon}{2}$.

1.24.b This sequence is not bounded (because of the Archimedean property) and so it cannot be Cauchy.

1.24.d This sequence is not Cauchy because it repeats the values 0 and 1 infinitely often. Thus beyond each 0 there is a 1, and beyond each 1 there is a 0. No matter how big N is, there will be n and m greater than N such that $x_n - x_m = 1 - 0 = 1$ which cannot be made less than ϵ if $0 < \epsilon < 1$.

1.31.a $\sup(S) = 1$ and $\inf(S) = -1$.

1.31.b $\sup(S) = \infty$ and $\inf(S) = -\infty$.

1.41 $\limsup x_n = 1$ and $\liminf x_n = -1$. The limit does not exist.

1.48 For example, $x_n = a + n$, $y_n = n$.

1.50 For example, $x_n = an$, $y_n = n$.

1.63 For example, let $y_j = \frac{1}{n}(1 + (-1)^n)$.

1.68 For example, if we let $(a_n, b_n) = \left(-\frac{1}{n}, \frac{1}{n}\right)$, then

$$\cap_1^\infty (a_n, b_n) = \{0\} \neq \emptyset.$$

1.79 The empty set $\emptyset$ is open because it is true (in a *vacuous* sense) that for each $x \in \emptyset$ there an $r > 0$ such that $(x - r, x + r) \subseteq \emptyset$. This is true because there is no element x in $\emptyset$, so that the claim is true for each (nonexistent) element of $\emptyset$.

1.81 For example, $\mathcal{O} = \{(-n, n) \mid n \in \mathbb{N}\}$ is such an open cover. The reader should take care not to confuse the set $\mathcal{O}$ with the union of its member intervals! We leave it to the reader to prove that $\mathcal{O}$ is an open cover and has the claimed property.

1.83 For example, let $\mathcal{O} = \left\{\left(\frac{1}{2n}, 2\right) \mid n \in \mathbb{N}\right\}$. We leave the necessary proofs to the reader.

1.92.c One example is $\emptyset$, the empty set. Can you give another example?

1.98 We can take $\delta = \frac{1}{200}$. The idea is to use the triangle inequality.

1.99.a True.

1.99.b False since the sequence is unbounded.

1.100 For example, we can use $x_n = y_n = (-1)^n$.

1.101 We have $\liminf x_n = 0$ and $\limsup x_n = 2$. Most errors with this question result from forgetting the meaning of $\limsup x_n = \lim_{n \to \infty} \sup(T_n)$, where T_n is the nth tail of the sequence.

1.102 For example, we can use $x_n = (-1)^n n$ and $y_n = (-1)^{n+1} n$.

1.103 Counterexample: Let $x_n = (1 + (-1)^n) n$. Then the subsequence x_{2n-1} is convergent.

1.104 For example, we can use $(a_n, b_n) = \left(0, \frac{1}{n}\right)$ for all n.

1.105 True.

1.106 For example, one can use $O_n = \left(\frac{1}{n}, 2\right)$.

1.107 False. For example, there is a sequence of rational numbers that converges to $\sqrt{2}$.

1.108 This is true since the set S is the union of countably many finite sets.

1.109 False. The set S is indexed by the set $\mathbb{Q}$, which is countable.

1.110 We can take $N > \frac{1}{\epsilon^2}$. Such N exist because of the Archimedean property of the real number system.

1.111 We give a counterexample. The constant sequence $x_n \equiv 1$ is convergent, and the sequence $y_n = (-1)^n$ is bounded. But $x_n y_n$ is divergent.

1.112 $\emptyset$.

1.113 For example, we can let $\mathcal{O} = \left\{ \left(\frac{1}{n}, 1 \right) \mid n \in \mathbb{N} \right\}$.

1.114 True, since $E^c = \bigcup_{n \in \mathbb{N}} \left(\frac{1}{n+1}, \frac{1}{n} \right) \cup (-\infty, 0) \cup ((1, \infty)$, a union of open sets.

SOLUTIONS FOR CHAPTER 2

2.4 Each real number $x \in \mathbb{R}$ is a cluster point of the set $\mathbb{Q}$ because the set $\mathbb{Q}$ is dense in in the set $\mathbb{R}$.

2.13.b The limit is na^{n-1}.

2.25 $c = na^{n-1}$. The reader should prove this claim.

2.45 $f(x) = x^2$ is uniformly continuous on $(0, 1)$. The reader should explain why.

2.58.b For example, let $f(x) = \tan x$ for all $x \in \left(-\frac{\pi}{2}, \frac{\pi}{2} \right)$.

2.59.b $\|f\|_{\sup} = \infty$.

2.64.a $\|f_n\|_{\sup} = e^{-1}$.

2.68.d The sequence of functions does not converge uniformly. The student should take care to distinguish this case from the one that precedes it.

2.71 The set is countable. Write $\mathbb{Q} = \{r_i \mid i \in \mathbb{N}\}$ and arrange the set of all possible functions $f_{a,b}$ as

$$\bigcup_{i \in \mathbb{N}} \{f_{r_i, r_j} \mid j \in \mathbb{N}\}.$$

Then invoke the fact that the union of countably many countable sets is countable.

2.72 True. Pick a suitable open interval for each rational number that lies in the given open set.

2.73 Counterexample:

$$f(x) = \begin{cases} \sin \frac{1}{x} & \text{if } 0 < x < 1, \\ 0 & \text{if } x = 0. \end{cases}$$

The indicated limit fails to exist and thus is not equal to any number L.

2.74 The set $\mathbb{R}$ is the set of all cluster points of the set $\mathbb{Q}$.

2.75 True. Apply the Intermediate Value theorem to $g(x) = f(x) - \sqrt{x}$ on $[0, 1]$.

2.76 For example, let $f_n(x) = \log x - \frac{1}{n}$.

2.77 For example, $f(x) = \sin\frac{1}{x}$ on $(0, 1)$. This was a homework problem.

2.78.a This is not a vector space because the difference between two such polynomials can have non-odd degree.

2.78.b Yes, this is a vector space.

2.79 $\|f\|_{\text{sup}} = 1$.

2.80.a True, since f_n converges pointwise on $\mathbb{R}$ because $f_n(x) \to 0$ for each fixed x as $n \to \infty$.

2.80.b False since $f_n \to 0$ pointwise but $\|f_n - 0\|_{\text{sup}} = \infty$ and does not approach 0 as $n \to \infty$.

2.81 You can use the derivative from elementary calculus for this exercise, to determine that $\|f_n\|_{\text{sup}} = \frac{1}{ne}$ for all $n \in \mathbb{N}$.

2.82.a True.

2.82.b False.

2.82.c False. It is very important to distinguish this case from the preceding one.

2.83 For example, let $x_n = \frac{2}{4n+1}$.

2.84 We can use $\delta = \frac{\epsilon}{2}$.

2.85 True, because this function can be extended to be continuous on the closed, finite interval $[0, 1]$.

2.86 We give a counterexample. Let

$$f(x) = \begin{cases} x & \text{if } x \leq 0, \\ x + 1 & \text{if } x > 0. \end{cases}$$

2.87 We can use $\delta = \epsilon^2$ because $|\sqrt{x} - \sqrt{a}| \leq \sqrt{|x - a|}$.

2.88.a True.

2.88.b True.

2.88.c False.

SOLUTIONS FOR CHAPTER 3

3.12 $\lim_{n\to\infty} \frac{2}{n} \sum_{k=1}^{n} \cos\left(1 + \frac{2k}{n}\right) = \int_1^3 \cos x \, dx = \sin - \sin 1$.

3.17 For example, we could choose $P = \{0, 0.95, 1.05, 2\}$.

3.33.a $f_n(x)$ converges to 1 if and only if $x \in \mathbb{Q} \cap [0, 1]$. Proofs are left to the reader.

3.33.b $f_n(x) \to 0$ for all $x \in [0, 1]$. Proofs are left to the reader.

3.35 $\lim_{n\to\infty} \int_1^2 1 + \frac{1}{(1+x^2)^n} \, dx \to 1$. Proofs are left to the reader.

3.45 For example, let f be the indicator function of a single point.

3.56 For example, let $g(x) = c \sin x$ for some constant $c \neq 0$.

3.59.a We find that $f(x) = 0$ for all $x \in \mathbb{R}$.

3.59.b $\|f_n\|_{\sup} = \frac{1}{2\sqrt{n}} \to 0$ as $n \to \infty$.

3.59.c Thus $f_n \to 0$ uniformly on $\mathbb{R}$.

3.60.a For the first interval the answer is *yes*. In fact $|f_n(x) - 0| = |x|^n \to 0$ as $n \to \infty$, provided that $-1 < x \leq 1$.

3.60.b For the second interval, the answer is *no* since the sequence $(-1)^n$ diverges.

3.61 We can pick $\delta = \epsilon$. Then $\|P\| < \delta$ implies that

$$|P(f, \{\bar{x}_i\}) - 3| < \epsilon.$$

3.62 This is the limit of a sequence of Riemann sums leading to $\int_1^3 \cos(x) \, dx = \sin(3) - \sin(1)$.

3.63 For example, we can let $P = \{0, .99, 1.01, 2\}$, a four-point partition of $[0, 2]$.

3.64 We give a counterexample. Let

$$f(x) = \begin{cases} 1 & \text{if } x \in \mathbb{Q} \cap [0, 1], \\ -1 & \text{if } x \in [0, 1] \setminus \mathbb{Q}. \end{cases}$$

3.65 $|T(f)| \leq \|f\|_{\sup} \int_0^1 \sin(x) \, dx = (1 - \cos 1)\|f\|_{\sup}$, so we can take $K = 1 - \cos 1$.

3.66 We can take $K = \int_{\frac{1}{e}}^{e} \frac{1}{x} \, dx = 2$.

3.67.a $\overline{\int_0^1} f(x) \, dx = \frac{1}{3}$ and $\underline{\int_0^1} f(x) \, dx = -\frac{1}{3}$.

3.67.b False.

3.68 True since $\mathcal{R}[a, b]$ is a vector space, so that $\frac{(f+g)+(f-g)}{2}$ and $\frac{(f+g)-(f-g)}{2}$ both lie in $\mathcal{R}[a, b]$.

3.69.a $\mathcal{C}[a, b]$ is a vector space.

3.69.b $\mathcal{P}_{10}(\mathbb{R})$ is not a vector space. For example, it has no zero vector.

3.70 $\|f_n\|_{\sup} = \frac{1}{ne}$. One can use the derivative from elementary calculus for this problem to determine the existence and value of the maximum value of f_n.

3.71 $\lim_{n \to \infty} \int_{\pi/4}^{\pi/2} \cos^n x \, dx = 0$ since the integrand converges uniformly to zero.

3.72 This is false because $T(f + g) \neq Tf + Tg$, if $g = -f$.

3.73 $\overline{\int_0^1} f(x) \, dx = \frac{1}{3}$ and $\underline{\int_0^1} f(x) \, dx = \frac{-1}{3}$.

SOLUTIONS FOR CHAPTER 4

4.25 $F'(x) = e^{-x^4}$.

4.32.c $\int_{-1}^1 f(x) \, dx = 2\sin(1)$.

4.43 $\lim_{x \to 0+} x^x = 1$.

4.45 $\lim_{h \to 0} \frac{P(x+3h)+P(x-3h)-2P(x)}{h^2} = 9P(x)$.

4.53 True. Let $f(x) = x - \ln(1 + x)$. We see that $f(0) = 0$ and $f'(x) > 0$ for all $x > 0$. Thus the Mean Value theorem tells us that f is increasing and $f(x) > 0$ for all $x > 0$.

4.54 For such an example we can use

$$F(x) = \begin{cases} x^2 \sin\left(\frac{1}{x^2}\right) & \text{if } 0 < x \leq 1, \\ 0 & \text{if } x = 0. \end{cases}$$

4.55 False. Use any step function with distinct values on two contiguous intervals to give a counterexample.

4.56 True. This is the first version of the Fundamental theorem of the calculus.

4.57 This is true for any function with the given properties–not only those functions which are derivatives.

4.58 For example, we can use

$$
f(x) = \begin{cases} 2x \sin \frac{1}{x^2} - \frac{2}{x} \cos \frac{1}{x^2} & \text{if } 0 < x \leq 1, \\ 0 & \text{if } x = 0. \end{cases}
$$

4.59 True: $\|T\| \leq 1$. The reader could show as an additional problem that $\|T\| = 1$.

4.60 $t = \frac{2}{3}$.

4.61 The limit is $16P''(x)$.

4.62 The limit is 0.

4.63 Yes, because $\|T\| \leq 1$.

SOLUTIONS FOR CHAPTER 5

5.3 Let $a_1 = 1$ and let $a_n = \sqrt{n} - \sqrt{n-1}$ for each $n > 1$.

5.9.a Divergent.

5.9.b Convergent.

5.12.a Divergent.

5.19.a Hint: Use the Ratio Test together with the definition of e as a limit.

5.19.b Convergent.

5.20 $\frac{1}{6}$.

5.43.a Converges uniformly by the M-test.

5.43.c Not uniformly convergent on $\left[0, \frac{\pi}{2}\right)$.

5.49 The sum is $\ln 2$.

5.50.e The interval of convergence is $I = (-e, e)$. The reader should prove divergence at the endpoints by applying the nth term test and using the hint.

5.54 $f^{(100)}(0) = 0$ and $f^{(101)}(0) = 100!$.

5.70.a Absolutely convergent.

5.70.b Conditionally convergent.

5.71 Let $x_k = \frac{(-1)^{k+1}}{k}$, for example.

5.72 True.

5.73 False. This set of series is not closed under subtraction, for example.

5.74.a Diverges.

5.74.b Converges (absolutely).

5.75 Let $x_k = \frac{1}{k^2}$ and $y_k = \frac{1}{k}$, for example.

5.76.a Diverges, by the ratio test.

5.76.b Converges, by the comparison test.

5.77 False, because the series is absolutely convergent.

5.78 For example, let $x_k = y_k = \frac{1}{2^k}$. Then

$$\sum_1^\infty x_k y_k = \frac{1}{3} \neq \left(\sum_1^\infty x_k \right) \left(\sum_1^\infty y_k \right) = 1.$$

The student should be able to write another such example.

5.79 $c_5 = x_1 y_4 + x_2 y_3 + x_3 y_2 + x_4 y_1$.

5.80 $\|x^{(n)}\|_1 = \frac{n+1}{2n}$.

5.81.a $M_k = e^{-2k}$, so the series $\sum_1^\infty M_k$ converges.

5.81.b $M_k \equiv 1$, so the series $\sum_1^\infty M_k$ diverges.

SOLUTIONS FOR CHAPTER 6

6.15 For example, let $f(x) = ix$ for all $x \in [a, b]$. There are infinitely many possible answers.

6.19.a $\widehat{f}(n) = \begin{cases} \frac{i}{2\pi n} \left(e^{-2\pi i n b} - e^{-2\pi i n a} \right), & n \in \mathbb{Z} \setminus \{0\}, \\ b - a, & n = 0. \end{cases}$

6.19.c $\widehat{f}(n) = \begin{cases} \frac{i}{2\pi n}, & n \neq 0, \\ \frac{1}{2}, & n = 0. \end{cases}$

6.19.d $\widehat{f}(n) = \begin{cases} \frac{i(-1)^n}{2\pi n}, & n \neq 0, \\ 0, & n = 0. \end{cases}$

6.19.e $\widehat{f}(n) = \begin{cases} \frac{(-1)^{|n|}}{2\pi^2 n^2}, & n \neq 0, \\ \frac{1}{12}, & n = 0. \end{cases}$

6.35 Hint: Write each trigonometric function using Euler's formula, and multiple using the multiplicative property of the exponential function.

6.44.a $\frac{\pi^2}{6}$.

6.51.a $\int_{\frac{1}{2}}^{1} f(x)\, dx = 1$.

6.51.c $\int_{0}^{1} f(x) \sin 2\pi x\, dx = 0$.

6.52.b $\int_{0}^{1} f(x) \sin 6\pi x\, dx = \frac{1}{9}$.

6.53.a $\frac{\sqrt{3}}{2}$.

6.53.b i.

6.53.c $\frac{3}{2}$.

6.54 $-\frac{i}{3}$.

6.55 False. The product is not linear in the second variable.

6.56.a $\widehat{f}(n)$ is both even and real-valued.

6.56.b $\widehat{f}(n)$ is both odd and imaginary-valued.

6.57.a $f(x) = \frac{1}{4} \cos 6\pi x + \frac{3}{4} \cos 2\pi x$.

6.57.b $\frac{3}{8}$.

6.58 True, because $\lim S_n$ is continuous, hence Riemann integrable.

6.59.a $\frac{\pi^4}{90}$.

6.59.b $\frac{8\pi^4}{729}$.

SOLUTIONS FOR CHAPTER 7

7.19 1.

7.21 6.

7.24 2.

7.27 For example, we can use $g = 1_{(p,b]}$ if $a \leq p < b$, or $g = 1_p$ if $p = b$.

7.30 For example, let $g = 1_{\{1\}}$.

7.41 This is false as the reader should be able to deduce from the fact that $\mathcal{BV}[0,1]$ is a vector space.

7.42 For example, let

$$f(x) = \begin{cases} \sin \frac{\pi}{x}, & x \neq 0, \\ 0, & x = 0. \end{cases}$$

7.43 6.

7.44 False.

7.45 True.

7.46 0.

7.47 We can choose $g(x) = 2 \cdot 1_{[1,2]}$, twice the indicator function of the subinterval $[1,2] \subset [0,2]$.

7.48 $\int_0^2 f \, dg = 3e$.

7.49 We can take $f = 1_{(a,b)}$ and $g = 1_{(b,c)}$. Then on the full interval $[a,c]$ both functions have a discontinuity at b. But on $[a,b]$ or the interval $[b,c]$, only one function or the other one has a discontinuity.

SOLUTIONS FOR CHAPTER 8

8.7.a The limit is $(0,e)$.

8.7.b The limit does not exist.

8.17.a Open.

8.17.c Neither.

8.17.e Closed.

8.42.a Not compact.

8.42.c Compact.

8.59 Any $\mathbf{y}$ perpendicular to $\mathbf{x}$ will suffice. For example: let $\mathbf{y} = (1,1,-1)$.

8.60 True. If not there exists a sequence for which $\|\mathbf{x}^{(k)}\| \to \infty$, and no subsequence converges.

8.61 This is a square box with vertices at $(\pm 1, 0)$ and $(0, \pm 1)$.

8.62 For example, let $D = \{(m, n) \mid m \in \mathbb{Z} \text{ and } n \in \mathbb{Z}\}$. Then the point $(0, 0)$ is an isolated point of D.

8.63 The set $S = \mathbb{Q}^{\times 2}$ is an example.

8.64 True. For example, this is true for every non-$\emptyset$ finite set.

8.65 False. The graph is not closed, lacking any points on the y-axis between 1 and -1.

8.66 False. For example, let $E_k = \mathbb{E}^n \setminus B_k(\mathbf{0})$.

8.67 True, because $\mathbf{x}_n \to \mathbf{x}_1$ as $n \to \infty$.

8.68 True, because g is continuous.

8.69.a This means that $x_2 = \pm x_1$, the graph of which consists of two intersecting straight lines, each of which is a connected set. Thus S is *connected*.

8.69.b The two open sets $A = \{\mathbf{x} \in \mathbb{E}^2 \mid x_1 > 0\}$ and $B = \{\mathbf{x} \in \mathbb{E}^2 \mid x_1 < 0\}$ separate T, so that T is not connected. There is no point in T corresponding to $x_1 = 0$.

8.70 False. For example, this would make $\mathbb{E}^n$ disconnected! For the two open separating sets, just take any one of the open balls for A and the union of all the others for B. But $\mathbb{E}^n$ is connected.

SOLUTIONS FOR CHAPTER 9

9.6.a The limit does not exist.

9.19 For example, let $\mathbf{f}(x) = (\cos x, \sin x)$.

9.26 Closed.

9.52 Here is a counterexample. Let A be the graph of $x_2 = \frac{1}{x_1}$ in $\mathbb{E}^2$, and let B be the x_1-axis.

9.53 True. This follows from the fact that any open set containing the point $(0, 1)$ must intersect the graph of $f|_{(0,\infty)}$ which is connected since the restriction of f is continuous on $(0, \infty)$.

9.54 Let
$$f(\mathbf{x}) = \begin{cases} \frac{x_1^2 x_2}{x_1^4 + x_2^2} & \text{if } \mathbf{x} \neq \mathbf{0}, \\ 0 & \text{if } \mathbf{x} = \mathbf{0}. \end{cases}$$

9.55 $\lim_{\mathbf{x} \to \mathbf{a}} \mathbf{f}(\mathbf{x})$ exists if and only if for each $\epsilon > 0$ there exists $\delta > 0$ such that $\mathbf{x}$ and $\mathbf{y}$ in $B_\delta(\mathbf{a}) \cap (D_f \setminus \{\mathbf{a}\})$ implies

$$\|\mathbf{f}(\mathbf{x}) - \mathbf{f}(\mathbf{y})\| < \epsilon.$$

9.56 True since $f \in C(\mathbb{E}^n, \mathbb{R})$ and $S = f^{-1}(\{0\})$, the pre-image of a closed set.

9.57 False, since the pre-image of an open set under a continuous function need only be *relatively open* in D.

9.58 This limit does not exist, since the limit along either axis is zero, but the limit along the line $x_2 = x_1$ is $\frac{1}{2}$.

9.59 True. Necessity follows directly from continuity at 0. Sufficiency is nearly the same as Exercise 9.33.b.

9.60 Here is a counterexample. $B_1(\mathbf{0})$ is a relatively closed subset of itself, but it is not closed so it cannot be compact, although it is bounded.

9.61 True, since A and B are connected being convex, and their union is connected since $A \cap B \neq \emptyset$.

9.62 False. Let $\mathbf{f}(x) = (\cos \pi x, \sin \pi x)$. Then $\|\mathbf{f}(x)\| \equiv 1$.

9.63 True, because the continuous image of a connected set is connected.

SOLUTIONS FOR CHAPTER 10

10.5 $\|A\| = \|B\| = \|A + B\| = 1$.

10.25.a $\mathbf{f}'(x)$ exists for all x.

10.26.a $\det \mathbf{f}'(\mathbf{x}) = x_1$.

10.26.c $\det \mathbf{f}'(\mathbf{x}) = x_1^2 \sin x_3$.

10.28 $D_{(1,-2)}\mathbf{f}(1, 2) = \begin{pmatrix} -5 \\ -10 \end{pmatrix}$.

10.44 $\begin{pmatrix} -3 & 3 & 6 \\ 4 & -4 & -3 \end{pmatrix}$.

10.48 $\frac{\partial z_i}{\partial x_j} = \sum_{k=1}^m \frac{\partial z_i}{\partial y_k} \frac{\partial y_k}{\partial x_j}$.

10.60.a $\{(x_1, x_2) \mid x_1 \neq 0\}$ and $0 < r < |x_1|$.

10.69 $x_1 x_2$.

10.89 $\|A\| = 1 = \|B\|$, and $\|A + B\| = 1$.

10.90 This is false, since $\|X + X\| = \|2X\| = 2\|X\|$.

10.91 This is false since the function $c_{1,1}(A) = a_{1,1}$ is a continuous function of A and $c_{1,1}(S)$ is not a connected subset of $\mathbb{E}^1$.

10.92 Let T have the matrix $\begin{pmatrix} 0 & 0 \\ 1 & 0 \end{pmatrix}$ in the standard basis.

10.93 This is false, since $\mathbf{f}'(\mathbf{x})$ is not linear on $\mathbf{x} \in D$. Rather $\mathbf{f}'(\mathbf{x})$ is a linear map from $\mathbb{E}^2 \to \mathbb{E}^3$ for each fixed $\mathbf{x} \in \mathbb{E}^2$. Moreover, D need not be a vector space, so that $c\mathbf{x_1} + \mathbf{x_2}$ need not be in D.

10.94 $[\mathbf{f}'(\mathbf{x})] = \begin{pmatrix} e^{x_1}\cos x_2 & -e^{x_1}\sin x_2 \\ e^{x_1}\sin x_2 & e^{x_1}\cos x_2 \end{pmatrix}$. Thus

$$D_{\mathbf{v}}\mathbf{f}\left(1, \frac{\pi}{6}\right) = \begin{pmatrix} \frac{e\sqrt{3}}{2} & -\frac{e}{2} \\ \frac{e}{2} & \frac{e\sqrt{3}}{2} \end{pmatrix}\begin{pmatrix} 2 \\ 1 \end{pmatrix} = \frac{e}{2}\begin{pmatrix} 2\sqrt{3}-1 \\ 2+\sqrt{3} \end{pmatrix}.$$

10.95 The matrix

$$[(\mathbf{f} \circ \mathbf{g})'(\mathbf{0})] = \begin{pmatrix} 1 & 2 \\ 3 & -1 \end{pmatrix}\begin{pmatrix} 1 & -1 & 0 \\ -2 & 3 & -2 \end{pmatrix} = \begin{pmatrix} -3 & 5 & -4 \\ 5 & -6 & 2 \end{pmatrix}.$$

10.96 This is true, because if we did have $f(a) = f(b)$ for some $a < b$ there would be an extreme point someplace in the open interval (a, b), because of the extreme value theorem and because f cannot be constant on $[a, b]$. For such an extreme point p and $r > 0$, f could not be injective on $(p - r, p + r)$.

10.97 By direct calculation, the Jacobian

$$\frac{\partial(f_1, f_2, f_3)}{\partial(x_1, x_2, x_3)} = x_1.$$

Thus $\mathbf{f}$ has nonvanishing Jacobian off the plane $x_1 = 0$, and for each such $\mathbf{x}$ we can take a ball of radius $r = \min(|x_1|, \pi)$.

10.98 The Jacobian

$$\frac{\partial f}{\partial x_3} = x_4.$$

Thus the equation

$$f(\mathbf{x}) = f(\mathbf{x_0})$$

has a local C^1-solution for x_3 in terms of the other three variables in a neighborhood of $\mathbf{x_0}$ provided $x_4 \neq 0$.

10.99 Let $\mathbf{f}(x) = (\cos x, \sin x)$ for all $x \in [0, 2\pi)$.

10.100 Denote $\mathbf{g}(x_1) = (x_1, x_2(x_1), x_3(x_1))$. Thus $\mathbf{f} \circ \mathbf{g}(x_1) \equiv 0$. Hence

$$[\mathbf{f}'(\mathbf{x_0})][g'(a)] = \begin{pmatrix} 1 & 1 & 3 \\ -1 & 0 & 2 \end{pmatrix}\begin{pmatrix} 1 \\ x_2'(a) \\ x_3'(a) \end{pmatrix} = \begin{pmatrix} 0 \\ 0 \end{pmatrix}.$$

This yields the equations

$$1 + x_2'(a) + 3x_3'(a) = 0,$$
$$-1 + 2x_3'(a) = 0$$

and the solutions $x_3'(a) = \frac{1}{2}$ and $x_2'(a) = -\frac{5}{2}$.

SOLUTIONS FOR CHAPTER 11

11.6 For example, we can let

$$\mathcal{P} = \{((0,0),(.95,1)),((.95,0),(1.05,1)),((1.05,1),(2,1))\}.$$

11.48 $\mathbf{g}'(\rho,\theta,\phi) = \rho^2 \sin\phi$ and the volume is $\frac{4}{3}\pi a^3$.

11.52 The volumes are identical.

11.58 For example, let f be the indicator function of the set of points for which both coordinates are rational.

11.59 Let f be the indicator of the points with both coordinates rational, and let g be the indicator of the points with both coordinates irrational.

11.60 For example, let f be the indicator function of the set of points having first coordinate rational.

11.61 $\|T\| = \prod_{i=1}^{n}(b_i - a_i)$.

11.62 The function f is not continuous anywhere on $[0,1]$.

11.63 No, S is not Jordan measurable since its boundary is not a Lebesgue null set.

11.64 True, because it is a compact subset of a Lebesgue null set.

11.65 $o(f,0) = 2$

11.66 The Jordan measure is zero.

11.67 We must have $f(x_1) = 0$ except on a Lebesgue null set. This follows from Lebesgue's theorem.

11.68.a

$$[\mathbf{f}'(\mathbf{x})] = \begin{pmatrix} -1 & 0 & y \\ 0 & -1 & x \\ 0 & 0 & -1 \end{pmatrix}.$$

11.68.b The volume is $\frac{32\pi}{3}$.

REFERENCES

1. Tom M. Apostol, *Mathematical Analysis: A Modern Approach to Advanced Calculus*, Addison-Wesley, Reading, MA, 1957.

2. C. Berge, *Principles of Combinatorics*, Academic Press, New York, 1971.

3. Carl B. Boyer, *The History of the Calculus*, Dover Publications, New York, 1949.

4. R. Creighton Buck, *Advanced Calculus*, McGraw-Hill, New York, 1956.

5. Lennart Carleson, On convergence and growth of partial sumas of Fourier series, *Acta Mathematica*, Vol. 116, 1966, pp. 135–157.

6. H. Dym and H. P. McKean, *Fourier Series and Integrals*, Academic Press, New York, 1972.

7. Bernard R. Gelbaum and John M. H. Olmsted, *Counterexamples in Analysis*, Holden-Day, San Francisco, 1964.

8. Casper Goffman and George Pedrick, *First Course in Functional Analysis*, Prentice-Hall, Englewood Cliffs, NJ, 1965.

9. Kenneth Hoffman, *Analysis in Euclidean Space*, Prentice-Hall, Englewood Cliffs, NJ, 1975.

10. Kenneth Hoffman and Ray Kunze, *Linear Algebra*, Prentice-Hall, Englewood Cliffs, NJ, 1971.

11. E. Kamke, *Theory of Sets*, Dover Publications, New York, 1950.

12. E. Landau, *Foundations of Analysis*, Chelsea Publishing Co., Broomall, PA, 1960.

Advanced Calculus: An Introduction to Linear Analysis. By Leonard F. Richardson
Copyright © 2008 John Wiley & Sons, Inc.

13. H. E. Lomeli and C. L. Garcia, Variations on a theorem of Korovkin, *American Math Monthly*, Vol. 113, No. 8, 2006, pp. 744–750.

14. G. D. Mostow, J. H. Sampson and J.-P. Meyer, *Fundamental Structures of Algebra*, McGraw Hill, New York, 1963.

15. Otto Neugebauer, *The Exact Sciences in Antiquity*, Brown University Press, Providence, RI, 1957.

16. John M. H. Olmsted, *The Real Number System*, Appleton-Century-Crofts, New York, 1962.

17. J. Dauben, Review: The Universal History of Numbers, *Notices of the American Mathematical Society*, Vol. 49, No. 1, January 2002, pp. 32–38.

18. Frigyes Riesz and Béla Sz.-Nagy, *Functional Analysis*, Frederick Ungar Publishing Co., New York, 1955.

19. Walter Rudin, *Principles of Mathematical Analysis*, McGraw-Hill, New York, 1964.

20. Michael Spivak, *Calculus on Manifolds*, W. A. Benjamin, New York, 1965.

INDEX

Advanced Calculus: An Introduction to Linear Analysis. By Leonard F. Richardson
Copyright © 2008 John Wiley & Sons, Inc.